Deciphering Object-Oriented Programming with C++

A practical, in-depth guide to implementing object-oriented design principles to create robust code

Dorothy R. Kirk

BIRMINGHAM—MUMBAI

Deciphering Object-Oriented Programming with C++

Copyright © 2022 Dorothy R. Kirk

Associate Group Product Manager: Gebin George
Content Development Editor: Rosal Colaco
Senior Editor: Tiksha Abhimanyu Lad
Technical Editor: Jubit Pincy
Copy Editor: Safis Editing
Project Manager: Prajakta Naik
Proofreader: Safis Editing
Indexer: Manju Arasan
Production Designer: Prashant Ghare
Marketing Coordinator: Sonakshi Bubbar

First published: September 2022
Production reference: 1020922

Published by Packt Publishing Ltd.
Livery Place
35 Livery Street
Birmingham
B3 2PB, UK.

ISBN 978-1-80461-390-0

www.packt.com

To my family, especially my two children.

For my oldest, a dedicated and brilliant physicist, astronomer, and astrophysicist.

For my youngest, a superlative writer and advocate for equality and our planet.

May they make our world (and universe) a better place.

– Dorothy R. Kirk

Contributors

About the author

Dorothy R. Kirk has specialized in object-oriented (OO) technologies since nearly their inception. She began as an early adopter of C++ at General Electric in research and development (R&D). After working on various projects, she was one of 10 charter members to start GE's Advanced Concepts Center to commercially utilize OO technologies. She later started her own OO training and consulting business, specializing in OOP using C++ and Java. She additionally programs in Python. She has developed and taught many OO courses and has clients spanning industries such as academia, finance, transportation, software, embedded systems, manufacturing, and medical imaging. She has also taught C++ and OO courses at Penn State University.

Ms. Kirk has earned a Bachelor of Science degree in Computer and Information Sciences from the University of Delaware and a Master of Science degree in Computer Science from Temple University.

Ms. Kirk is married and has two grown children; she and her family live on a horse farm in Pennsylvania.

I want to thank those who have consistently supported my endeavors, including my husband, children, parents, sister, extended family, and close friends. I also want to thank the many people at Packt who have contributed to this project, including Tiksha Abhimanyu Lad, Rosal Colaco, and Jubit Pincy, and especially Alok Dhuri and Prajakta Naik, who have worked with me often and tirelessly.

About the reviewers

Vinícius G. Mendonça is a professor at PUCPR and a mentor at Apple Developer Academy. He has a master's degree in Computer Vision and Image Processing (PUCPR) and a specialization degree in Game Development (Universidade Positivo). He is also one of the authors of the book *Learn OpenCV 4 by Building Projects*, also by Packt Publishing.

He has been in this field since 1996. His former experience includes designing and programming a multithreaded framework for PBX tests at Siemens, coordination of Aurélio Dictionary software (including its apps for Android, IOS, and Windows phones), and coordination of an augmented reality educational activity for Positivo's *Mesa Alfabeto*, presented at CEBIT. Currently, he works with server-side Node.js at a company called Tenet Tech.

> *First of all, I would like to thank Thais, my spouse and the love of my life, for being a wonderful partner and for supporting me every day. I'd also like to thank my daughters: Mariana, Alice, Laura, Helena, and Renata – you're the best, girls! Also, my compliments to my stepson, Bruno, who makes me proud, since he is also becoming a skilled programmer and listens to my technical mumbo jumbo all day. You are my reason to live, and I love you all.*

Yacob Cohen-Arazi (**Kobi**), a husband and father to three kids, is a Principal Software Engineer with over 20 years of experience, currently working at Qualcomm in San Diego, California. He now works on the next generation of 5G cellular and radio access networks, and in his previous roles, he also worked on automotive, machine learning, Wi-Fi, and 4G domains. Kobi is an expert in C++ and enjoys sharing his knowledge with the community. He has led the San Diego C++ Meetup for the past 3 years, where he presents various topics on advanced, modern C++ constructs. In his free time, Kobi likes cycling, swimming, and spending time with his family. He holds a bachelor's degree in Computer Science from The Academic College of Tel Aviv-Yaffo.

Table of Contents

Part 1: C++ Building Block Essentials

1

Understanding Basic C++ Assumptions   3

2

Adding Language Necessities   27

3

Indirect Addressing – Pointers   47

4

Indirect Addressing – References   83

Part 2: Implementing Object-Oriented Concepts in C++

5

Exploring Classes in Detail   105

6

Implementing Hierarchies with Single Inheritance   167

7

Utilizing Dynamic Binding through Polymorphism   203

8

Mastering Abstract Classes   233

9

Exploring Multiple Inheritance   253

10

Implementing Association, Aggregation, and Composition   279

Part 3: Expanding Your C++ Programming Repertoire

11

Handling Exceptions   305

12

Friends and Operator Overloading   327

13

Working with Templates   347

14

Understanding STL Basics   365

15

Testing Classes and Components   393

Part 4: Design Patterns and Idioms in C++

16

Using the Observer Pattern   415

20

Removing Implementation Details Using the pImpl Pattern 495

Part 5: Considerations for Safer Programming in C++

21

Making C++ Safer 517

Assessments 545

Preface

Companies choose to utilize C++ for its speed; however, object-oriented (OO) software design leads to more robust code that is easier to modify and maintain. Understanding how to utilize C++ as an object-oriented language is, therefore, essential. Programming in C++ won't guarantee object-oriented programming (OOP) – you must understand OO concepts and how they map to C++ language features and programming techniques. Additionally, programmers will want to harness additional skills beyond OOP to make code more generic and robust, as well as employ well-tested, creative solutions that can be found in popular design patterns. It is also critical for programmers to understand language features and conventions that can make C++ a safer language to use.

A programmer who learns how to use C++ as an object-oriented language, following safe programming conventions, will become a valuable C++ developer – a developer whose code is easy to maintain, modify, and be understood by others.

This book has step-by-step explanations of essential OO concepts, paired with practical examples in code and often with diagrams so you can truly understand how and why things work. Self-assessment questions are available to test your skills.

This book first provides necessary building blocks of skills (which may not be object-oriented) that provide the essential foundation on which to build OOP skills. Next, OO concepts will be described and paired with language features as well as coding techniques so that you can understand how to use C++ as an OOP language successfully. Additionally, more advanced skills are added to augment the programmer's repertoire, including friend function/classes, operator overloading, templates (to build more generic code), exception handling (to build robust code), Standard Template Library (STL) basics, as well as design patterns and idioms. The book wraps up by re-examining programming constructs presented throughout the book, paired with conventions that lead to safer programming in C++. The end goal is to enable you to produce robust code that is easy to maintain and understand by others.

By the end of this book, you will understand both essential and advanced OO concepts and how to implement these concepts in C++. You will have a versatile toolkit of C++ programming skills. You will additionally understand ways to make safer, more robust, and easily maintainable code, as well as understand how to employ well-tested design patterns as part of your programming repertoire.

Who this book is for?

Programmers wanting to utilize C++ for OOP will find this book essential to understand how to implement OO designs in C++ through both language features and refined programming techniques, while creating robust and easily maintainable code. This OOP book assumes prior programming experience; however, if you have limited or no prior C++ experience, the early chapters will help you learn essential C++ skills to serve as the basis for the many OOP sections, advanced features, design patterns, and conventions to promote safe programming in C++.

What this book covers

Chapter 1, Understanding Basic C++ Assumptions, provides a concise review of basic language features that are assumed knowledge within the book. Existing programmers can quickly grasp the language basics reviewed in this first chapter.

Chapter 2, Adding Language Necessities, reviews non-OO features that are critical C++ building blocks: const qualifiers, function prototyping (default values), and function overloading.

Chapter 3, Indirect Addressing – Pointers, reviews pointers in C++, including memory allocation/deallocation, pointer usage/dereferencing, usage in function arguments, void pointers, and introduces the concept of smart pointers.

Chapter 4, Indirect Addressing – References, introduces references as an alternative to pointers, including initialization, function arguments/return values, and const qualification.

Chapter 5, Exploring Classes in Detail, introduces OOP by first exploring OO and the concepts of encapsulation and information hiding, and then covers class features in detail: member functions, the this pointer, access labels and regions, constructors, destructor, and qualifiers on data members and member functions (const, static, and inline).

Chapter 6, Implementing Hierarchies with Single Inheritance, details generalization and specialization using single inheritance. This chapter covers inheriting members, the use of base class constructors, inherited access regions, order of construction/destruction, final classes, as well as public versus private and protected base classes, and how this changes the meaning of inheritance.

Chapter 7, Utilizing Dynamic Binding through Polymorphism, describes the OO concept of polymorphism and then distinguishes operation from method, details virtual functions and runtime binding of methods to operations (including how the v-table works), and differentiates the use of virtual, override, and final.

Chapter 8, Mastering Abstract Classes, explains the OO concept of abstract classes, their implementation using pure virtual functions, the OO concept of an interface and how to implement it, as well as up and down casting within a public inheritance hierarchy.

Chapter 9, Exploring Multiple Inheritance, details how to use multiple inheritance as well as its controversy in OO designs. This chapter covers virtual base classes, diamond-shaped hierarchies, and when to consider alternate designs by examining the OO concept of a discriminator.

Chapter 10, Implementing Association, Aggregation, and Composition, describes the OO concepts of association, aggregation, and composition and how to implement each concept using pointers, sets of pointers, containment, and sometimes references.

Chapter 11, Handling Exceptions, explains how to try, throw, and catch exceptions by considering many exception scenarios. This chapter shows how to extend an exception handling hierarchy.

Chapter 12, Friends and Operator Overloading, explains the proper use of friend functions and classes, and examines operator overloading (which may use friends) to allow an operator to work with user defined types in the same way it works with standard types.

Chapter 13, Working with Templates, details template functions and classes to genericize certain types of code to work with any data type. This chapter also shows how operator overloading can make selected code more generic for any type to further support the use of templates.

Chapter 14, Understanding STL Basics, introduces the Standard Template Library in C++ and demonstrates how to use common containers, such as list, iterator, deque, stack, queue, priority_queue, and map. Additionally, STL algorithms and functors are introduced.

Chapter 15, Testing Classes and Components, illustrates OO testing methods using the canonical class form and drivers to test classes, and shows how to test classes related through inheritance, association, and aggregation. This chapter also shows how to test classes that utilize exception handling.

Chapter 16, Using the Observer Pattern, introduces design patterns overall and then explains the Observer pattern, with an in-depth example illustrating the components of the pattern.

Chapter 17, Applying the Factory Pattern, introduces the Factory Method pattern and showcases its implementation with and without an Object Factory. It also compares an Object Factory to an Abstract Factory.

Chapter 18, Applying the Adapter Pattern, examines the Adapter pattern, providing strategies and examples utilizing inheritance versus association to implement the pattern. Additionally, this chapter demonstrates a wrapper class as a simple Adapter.

Chapter 19, Using the Singleton Pattern, examines in detail the Singleton pattern with a sophisticated paired-class implementation. Singleton registries are also introduced.

Chapter 20, Removing Implementation Details Using the pImpl Pattern, describes the pImpl pattern, which is used to reduce compile-time dependencies within code. A detailed implementation is explored using unique pointers. Performance issues are explored relating to the pattern.

Chapter 21, *Making C++ Safer*, revisits topics covered throughout the book, with the intention of identifying core programming guidelines that can be used to make C++ a safer language for the development of robust software.

Assessments contains all the answers to the questions from every chapter.

To get the most out of this book

It is assumed that you have a current C++ compiler available. You will want to try the many online code examples! You can use any C++ compiler; however, C++17 or higher is recommended. The code presented will be C++20 compliant. Minimally, please download g++ from `https://gcc.gnu.org`.

Software/hardware covered in the book	Operating system requirements
Any C++ compiler, minimally C++17 C++20 (recommended)	Any: Windows, macOS (OS X), or Unix/Linux
Minimally download https://gcc.gnu.org.	Works for any aforementioned OS

Please keep in mind that though there is an ISO standard for C++, some compilers vary and interpret the standard with very slight differences.

If you are using the digital version of this book, we advise you to type the code yourself or access the code from the book's GitHub repository (a link is available in the next section). Doing so will help you avoid any potential errors related to the copying and pasting of code.

Trying the coding examples as you read the book is highly recommended. Completing the Assessments will further enhance your grasp of each new concept.

Download the example code files

You can download the example code files for this book from GitHub at `https://github.com/PacktPublishing/Deciphering-Object-Oriented-Programming-with-CPP`. If there's an update to the code, it will be updated in the GitHub repository.

We also have other code bundles from our rich catalog of books and videos available at `https://github.com/PacktPublishing/`. Check them out!

Code in Action

The Code in Action videos for this book can be viewed at `https://bit.ly/3pylFkV`.

Download the color images

We also provide a PDF file that has color images of the screenshots and diagrams used in this book. You can download it here: `https://packt.link/ZvNhC`.

Conventions used

There are a number of text conventions used throughout this book.

`Code in text`: Indicates code words in text, database table names, folder names, filenames, file extensions, pathnames, dummy URLs, user input, and Twitter handles. Here is an example: "With that in mind, let's take a look at our Adapter class, `Humanoid`."

A block of code is set as follows:

```
class Humanoid: private Person    // Humanoid is abstract
{
protected:
    void SetTitle(const string &t) { ModifyTitle(t); }
public:
    Humanoid() = default;
    Humanoid(const string &, const string &,
             const string &, const string &);
    // class definition continues
```

When we wish to draw your attention to a particular part of a code block, the relevant lines or items are set in bold:

```
    const string &GetSecondaryName() const
        { return GetFirstName(); }
    const string &GetPrimaryName() const
        { return GetLastName(); }
```

Any command-line input or output is written as follows:

```
Orkan Mork McConnell
Nanu nanu
Romulan Donatra Jarok
jolan'tru
Earthling Eve Xu
Hello
```

```
Earthling Eve Xu
Bonjour
```

> **Tips or important notes**
> Appear like this.

Get in touch

Feedback from our readers is always welcome.

General feedback: If you have questions about any aspect of this book, email us at customercare@ packtpub.com and mention the book title in the subject of your message.

Errata: Although we have taken every care to ensure the accuracy of our content, mistakes do happen. If you have found a mistake in this book, we would be grateful if you would report this to us. Please visit www.packtpub.com/support/errata and fill in the form.

Piracy: If you come across any illegal copies of our works in any form on the internet, we would be grateful if you would provide us with the location address or website name. Please contact us at copyright@packt.com with a link to the material.

If you are interested in becoming an author: If there is a topic that you have expertise in and you are interested in either writing or contributing to a book, please visit authors.packtpub.com.

Share Your Thoughts

Once you've read *Deciphering Object-Oriented Programming with C++*, we'd love to hear your thoughts! Scan the QR code below to go straight to the Amazon review page for this book and share your feedback.

https://packt.link/r/1-804-61390-8

Your review is important to us and the tech community and will help us make sure we're delivering excellent quality content.

Part 1:
C++ Building Block
Essentials

The goal of this part is to ensure that you have a strong background in non-OO C++ skills with which to build the forthcoming OOP skills in C++. This is the shortest section of the book, designed to quickly get you up to speed in preparation for OOP and more advanced book chapters.

The first chapter quickly reviews the basic skills that you are assumed to have to progress through the book: basic language syntax, looping constructs, operators, function usage, user defined type basics (`struct`, `typedef`, `class` basics, `using` statement, `enum`, strongly-typed `enum`), and `namespace` basics. The next chapter discusses `const` qualified variables, function prototyping, prototyping with default values, and function overloading.

The subsequent chapter covers indirect addressing with pointers by introducing `new()` and `delete()` to allocate basic types of data, dynamically allocating arrays of 1, 2, and N dimensions, managing memory with `delete`, passing parameters as arguments to functions, using void pointers, and an overview of smart pointers. This section concludes with a chapter featuring indirect addressing with references that will take you through a review of reference basics, references to existing objects, and as arguments to functions.

Although this book will gradually progress to use smart pointers as a preference (and recommends smart pointers for safety), gaining proficiency with native C++ pointers will be an important skill to have. This skill will be crucial for modifying and deciphering existing code with native pointers, as well as for clearly understanding the potential misuses and pitfalls of native pointers.

This part comprises the following chapters:

- *Chapter 1, Understanding Basic C++ Assumptions*
- *Chapter 2, Adding Language Necessities*
- *Chapter 3, Indirect Addressing – Pointers*
- *Chapter 4, Indirect Addressing – References*

1

Understanding Basic C++ Assumptions

This chapter will briefly identify the basic language syntax, constructs, and features of C++, which you are assumed to have from familiarity either with the basic syntax of C++, C, Java, or similar languages. These core language features will be reviewed concisely. If these basic syntax skills are not familiar to you after completing this chapter, please first take the time to explore a more basic syntax-driven C++ text before continuing with this book. The goal of this chapter is not to teach each of the assumed skills in detail but to briefly provide a synopsis of each basic language feature to allow you to quickly recall a skill that should already be in your programming repertoire.

In this chapter, we will cover the following main topics:

- Basic language syntax
- Basic I/O
- Control structures, statements, and looping
- Operators
- Function basics
- User defined type basics
- Namespace basics

By the end of this chapter, you'll have a succinct review of the very basic C++ language skills in which you're assumed to be proficient. These skills will be necessary in order to move forward with the next chapter successfully. Because most of these features do not utilize OO features of C++, I will refrain from using OO terminology (as much as possible) and will instead introduce appropriate OO terminology when we move into the OO sections of this book.

Technical requirements

Please ensure that you have a current C++ compiler available; you'll want to try many of the online code examples. Minimally, please download g++ from `https://gcc.gnu.org`.

Online code for full program examples can be found in the following GitHub URL: `https://github.com/PacktPublishing/Deciphering-Object-Oriented-Programming-with-CPP/tree/main/Chapter01`. Each full program example can be found in the GitHub under the appropriate chapter heading (subdirectory) in a file that corresponds to the chapter number, followed by a dash, followed by the example number in the chapter at hand. For example, the first full program in *Chapter 1*, *Understanding Basic C++ Assumptions*, can be found in the subdirectory `Chapter01` in a file named `Chp1-Ex1.cpp` under the aforementioned GitHub directory.

The **Code in Action (CiA)** video for this chapter can be viewed at: `https://bit.ly/3PtOYjf`.

Reviewing basic C++ language syntax

In this section, we will briefly review basic C++ syntax. We'll assume that you are either a C++ programmer with non-OO programming skills, or that you've programmed in C, Java, or a similar strongly typed checked language with related syntax. You may also be a long-standing professional programmer who is able to pick up another language's basics quickly. Let's begin our brief review.

Comment styles

Two styles of comments are available in C++:

- The `/*    */` style provides for comments spanning multiple lines of code. This style may not be nested with other comments of this same style.
- The `//` style of comment provides for a simple comment to the end of the current line.

Using the two comment styles together can allow for nested comments, which can be useful when debugging code.

Variable declarations and standard data types

Variables may be of any length, and may consist of letters, digits, and underscores. Variables are case sensitive and must begin with a letter or an underscore. Standard data types in C++ include the following:

- `int`: To store whole numbers
- `float`: To store floating point values
- `double`: To store double precision floating point values

- `char`: To store a single character
- `bool`: For boolean values of `true` or `false`

Here are a few straightforward examples using the aforementioned standard data types:

```
int x = 5;
int a = x;

float y = 9.87;
float y2 = 10.76f;  // optional 'f' suffix on float literal
float b = y;

double yy = 123456.78;
double c = yy;

char z = 'Z';
char d = z;

bool test = true;
bool e = test;
bool f = !test;
```

Reviewing the previous fragment of code, note that a variable can be assigned a literal value, such as `int x = 5;` or that a variable may be assigned the value or contents of another variable, such as `int a = x;`. These examples illustrate this ability with various standard data types. Note that for the `bool` type, the value can be set to `true` or `false`, or to the opposite of one of those values using `!` (not).

Variables and array basics

Arrays can be declared of any data type. The array name represents the starting address of the contiguous memory associated with the array's contents. Arrays are zero-based in C++, meaning they are indexed starting with array `element [0]` rather than array `element [1]`. Most importantly, range checking is not performed on arrays in C++; if you access an element outside the size of an array, you are accessing memory belonging to another variable, and your code will likely fault very soon.

Let's review some simple array declarations (some with initialization), and an assignment:

```
char name[10] = "Dorothy"; // size is larger than needed
float grades[20];  // array is not initialized; caution!
```

```
grades[0] = 4.0;  // assign a value to one element of array
float scores[] = {3.3, 4.3, 4.0, 3.7}; // initialized array
```

Notice that the first array, name, contains 10 char elements, which are initialized to the seven characters in the string literal "Dorothy", followed by the null character ('\0'). The array currently has two unused elements at the end. The elements in the array can be accessed individually using name[0] through name[9], as arrays in C++ are zero-based. Similarly, the array above, which is identified by the variable grades, has 20 elements, none of which are initialized. Any array value accessed prior to initialization or assignment can contain any value; this is true for any uninitialized variable. Notice that just after the array grades is declared, its 0th element is assigned a value of 4.0. Finally, notice that the array of float, scores, is declared and initialized with values. Though we could have specified an array size within the [] pair, we did not – the compiler is able to calculate the size based upon the number of elements in our initialization. Initializing an array when possible (even using zeros), is always the safest style to utilize.

Arrays of characters are often conceptualized as strings. Many standard string functions exist in libraries such as <cstring>. Arrays of characters should be null-terminated if they are to be treated as strings. When arrays of characters are initialized with a string of characters, the null character is added automatically. However, if characters are added one by one to the array via assignment, it would then be the programmer's job to add the null character ('\0') as the final element in the array.

In addition to strings implemented using arrays of characters (or a pointer to characters), there is a safer data type from the C++ Standard Library, std::string. We will understand the details of this type once we master classes in *Chapter 5, Exploring Classes in Detail*; however, let us introduce string now as an easier and less error-prone way to create strings of characters. You will need to understand both representations; the array of char (and pointer to char) implementations will inevitably appear in C++ library and other existing code. Yet you may prefer string in new code for its ease and safety.

Let's see some basic examples:

```
// size of array can be calculated by initializer
char book1[] = "C++ Programming";
char book2[25];  // this string is uninitialized; caution!
// use caution as to not overflow destination (book2)
strcpy(book2, "OO Programming with C++");
strcmp(book1, book2);
length = strlen(book2);
string book3 = "Advanced C++ Programming";  // safer usage
string book4("OOP with C++'); // alt. way to init. string
string book5(book4); // create book5 using book4 as a basis
```

Here, the first variable `book1` is declared and initialized to a string literal of `"C++ Programming"`; the size of the array will be calculated by the length of the quoted string value plus one for the null character (`'\0'`). Next, variable `book2` is declared to be an array of 25 characters in length, but is not initialized with a value. Next, the function `strcpy()` from `<cstring>` is used to copy the string literal `"OO Programming with C++"` into the variable `book2`. Note that `strcpy()` will automatically add the null-terminating character to the destination string. On the next line, `strcmp()`, also from `<cstring>`, is used to lexicographically compare the contents of variables `book1` and `book2`. This function returns an integer value, which can be captured in another variable or used in a comparison. Lastly, the function `strlen()` is used to count the number of characters in `book2` (excluding the null character).

Lastly, notice that `book3` and `book4` are each of type `string`, illustrating two different manners to initialize a string. Also notice that `book5` is initialized using `book4` as a basis. As we will soon discover, there are many safety features built into the `string` class to promote safe string usage. Though we have reviewed examples featuring two of several manners to represent strings (a native array of characters versus the string class), we will most often utilize `std::string` for its safety. Nonetheless, we have now seen various functions, such as `strcpy()` and `strlen()`, that operate on native C++ strings (as we will inevitably come across them in existing code). It is important to note that the C++ community is moving away from native C++ strings – that is, those implemented using an array of (or pointer to) characters.

Now that we have successfully reviewed basic C++ language features such as comment styles, variable declarations, standard data types, and array basics, let's move forward to recap another fundamental language feature of C++: basic keyboard input and output using the `<iostream>` library.

Recapping basic I/O

In this section, we'll briefly review simple character-based input and output with the keyboard and monitor. Simple manipulators will also be reviewed to both explain the underlying mechanics of I/O buffers and to provide basic enhancements and formatting.

The iostream library

One of the easiest mechanisms for input and output in C++ is the use of the `<iostream>` library. The header file `<iostream>` contains definitions of data types **istream** and **ostream**. Instances of these data types, `cin`, `cout`, `cerr`, and `clog`, are incorporated by including the `std` namespace. The `<iostream>` library facilitates simple I/O and can be used as follows:

- `cin` can be used in conjunction with the extraction operator `>>` for buffered input
- `cout` can be used in conjunction with the insertion operator `<<` for buffered output
- `cerr` (unbuffered) and `clog` (buffered) can also be used in conjunction with the insertion operator, but for errors

Let's review an example showcasing simple I/O:

https://github.com/PacktPublishing/Deciphering-Object-Oriented-Programming-with-CPP/blob/main/Chapter01/Chp1-Ex1.cpp

```cpp
#include <iostream>
using namespace std;   // we'll limit the namespace shortly

int main()
{
    char name[20];   // caution, uninitialized array of char
    int age = 0;
    cout << "Please enter a name and an age: ";
    cin >> name >> age; // caution, may overflow name var.
    cout << "Hello " << name;
    cout << ". You are " << age << " years old." << endl;
    return 0;
}
```

First, we include the <iostream> library and indicate that we're using the std namespace to gain usage of cin and cout (more on namespaces later in this chapter). Next, we introduce the main() function, which is the entry point in our application. Here, we declare two variables, name and age, neither of which is initialized. Next, we prompt the user for input by placing the string "Please enter a name and an age: " in the buffer associated with cout. When the buffer associated with cout is flushed, the user will see this prompt on the screen.

The keyboard input string is then placed in the buffer associated with cout using the extraction operator <<. Conveniently, one mechanism that automatically flushes the buffer associated with cout is the use of cin to read keyboard input into variables, such as seen on the next line, where we read the user input into the variables name and age, respectively.

Next, we print out a greeting of "Hello" to the user, followed by the name entered, followed by an indication of their age, gathered from the second piece of user input. The endl at the end of this line both places a newline character '\n' into the output buffer and ensures that the output buffer is flushed – more of that next. The return 0; declaration simply returns a program exit status to the programming shell, in this case, the value 0. Notice that the main() function indicates an int for a return value to ensure this is possible.

Basic iostream manipulators

Often, it is desirable to be able to manipulate the contents of the buffers associated with `cin`, `cout`, and `cerr`. Manipulators allow the internal state of these objects to be modified, which affects how their associated buffers are formatted and manipulated. Manipulators are defined in the `<iomanip>` header file. Common manipulator examples include the following:

- `endl`: Places a newline character (`'\n'`) in the buffer associated with `cout` then flushes the buffer
- `flush`: Clears the contents of the output stream
- `setprecision(int)`: Defines the precision (number of digits) used to output floating point numbers
- `setw(int)`: Sets the width for input and output
- `ws`: Removes whitespace characters from the buffer

Let's see a simple example:

https://github.com/PacktPublishing/Deciphering-Object-Orient-ed-Programming-with-CPP/blob/main/Chapter01/Chp1-Ex2.cpp

```cpp
#include <iostream>
#include <iomanip>
using namespace std;    // we'll limit the namespace shortly
int main()
{
    char name[20];      // caution; uninitialized array
    float gpa = 0.0;    // grade point average
    cout << "Please enter a name and a gpa: ";
    cin >> setw(20) >> name >> gpa;   // won't overflow name
    cout << "Hello " << name << flush;
    cout << ". GPA is: " << setprecision(3) << gpa << endl;
    return 0;
}
```

In this example, first, notice the inclusion of the `<iomanip>` header file. Also, notice that `setw(20)` is used to ensure that we do not overflow the name variable, which is only 20 characters long; `setw()` will automatically deduct one from the size provided to ensure there is room for the null character. Notice that `flush` is used on the second output line – it's not necessary here to flush the output buffer; this manipulator merely demonstrates how a `flush` may be applied. On the last output line with cout, notice that setprecision(3) is used to print the floating point gpa. Three points of precision account for the decimal point plus two places to the right of the decimal. Finally, notice that we add the endl manipulator to the buffer associated with cout. The endl manipulator will first insert a newline character ('\n') into the buffer and then flush the buffer. For performance, if you don't need a buffer flush to immediately see the output, using a newline character alone is more efficient.

Now that we have reviewed simple input and output using the <iostream> library, let's move forward by briefly reviewing control structures, statements, and looping constructs.

Revisiting control structures, statements, and looping

C++ has a variety of control structures and looping constructs that allow for non-sequential program flow. Each can be coupled with simple or compound statements. Simple statements end with a semicolon; more compound statements are enclosed in a block of code using a pair of brackets { }. In this section, we will be revisiting various types of control structures (if, else if, and else), and looping constructs (while, do while, and for) to recap simple methods for non-sequential program flow within our code.

Control structures – if, else if, and else

Conditional statements using if, else if, and else can be used with simple statements or a block of statements. Note that an if clause can be used without a following else if or else clause. Actually, else if is really a condensed version of an else clause with a nested if clause inside of it. Practically speaking, developers flatten the nested use into else if format for readability and to save excess indenting. Let's see an example:

https://github.com/PacktPublishing/Deciphering-Object-Oriented-Programming-with-CPP/blob/main/Chapter01/Chp1-Ex3.cpp

```cpp
#include <iostream>
using namespace std;    // we'll limit the namespace shortly

int main()
{
    int x = 0;
    cout << "Enter an integer: ";
    cin >> x;
    if (x == 0)
        cout << "x is 0" << endl;
    else if (x < 0)
        cout << "x is negative" << endl;
```

```
    else
    {
        cout << "x is positive";
        cout << "and ten times x is: " << x * 10 << endl;
    }
    return 0;
}
```

Notice that in the preceding else clause, multiple statements are bundled into a block of code, whereas in the if and else if conditions, only a single statement follows each condition. As a side note, in C++, any non-zero value is considered to be true. So, for example, testing if (x) would imply that x is not equal to zero – it would not be necessary to write if (x !=0), except possibly for readability.

It is worth mentioning that in C++, it is wise to adopt a set of consistent coding conventions and practices (as do many teams and organizations). As a straightforward example, the placement of brackets may be specified in a coding standard (such as starting the { on the same line as the keyword else, or on the line below the keyword else with the number of spaces it should be indented). Another convention may be that even a single statement following an else keyword be included in a block using brackets. Following a consistent set of coding conventions will allow your code to be more easily read and maintained by others.

Looping constructs – while, do while, and for loops

C++ has several looping constructs. Let's take a moment to review a brief example for each style, starting with the while and do while loop constructs:

https://github.com/PacktPublishing/Deciphering-Object-Oriented-Programming-with-CPP/blob/main/Chapter01/Chp1-Ex4.cpp

```
#include <iostream>
using namespace std;    // we'll limit the namespace shortly

int main()
{
    int i = 0;
    while (i < 10)
    {
        cout << i << endl;
        i++;
    }
```

```
    i = 0;
    do
    {
        cout << i << endl;
        i++;
    } while (i < 10);
    return 0;
}
```

With the while loop, the condition to enter the loop must evaluate to true prior to each entry of the loop body. However, with the do while loop, the first entry to the loop body is guaranteed – the condition is then evaluated before another iteration through the loop body. In the preceding example, both the while and do while loops are executed 10 times, each printing values 0-9 for variable i.

Next, let's review a typical for loop. The for loop has three parts within the (). First, there is a statement that is executed exactly once and is often used to initialize a loop control variable. Next, separated on both sides by semicolons in the center of the () is an expression. This expression is evaluated each time before entering the body of the loop. The body of the loop is only entered if this expression evaluates to true. Lastly, the third part within the () is a second statement. This statement is executed immediately after executing the body of the loop and is often used to modify a loop control variable. Following this second statement, the center expression is re-evaluated. Here is an example:

https://github.com/PacktPublishing/Deciphering-Object-Oriented-Programming-with-CPP/blob/main/Chapter01/Chp1-Ex5.cpp

```cpp
#include <iostream>
using namespace std;    // we'll limit the namespace shortly

int main()
{
    // though we'll prefer to declare i within the loop
    // construct, let's understand scope in both scenarios
    int i;
    for (i = 0; i < 10; i++)
        cout << i << endl;
    for (int j = 0; j < 10; j++)   // preferred declaration
        cout << j << endl;         // of loop control variable
    return 0;
}
```

Here, we have two for loops. Prior to the first loop, variable i is declared. Variable i is then initialized with a value of 0 in statement 1 between the loop parentheses (). The loop condition is tested, and if true, the loop body is then entered and executed, followed by statement 2 being executed prior to the loop condition being retested. This loop is executed 10 times for i values 0 through 9. The second for loop is similar, with the only difference being variable j is both declared and initialized within statement 1 of the loop construct. Note that variable j only has scope for the for loop itself, whereas variable i has scope of the entire block in which it is declared, from its declaration point forward.

Let's quickly see an example using nested loops. The looping constructs can be of any type, but in the following, we'll review nested for loops:

https://github.com/PacktPublishing/Deciphering-Object-Oriented-Programming-with-CPP/blob/main/Chapter01/Chp1-Ex6.cpp

```cpp
#include <iostream>
using namespace std;     // we'll limit the namespace shortly

int main()
{
    for (int i = 0; i < 10; i++)
    {
        cout << i << endl;
        for (int j = 0; j < 10; j++)
            cout << j << endl;
        cout << "\n";
    }
    return 0;
}
```

Here, the outer loop will execute ten times with i values of 0 through 9. For each value of i, the inner loop will execute ten times, with j values of 0 through 9. Remember, with for loops, the loop control variable is automatically incremented with the i++ or j++ within the loop construct. Had a while loop been used, the programmer would need to remember to increment the loop control variable in the last line of the body of each such loop.

Now that we have reviewed control structures, statements, and looping constructs in C++, we can move onward by briefly recalling C++'s operators.

Reviewing C++ operators

Unary, binary, and ternary operators exist in C++. C++ allows operators to have different meanings based on the context of usage. C++ also allows programmers to redefine the meaning of selected operators when used in the context of at least one user defined type. The operators are listed in the following concise list. We'll see examples of these operators throughout the remainder of this section and throughout the course. Here is a synopsis of the binary, unary, and ternary operators in C++:

Binary operators			
+	addition	+=	plus equal
-	subtraction	-=	minus equal
*	multiplication, pointer indirection	*=	times equal
/	division	/=	div equal
%	Modulus (remainder from an integer divide)		
&	address-of, reference, bitwise AND		
|	bitwise OR		
^	bitwise XOR		
<<	left shift, insertion operator		
>>	right shift, extraction operator		
==	equality (comparison)		
!=	inequality		
<	less than	<=	less than or equals
>	greater than	>=	greater than or equals
&&	logical AND		
||	logical OR		
Unary operators			
+	unary plus		
-	unary minus		
~	one's complement		
!	not		
--	decrement		
++	increment		
Ternary operator			
?:	conditional expression		

Figure 1.1 – C++ operators

In the aforementioned binary operator list, notice how many of the operators have "*shortcut*" versions when paired with the assignment operator =. For example, a = a * b can be written equivalently using a shortcut operator a *= b. Let's take a look at an example that incorporates an assortment of operators, including the usage of a shortcut operator:

```
score += 5;
score++;


if (score == 100)
    cout << "You have a perfect score!" << endl;
else
    cout << "Your score is: " << score << endl;


// equivalent to if - else above, but using ?: operator
(score == 100)? cout << "You have a perfect score" << endl:
                cout << "Your score is: " << score << endl;
```

In the previous code fragment, notice the use of the shortcut operator +=. Here, the statement score += 5; is equivalent to score = score + 5;. Next, the unary increment operator ++ is used to increment score by 1. Then we see the equality operator == to compare score with a value of 100. Finally, we see an example of the ternary operator ?: to replace a simple if - else statement. It is instructive to note that ?: is not preferred by some programmers, yet it is always interesting to review an example of its use.

Now that we have very briefly recapped the operators in C++, let's revisit function basics.

Revisiting function basics

A function identifier must begin with a letter or underscore and may also contain digits. The function's return type, argument list, and return value are optional. The basic form of a C++ function is as follows:

```
<return type> FunctionName (<argumentType argument1, ...>)
{
    expression 1...N;
    <return value/expression;>
}
```

Let's review a simple function:

https://github.com/PacktPublishing/Deciphering-Object-Orient-ed-Programming-with-CPP/blob/main/Chapter01/Chp1-Ex7.cpp

```cpp
#include <iostream>
using namespace std;    // we'll limit the namespace shortly

int Minimum(int a, int b)
{
    if (a < b)
        return a;
    else
        return b;
}

int main()
{
    int x = 0, y = 0;
    cout << "Enter two integers: ";
    cin >> x >> y;
    cout << "The minimum is: " << Minimum(x, y) << endl;
    return 0;
}
```

In the preceding simple example, first, a function Minimum() is defined. It has a return type of int and it takes two integer arguments: formal parameters a and b. In the main() function, Minimum() is called with actual parameters x and y. The call to Minimum() is permitted within the cout statement because Minimum() returns an integer value; this value is passed along to the extraction operator (<<) in conjunction with printing. In fact, the string "The minimum is: " is first placed into the buffer associated with cout, followed by the return value from calling function Minimum(). The output buffer is then flushed by endl (which first places a newline character in the buffer before flushing).

Notice that the function is first defined in the file and then called later in the file in the `main()` function. Strong type checking is performed on the call to the function by comparing the parameter types and their usage in the call to the function's definition. What happens, however, when the function call precedes its definition? Or if the call to the function is in a separate file from its definition?

In these cases, the default action is for the compiler to assume a certain *signature* to the function, such as an integer return type, and that the formal parameters will match the types of arguments in the function call. Often, the default assumptions are incorrect; when the compiler then encounters the function definition later in the file (or when another file is linked in), an error will be raised indicating that the function call and definition do not match.

These issues have historically been solved with a forward declaration of a function included at the top of a file where the function will be called. Forward declarations consist of the function return type, function name and types, and the number of parameters. In C++, a forward declaration has been improved upon and is instead known as a function prototype. Since there are many interesting details surrounding function prototyping, this topic will be covered in reasonable detail in the next chapter.

> **Important note**
>
> The specifier `[[nodiscard]]` can optionally be added to precede the return type of a function. This specifier is used to indicate that the return value from a function must not be ignored – that is, it must be captured in a variable or utilized in an expression. Should the function's return value consequently be ignored, a compiler warning will be issued. Note that the `nodiscard` qualifier can be added to the function prototype and optionally to the definition (or required in a definition if there is no prototype). Ideally, `nodiscard` should appear in both locations.

As we move to the object-oriented sections in this book (*Chapter 5*, *Exploring Classes in Detail*, and beyond), we will learn that there are many more details and quite interesting features relating to functions. Nonetheless, we have sufficiently recalled the basics needed to move forward. Next, let's continue our C++ language review with user defined types.

Reviewing user defined type basics

C++ offers several mechanisms to create user defined types. Bundling together like characteristics into one data type (later, we'll also add relevant behaviors) will form the basis for an object-oriented concept known as encapsulation explored in a later section of this text. For now, let's review the basic mechanisms to bundle together only data in `struct`, `class`, and `typedef` (to a lesser extent). We will also review enumerated types to represent lists of integers more meaningfully.

struct

A C++ structure in its simplest form can be used to collect common data elements together in a single unit. Variables may then be declared of the composite data type. The dot operator is used to access specific members of each structure variable. Here is a structure used in its most simple fashion:

https://github.com/PacktPublishing/Deciphering-Object-Orient-ed-Programming-with-CPP/blob/main/Chapter01/Chp1-Ex8.cpp

```cpp
#include <iostream>
using namespace std;    // we'll limit the namespace shortly

struct student
{
    string name;
    float semesterGrades[5];
    float gpa;
};

int main()
{
    student s1;
    s1.name = "George Katz";
    s1.semesterGrades[0] = 3.0;
    s1.semesterGrades[1] = 4.0;
    s1.gpa = 3.5;
    cout << s1.name << " has GPA: " << s1.gpa << endl;
    return 0;
}
```

Stylistically, type names are typically in lowercase when using structs. In the preceding example, we declare the user defined type student using a struct. Type student has three fields or data members: name, semesterGrades, and gpa. In the main() function, a variable s1 of type student is declared; the dot operator is used to access each of the variable's data members. Since structs are typically not used for OO programming in C++, we're not going to yet introduce significant OO terminology relating to their use. It's worthy to note that in C++, the tag student also becomes the type name (unlike in C, where a variable declaration would need the word struct to precede the type).

typedef and "using" alias declaration

A `typedef` declaration can be used to provide a more mnemonic representation for data types. In C++, the relative need for a `typedef` has been eliminated in usage with a `struct`. Historically, a `typedef` in C allowed the bundling together of the keyword `struct` and the structure tag to create a user defined type. However, in C++, as the structure tag automatically becomes the type, a `typedef` then becomes wholly unnecessary for a `struct`. Typedefs can still be used with standard types for enhanced readability in code, but in this fashion, the `typedef` is not being used to bundle together like data elements, such as with a `struct`. As a related historical note, `#define` (a preprocessor directive and macro replacement) was once used to create more mnemonic types, but `typedef` (and `using`) are certainly preferred. It's worthy to note when viewing older code.

A `using` statement can be used as an alternative to a simple `typedef` to create an alias for a type, known as an **alias-declaration**. The `using` statement can also be used to simplify more complex types (such as providing an alias for complex declarations when using the Standard Template Library or declaring function pointers). The current trend is to favor a `using` alias-declaration to a `typedef`.

Let's take a look at a simple `typedef` compared to a simple `using` alias-declaration:

```
typedef float dollars;
using money = float;
```

In the previous declaration, the new type `dollars` can be used interchangeably with the type `float`. Likewise, the new alias `money` can also be used interchangeably with type `float`. It is not productive to demonstrate the archaic use of `typedef` with a structure, so let's move on to the most used user defined type in C++, the `class`.

class

A `class` in its simplest form can be used nearly like a `struct` to bundle together related data into a single data type. In *Chapter 5*, *Exploring Classes in Detail*, we'll see that a `class` is typically also used to bundle related functions together with the new data type. Grouping together data and behaviors relevant to that data is the basis of encapsulation. For now, let's see a `class` in its simplest form, much like a `struct`:

https://github.com/PacktPublishing/Deciphering-Object-Oriented-Programming-with-CPP/blob/main/Chapter01/Chp1-Ex9.cpp

```cpp
#include <iostream>
using namespace std;   // we'll limit the namespace shortly

class Student
{
```

```
public:
    string name;
    float semesterGrades[5];
    float gpa;
};

int main()
{
    Student s1;
    s1.name = "George Katz";
    s1.semesterGrades[0] = 3.0;
    s1.semesterGrades[1] = 4.0;
    s1.gpa = 3.5;
    cout << s1.name << " has GPA: " << s1.gpa << endl;
    return 0;
}
```

Notice that the previous code is very similar to that used in the `struct` example. The main difference is the keyword `class` instead of the keyword `struct` and the addition of the access label `public:` at the beginning of the class definition (more on that in *Chapter 5, Exploring Classes in Detail*). Stylistically, the capitalization of the first letter in the data type, such as `Student`, is typical for classes. We'll see that classes have a wealth more features and are the building blocks for OO programming. We'll introduce new terminology such as *instance*, to be used rather than *variable*. However, this section is only a review of skills assumed, so we'll need to wait to get to the exciting OO features of the language. Spoiler alert: all the wonderful things classes will be able to do also applies to structs; however, we'll see that structs stylistically won't be used to exemplify OO programming.

enum and strongly-typed enum

Traditional enumerated types may be used to mnemonically represent lists of integers. Unless otherwise initialized, integer values in the enumeration begin with zero and increase by one throughout the list. Two enumerated types may not utilize the same enumerator names.

Strongly-typed enumerated types improve upon traditional enumerated types. Strongly-typed enums default to represent lists of integers, but may be used to represent any integral type, such as `int`, `short int`, `long int`, `char`, or `bool`. The enumerators are not exported to the surrounding scope, so enumerators may be reused between types. Strongly-typed enums allow forward declarations of their type (allowing such uses as these types as arguments to functions before the enumerator declaration).

Let's now see an example of traditional enums and strongly-typed enums:

https://github.com/PacktPublishing/Deciphering-Object-Orient-ed-Programming-with-CPP/blob/main/Chapter01/Chp1-Ex10.cpp

```cpp
#include <iostream>
using namespace std;    // we'll limit the namespace shortly

// traditional enumerated types
enum day {Sunday, Monday, Tuesday, Wednesday, Thursday,
          Friday, Saturday};
enum workDay {Mon = 1, Tues, Wed, Thurs, Fri};

// strongly-typed enumerated types can be a struct or class
enum struct WinterHoliday {Diwali, Hanukkah, ThreeKings,
  WinterSolstice, StLucia, StNicholas, Christmas, Kwanzaa};
enum class Holiday : short int {NewYear = 1, MLK, Memorial,
  Independence, Labor, Thanksgiving};

int main()
{
    day birthday = Monday;
    workDay payday = Fri;
    WinterHoliday myTradition = WinterHoliday::StNicholas;
    Holiday favorite = Holiday::NewYear;

    cout << "Birthday is " << birthday << endl;
    cout << "Payday is " << payday << endl;
    cout << "Traditional Winter holiday is " <<
            static_cast<int> (myTradition) << endl;
    cout << "Favorite holiday is " <<
            static_cast<short int> (favorite) << endl;

    return 0;
}
```

In the previous example, the traditional enumerated type day has values of 0 through 6, starting with Sunday. The traditional enumerated type workDay has values of 1 through 5, starting with Mon. Notice the explicit use of Mon = 1 as the first item in the enumerated type has been used to override the default starting value of 0. Interestingly, we may not repeat enumerators between two enumerated types. For that reason, you will notice that Mon is used as an enumerator in workDay because Monday has already been used in the enumerated type day. Now, when we create variables such as birthday or payday, we can use meaningful enumerated types to initialize or assign values, such as Monday or Fri. As meaningful as the enumerators may be within the code, please note that the values when manipulated or printed will be their corresponding integer values.

Moving forward to consider the strongly-typed enumerated types in the previous example, the enum for WinterHoliday is defined using a struct. Default values for the enumerators are integers, starting with the value of 0 (as with the traditional enums). However, notice that the enum for Holiday specifies the enumerators to be of type short int. Additionally, we choose to start the first item in the enumerated type with value 1, rather than 0. Notice when we print out the strongly-typed enumerators that we must cast the type using a static_cast to the type of the enumerator. This is because the insertion operator knows how to handle selected types, but these types do not include strongly-typed enums; therefore, we cast our enumerated type to a type understood by the insertion operator.

Now that we have revisited simple user defined types in C++, including struct, typedef (and using an alias), class, and enum, we are ready to move onward to reviewing our next language necessity, the namespace.

Recapping namespace basics

The namespace utility was added to C++ to add a scoping level beyond global scope to applications. This feature can be used to allow two or more libraries to be utilized without concern that they may contain duplicative data types, functions, or identifiers. The programmer needs to activate the desired namespace in each relevant portion of their application with the keyword using. Programmers can also create their own namespaces (usually for creating reusable library code) and activate each namespace as applicable. In the previous examples, we've seen the simple use of the std namespace to include cin and cout, which are instances of istream and ostream (whose definitions are found in <iostream>). Let's review how we can create namespaces ourselves:

https://github.com/PacktPublishing/Deciphering-Object-Oriented-Programming-with-CPP/blob/main/Chapter01/Chp1-Ex11.cpp

```cpp
#include <iostream>
// using namespace std; // Do not open entire std namespace
using std::cout;    // Instead, activate individual elements
using std::endl;    // within the namespace as needed
```

```cpp
namespace DataTypes
{
    int total;
    class LinkList
    {   // full class definition …
    };
    class Stack
    {   // full class definition …
    };
};


namespace AbstractDataTypes
{
    class Stack
    {   // full class definition …
    };
    class Queue
    {   // full class description …
    };
};


// Add entries to the AbstractDataTypes namespace
namespace AbstractDataTypes
{
    int total;
    class Tree
    {   // full class definition …
    };
};


int main()
{
    using namespace AbstractDataTypes; //activate namespace
    using DataTypes::LinkList;     // activate only LinkList
    LinkList list1;     // LinkList is found in DataTypes
```

```
    Stack stack1;      // Stack is found in AbstractDataTypes
    total = 5;         // total from active AbstractDataTypes
    DataTypes::total = 85;// specify non-active mbr., total
    cout << "total " << total << "\n";
    cout << "DataTypes::total " << DataTypes::total;
    cout << endl;
    return 0;
}
```

In the second line of the preceding code (which is commented out), we notice the keyword using applied to indicate that we'd like to use or activate the entire std namespace. Preferably, on the following two lines of code, we can instead activate only the elements in the standard namespace that we will be needing, such as std::cout or std::endl. We can utilize using to open existing libraries (or individual elements within those libraries) that may contain useful classes; the keyword using activates the namespace to which a given library may belong. Next in the code, a user specified namespace is created called DataTypes, using the namespace keyword. Within this namespace exists a variable total, and two class definitions: LinkList and Stack. Following this namespace, a second namespace, AbstractDataTypes, is created and includes two class definitions: Stack and Queue. Additionally, the namespace AbstractDataTypes is augmented by a second occurrence of the *namespace* definition in which a variable total and a class definition for Tree are added.

In the main() function, first, the AbstractDataTypes namespace is opened with the using keyword. This activates all names in this namespace. Next, the keyword using is combined with the scope resolution operator (::) to only activate the LinkList class definition from the DataTypes namespace. Had there also been a LinkList class within the AbstractDataType namespace, the initial visible LinkList would now be hidden by the activation of DataTypes::LinkList.

Next, a variable of type LinkList is declared, whose definition comes from the DataTypes namespace. A variable of type Stack is next declared; though both namespaces have a Stack class definition, there is no ambiguity since only one Stack has been activated. Next, we use cin to read into total, which is active from the AbstractDataTypes namespace. Lastly, we use the scope resolution operator to explicitly read into DataTypes::total, a variable that would otherwise be hidden. One caveat to note: should two or more namespaces contain the same identifier, the one last opened will preside, hiding all previous occurrences.

It is considered good practice to activate only the elements of a namespace we wish to utilize. From the aforementioned example, we can see potential ambiguity that can otherwise arise.

Summary

In this chapter, we reviewed core C++ syntax and non-OO language features to refresh your existing skill set. These features include basic language syntax, basic I/O with `<iostream>`, control structures/statements/looping, operator basics, function basics, simple user defined types, and namespaces. Most importantly, you are now ready to move to the next chapter, in which we will expand on some of these ideas with additional language necessities such as `const` qualified variables, understanding and using prototypes (including with default values), and function overloading.

The ideas in the next chapter begin to move us closer to our goal for OO programming, as many of these aggregate skills are used often and matter of factly as we move deeper into the language. It is important to remember that in C++, you can do anything, whether you mean to do so or not. There is great power in the language, and having a solid base for its many nuances and features is crucial. Over the next couple of chapters, the solid groundwork will be laid with an arsenal of non-OO C++ skills, so that we may realistically engage OO programming in C++ with a high level of understanding and success.

Questions

1. Describe a situation in which `flush`, rather than `endl`, may be useful for clearing the contents of the buffer associated with `cout`.

2. The unary operator `++` can be used as a pre- or post-increment operator, such as `i++` or `++i`. Can you describe a situation in which choosing a pre- versus post-increment for `++` would have different consequences in the code?

3. Create a simple program using a `struct` or `class` to make a user defined type for `Book`. Add data members for title, author, and number of pages. Create two variables of type `Book` and use the dot operator `.` to fill in the data members for each such instance. Use `iostreams` to both prompt the user for input values, and to print each `Book` instance when complete. Use only features covered in this chapter.

2
Adding Language Necessities

This chapter will introduce necessary non-OO features of C++ that are critical building blocks for C++'s object-oriented features. The features presented in this chapter represent topics that you will see matter-of-factly used from this point onward in the book. C++ is a language shrouded in areas of gray; from this chapter forward, you will become versed in not only language features, but in language nuances. The goal of this chapter will be to begin enhancing your skills from those of an average C++ programmer to one who is capable of operating among language subtleties successfully while creating maintainable code.

In this chapter, we will cover the following main topics:

- The `const` qualifier
- Function prototyping
- Function overloading

By the end of this chapter, you will understand non-OO features such as the `const` qualifier, function prototyping (including using default values), and function overloading (including how standard type conversion affects overloaded function choices and may create potential ambiguities). Many of these seemingly straightforward topics include an assortment of interesting details and nuances. These skills will be necessary in order to move forward with the next chapters in the book successfully.

Technical requirements

Online code for full program examples can be found in the following GitHub URL: `https://github.com/PacktPublishing/Deciphering-Object-Oriented-Programming-with-CPP/tree/main/Chapter02`. Each full program example can be found in the GitHub under the appropriate chapter heading (subdirectory) in a file that corresponds to the chapter number, followed by a dash, followed by the example number in the chapter at hand. For example, the first full program in *Chapter 2, Adding Language Necessities*, can be found in the subdirectory `Chapter02` in a file named `Chp2-Ex1.cpp` under the aforementioned GitHub directory.

The CiA video for this chapter can be viewed at: `https://bit.ly/3CM65dF`.

Using the const and constexpr qualifiers

In this section, we will add the `const` and `constexpr` qualifiers to variables, and discuss how they can be added to functions in both their input parameters and as return values. These qualifiers will be used quite liberally as we move forward in the C++ language. The use of `const` and `constexpr` can enable values to be initialized, yet never again modified. Functions can advertise that they will not modify their input parameters, or that their return value may only be captured (but not modified) by using `const` or `constexpr`. These qualifiers help make C++ a more secure language. Let's take a look at `const` and `constexpr` in action.

const and constexpr variables

A `const` qualified variable is a variable that must be initialized, and may never be assigned a new value. It is seemingly a paradox to pair the usage of `const` and a variable together – `const` implies not to change, yet the concept of a variable is to inherently hold different values. Nonetheless, it is useful to have a strongly type-checked variable whose one and only value can be determined at run time. The keyword `const` is added to the variable declaration.

Similarly, a variable declared using `constexpr` is a constant qualified variable – one that may be initialized and never assigned a new value. The usage of `constexpr` is becoming preferred whenever its use is possible.

In some situations, the value of a constant is not known at compile time. An example might be if user input or the return value of a function is used to initialize a constant. A `const` variable may be easily initialized at runtime. A `constexpr` can often, but not always be initialized at runtime. We will consider various scenarios in our example.

Let's consider a few examples in the following program. We will break this program into two segments for a more targeted explanation, however, the full program example can be found in its entirety at the following link:

`https://github.com/PacktPublishing/Deciphering-Object-Oriented-Programming-with-CPP/blob/main/Chapter02/Chp2-Ex1.cpp`

```
#include <iostream>
#include <iomanip>
#include <cstring> // though, we'll prefer std::string,
// char [ ] demos the const qualifier easily in cases below
```

```
using std::cout;       // preferable to: using namespace std;
using std::cin;
using std::endl;
using std::setw;


// simple const variable declaration and initialization
// Convention will capitalize those known at compile time
// (those taking the place of a former macro #define)
const int MAX = 50;
// simple constexpr var. declaration and init. (preferred)
constexpr int LARGEST = 50;


constexpr int Minimum(int a, int b)
// function definition w formal parameters
{
    return (a < b)? a : b;    // conditional operator ?:
}
```

In the previous program segment, notice how we declare a variable with the const qualifier preceding the data type. Here, `const int MAX = 50;` simply initializes MAX to 50. MAX may not be modified via assignment later in the code. Out of convention, simple `const` and `constexpr` qualified variables (taking the place of once used `#define` macros) are often capitalized, whereas values that are calculated (or might be calculated) are declared using typical naming conventions. Next, we introduce a constant variable using `constexpr int LARGEST = 50;` to declare a variable that likewise cannot be modified. This option is becoming the preferred usage but is not always possible to utilize.

Next, we have the definition for function `Minimum();` notice the use of the ternary conditional operator `? :` in this function body. Also notice that this function's return value is qualified with `constexpr` (we will examine this shortly). Next, let's examine the body of the `main()` function as we continue with the remainder of this program:

```
int main()
{
    int x = 0, y = 0;
    // Since 'a', 'b' could be calculated at runtime
    // (such as from values read in), we will use lowercase
    constexpr int a = 10, b = 15;// both 'a', 'b' are const
    cout << "Enter two <int> values: ";
```

```
    cin >> x >> y;
    // const variable initialized w return val. of a fn.
    const int min = Minimum(x, y);
    cout << "Minimum is: " << min << endl;
    // constexpr initialized with return value of function
    constexpr int smallest = Minimum(a, b);
    cout << "Smallest of " << a << " " << b << " is: "
         << smallest << endl;


    char bigName[MAX] = {""};   // const used to size array
    char largeName[LARGEST] = {""}; // same for constexpr
    cout << "Enter two names: ";
    cin >> setw(MAX) >> bigName >> setw(LARGEST) >>
          largeName;
    const int namelen = strlen(bigName);
    cout << "Length of name 1: " << namelen << endl;
    cout << "Length of name 2: " << strlen(largeName) <<
          endl;
    return 0;
}
```

In main(), let's consider the sequence of code in which we prompt the user to "Enter two values: " into variables x and y, respectively. Here, we call a function Minimum(x,y) and pass as actual parameters our two values x and y, which were just read in using cin and the extraction operator >>. Notice that alongside the const variable declaration of min, we initialize min with the return value of the function call Minimum(). It is important to note that setting min is bundled as a single declaration and initialization. Had this been broken into two lines of code – a variable declaration followed by an assignment – the compiler would have flagged an error. Variables qualified with const may only be initialized with a value, and never assigned a value after declaration.

Next, we initialize smallest with the return value of function Minimum(a, b);. Notice that the parameters a and b are literal values that can be determined at compile time. Also notice the return value of the Minimum() function has been qualified with constexpr. This qualification is necessary in order for constexpr smallest to be initialized with the function's return value. Note that had we tried to pass x and y to Minimum() to set smallest, we would get an error, as the values of x and y are not literal values.

In the last sequence of code in the previous example, notice that we use MAX (defined in the earlier segment of this full program example) to define a size for the fixed-sized array bigName in the declaration char bigName[MAX];. We similarly use LARGEST to define a size for the fixed-size array `largeName`. Here we see that either the `const` or `constexpr` can be used to size an array in this manner. We then further use `MAX` in `setw(MAX)` and `LARGEST` in `setw(LARGEST)` to ensure that we do not overflow `bigName` or `largeName`, while reading keyboard input with `cin` using the extraction operator `>>`. Finally, we initialize variable `const int namelen` with the return value of function `strlen(bigname)` and print this value out using `cout`. Note that because `strlen()` is not a function whose value is qualified with `constexpr`, we cannot use this return value to initialize a `constexpr`.

The output to accompany the aforementioned full program example is as follows:

```
Enter two <int> values: 39 17
Minimum is: 17
Smallest of 10 15 is: 10
Enter two names: Gabby Dorothy
Length of name 1: 5
Length of name 2: 7
```

Now that we have seen how to `const` and `constexpr` qualify variables, let's consider constant qualification with functions.

const qualification with functions

The keywords `const` and `constexpr` can also be used in conjunction with functions. These qualifiers can be used among parameters to indicate that the parameters themselves will not be modified. This is a useful feature—the caller of the function will understand that the function will not modify input parameters qualified in these manners. However, because non-pointer (and non-reference) variables are passed *by value* to functions as copies of the actual parameters on the stack, `const` or `constexpr` qualifying these inherent copies of parameters does not serve a purpose. Hence, `const` or `constexpr` qualifying parameters that are of standard data types is not necessary.

The same principle applies to return values from functions. A return value from a function can be `const` or `constexpr` qualified; however, unless a pointer (or reference) is returned, the item passed back on the stack as the return value is a copy. For this reason, `const` qualified return values are more meaningful when the return type is a pointer to a constant object (which we will cover in *Chapter 3, Indirect Addressing: Pointers*, and beyond). Note that a `constexpr` qualified return value is required of a function whose return value will be used to initialize a `constexpr` variable, as we have seen in our previous example. As one final use of `const`, we can utilize this keyword when we move onto OO details for a class to specify that a particular member function will not modify any data members of that class. We will look at this scenario in *Chapter 5, Exploring Classes in Detail*.

Now that we understand the use of the `const` and `constexpr` qualifiers for variables and have seen potential uses of `const` and `constexpr` in conjunction with functions, let us move onward to the next language feature in this chapter: function prototypes.

Working with function prototypes

In this section, we will examine the mechanics of function prototyping, such as necessary placement in files and across multiple files for greater program flexibility. We will also add optional names to prototype arguments, as well as understand how and why we may choose to add default values to C++ prototypes. Function prototypes ensure C++ code is strongly type-checked.

Prior to proceeding to function prototypes, let's take a moment to review some necessary programming terms. A **function definition** refers to the body of code comprising a function, whereas a declaration of a function (also known as a **forward declaration**) merely introduces a function name with its return type and argument types. Forward declarations allow the compiler to perform strong type checking between the function call and its definition by instead comparing the call with the forward declaration. Forward declarations are useful because function definitions do not always appear in a file prior to a function call; sometimes, function definitions appear in a separate file from their calls.

Defining function prototypes

A **function prototype** is a forward declaration of a function that describes how a function should be correctly invoked. A prototype ensures strong type checking between a function call and its definition. A simple function prototype consists of the following:

- Function's return type
- Function's name
- Function's type and number of arguments

A function prototype allows a function call to precede the function's definition or allows for calls to functions that exist in separate files. As we learn about more C++ language features, such as exceptions, we will see that additional elements contribute to a function's extended prototype (and extended signature). For now, let's examine a simple example:

https://github.com/PacktPublishing/Deciphering-Object-Oriented-Programming-with-CPP/blob/main/Chapter02/Chp2-Ex2.cpp

```
#include <iostream>
using std::cout;      // preferred to: using namespace std;
using std:: endl;
```

```cpp
[[nodiscard]] int Minimum(int, int);    // fn. prototype

int main()
{
    int x = 5, y = 89;
    // function call with actual parameters
    cout << Minimum(x, y) << endl;
    return 0;
}


[[nodiscard]] int Minimum(int a, int b)  // fn. definition
                                 // with formal parameters
{
    return (a < b)? a : b;
}
```

Notice that we prototype int Minimum(int, int); near the beginning of the aforementioned example. This prototype lets the compiler know that any calls to Minimum() should take two integer arguments and should return an integer value (we'll discuss type conversions later in this section).

Also notice the use of [[nodiscard]] preceding the return type of the function. This indicates that the programmer should store the return value or otherwise utilize the return value (such as in an expression). The compiler will issue a warning if the return value of this function is ignored.

Next, in the main() function, we call the function Minimum(x, y). At this point, the compiler checks that the function call matches the aforementioned prototype with respect to type and number of arguments, and return type. Namely, that the two arguments are integers (or could easily be converted to integers) and that the return type is an integer (or could easily be converted to an integer). The return value will be utilized as a value to print using cout. Lastly, the function Minimum() is defined in the file. Should the function definition not match the prototype, the compiler will raise an error.

The existence of the prototype allows the call of a given function to be fully type-checked prior to the function's definition being seen by the compiler. The current example is of course contrived to demonstrate this point; we could have instead switched the order in which Minimum() and main() appear in the file. However, imagine that the definition of Minimum() was contained in a separate file (the more typical scenario). In this case, the prototype will appear at the top of the file that will call this function (along with header file inclusions) so that the function call can be fully type-checked against the prototype.

In the aforementioned multiple file scenario, the file containing the function definition will be separately compiled. It will then be the linker's job to ensure that when multiple files are linked together, the function definition and all prototypes match so that the linker can resolve any references to such function calls. Should the prototypes not match the function definition, the linker will not be able to link the various sections of code together into one compiled unit.

Let's take a look at this example's output:

```
5
```

Now that we understand function prototype basics, let's see how we can add optional argument names to function prototypes.

Naming arguments in function prototypes

Function prototypes may optionally contain names that may differ from those in either the formal or actual parameter lists. Argument names are ignored by the compiler, yet can often enhance readability. Let's revisit our previous example, adding optional argument names in the function prototype:

```
https://github.com/PacktPublishing/Deciphering-Object-Orient-
ed-Programming-with-CPP/blob/main/Chapter02/Chp2-Ex3.cpp
```

```cpp
#include <iostream>
using std::cout;    // preferred to: using namespace std;
using std::endl;

// function prototype with optional argument names
[[nodiscard]] int Minimum(int arg1, int arg2);

int main()
{
    int x = 5, y = 89;
    cout << Minimum(x, y) << endl;      // function call
    return 0;
}

[[nodiscard]] int Minimum(int a, int b) // fn. definition
{
    return (a < b)? a : b;
}
```

This example is nearly identical to the one preceding it. However, notice that the function prototype contains named arguments `arg1` and `arg2`. These identifiers are immediately ignored by the compiler. As such, these named arguments do not need to match either the formal or actual parameters of the function and are optionally present merely to enhance readability.

The output to accompany this example is the same as the previous example:

```
5
```

Next, let's move forward with our discussion by adding a useful feature to function prototypes: default values.

Adding default values to function prototypes

Default values may be specified in function prototypes. These values will be used in the absence of actual parameters in the function call and will serve as the actual parameters themselves. Default values adhere to the following criteria:

- Default values must be specified from right to left in the function prototype, without omitting any values.

- Actual parameters are substituted from left to right in the function call; hence the right to left order for default value specification in the prototype is significant.

A function prototype may have all, some, or none of its values filled with default values, as long as the default values adhere to the aforementioned specifications.

Let's see an example using default values:

https://github.com/PacktPublishing/Deciphering-Object-Oriented-Programming-with-CPP/blob/main/Chapter02/Chp2-Ex4.cpp

```cpp
#include <iostream>
using std::cout;      // preferred to: using namespaces std;
using std::endl;

// fn. prototype with one default value
[[nodiscard]] int Minimum(int arg1, int arg2 = 100000);

int main()
{
    int x = 5, y = 89;
    cout << Minimum(x) << endl; // function call with only
```

```
                                        // one argument (uses default)
    cout << Minimum(x, y) << endl; // no default vals used
    return 0;
}


[[nodiscard]] int Minimum(int a, int b) // fn. definition
{
    return (a < b)? a : b;
}
```

In this example, notice that a default value is added to the rightmost argument in the function prototype for int Minimum(int arg1, int arg2 = 100000);. This means that when Minimum() is called from main(), it may be called with either one argument, Minimum(x), or with two arguments, Minimum(x, y). When Minimum() is called with a single argument, the single argument is bound to the leftmost argument in the formal parameters of the function, and the default value is bound to the next sequential argument in the formal parameter list. However, when Minimum() is called with two arguments, both of the actual parameters are bound to the formal parameters in the function; the default value is not used.

Here is the output for this example:

```
5
5
```

Now that we have a handle on default values within a function prototype, let's expand on this idea by using different default values with prototypes in various program scopes.

Prototyping with different default values in different scopes

Functions may be prototyped in different scopes with different default values. This allows functions to be built generically and customized through prototypes within multiple applications or for use in multiple sections of code.

Here is an example illustrating multiple prototypes for the same function (in different scopes) using different default values:

```
https://github.com/PacktPublishing/Deciphering-Object-Orient-
ed-Programming-with-CPP/blob/main/Chapter02/Chp2-Ex5.cpp
```

```
#include <iostream>
using std::cout;      // preferred to: using namespace std;
using std::endl;
```

```cpp
// standard function prototype
[[nodiscard]] int Minimum(int, int);

void Function1(int x)
{
    // local prototype with default value
    [[nodiscard]] int Minimum(int arg1, int arg2 = 500);
    cout << Minimum(x) << endl;
}

void Function2(int x)
{
    // local prototype with default value
    [[nodiscard]] int Minimum(int arg1, int arg2 = 90);
    cout << Minimum(x) << endl;
}

[[nodiscard]] int Minimum(int a, int b)  // fn. definition
{
    return (a < b)? a : b;
}

int main()
{
    Function1(30);
    Function2(450);
    return 0;
}
```

In this example, notice that int Minimum(int, int); is prototyped near the top of the file. Then notice that Minimum() is re-prototyped in the more local scope of Function1() as int Minimum(int arg1, int arg2 = 500);, specifying a default value of 500 for its rightmost argument. Likewise, in the scope of Function2(), function Minimum() is re-prototyped as int Minimum(int arg1, int arg2 = 90);, specifying a default value of 90 in the rightmost argument. When Minimum() is called from within Function1() or Function2(), the local prototypes in each of these function scopes, respectively, will be used – each with their own default values.

In this fashion, specific areas of a program may be easily customized with default values that may be meaningful within a specific portion of an application. However, be sure to *only* employ re-prototyping of a function with individualized default values within the scope of a calling function to ensure that this customization can be easily contained within the safety of a very limited scope. Never re-prototype a function in global scope with differing default values – this could lead to unexpected and error-prone results.

The output for the example is as follows:

```
30
90
```

Having now explored function prototypes with respect to default usage in single and multiple files, using default values in prototypes, and re-prototyping functions in different scopes with individual default values, we are now able to move forward with the last major topic in this chapter: function overloading.

Understanding function overloading

C++ allows for two or more functions that share a similar purpose, yet differ in the types or number of arguments they take, to co-exist with the same function name. This is known as **function overloading**. This allows more generic function calls to be made, leaving the compiler to choose the correct version of the function based on the type of the variable (object) using the function. In this section, we will add default values to the basics of function overloading to provide flexibility and customization. We will also learn how standard type conversions may impact function overloading, and potential ambiguities that may arise (as well as how to resolve those types of uncertainties).

Learning the basics of function overloading

When two or more functions with the same name exist, the differentiating factor between these similar functions will be their signature. By varying a function's signature, two or more functions with otherwise identical names may exist in the same namespace. Function overloading depends on the signature of a function as follows:

- The **signature of a function** refers to a function's name, plus its type and number of arguments.

- A function's return type is not included as part of its signature.

- Two or more functions with the same purpose may share the same name, provided that their signatures differ.

A function's signature helps provide an internal, "mangled" name for each function. This encoding scheme guarantees that each function is uniquely represented internally to the compiler.

Let's take a few minutes to understand a slightly larger example that will incorporate function overloading. To simplify the explanation, this example is broken into three segments; nonetheless, the full program can be found in its entirety at the following link:

https://github.com/PacktPublishing/Deciphering-Object-Orient-ed-Programming-with-CPP/blob/main/Chapter02/Chp2-Ex6.cpp

```cpp
#include <iostream>
#include <cmath>
using std::cout;     // preferred to: using namespace std;
using std::endl;


constexpr float PI = 3.14159;


class Circle     // simple user defined type declarations
{
public:
    float radius;
    float area;
};


class Rectangle
{
public:
    float length;
    float width;
    float area;
};


void Display(Circle);       // 'overloaded' fn. prototypes
void Display(Rectangle);    // since they differ in signature
```

At the beginning of this example, notice that we include the math library with #include <cmath>, to provide access to basic math functions, such as pow(). Next, notice the class definitions for Circle and Rectangle, each with relevant data members (radius and area for Circle; length, width, and area for Rectangle). Once these types have been defined, prototypes for two overloaded Display() functions are shown. Since the prototypes for the two display functions utilize user defined types Circle and Rectangle, it is important that Circle and

Rectangle have both previously been defined. Now, let's examine the body of the main() function as we continue with the next segment of this program:

```cpp
int main()
{
    Circle myCircle;
    Rectangle myRect;
    Rectangle mySquare;

    myCircle.radius = 5.0;
    myCircle.area = PI * pow(myCircle.radius, 2.0);

    myRect.length = 2.0;
    myRect.width = 4.0;
    myRect.area = myRect.length * myRect.width;

    mySquare.length = 4.0;
    mySquare.width = 4.0;
    mySquare.area = mySquare.length * mySquare.width;

    Display(myCircle);    // invoke: void display(Circle)
    Display(myRect);      // invoke: void display(Rectangle)
    Display(mySquare);
    return 0;
}
```

Now, in the main() function, we declare a variable of type Circle and two variables of type Rectangle. We then proceed to load the data members for each of these variables in main() using the dot operator (.) with appropriate values. Next in main(), there are three calls to Display(). The first function call, Display(myCircle), will call the version of Display() that takes a Circle as a formal parameter because the actual parameter passed to this function is in fact of user defined type Circle. The next two function calls, Display(myRect) and Display(mySquare), will call the overloaded version of Display() that takes a Rectangle as a formal parameter because the actual parameters passed in each of these two calls are of type Rectangle themselves. Let's complete this program by examining both function definitions for Display():

```cpp
void Display (Circle c)
{
    cout << "Circle with radius " << c.radius;
```

```
      cout << " has an area of " << c.area << endl;
}


void Display (Rectangle r)
{
    cout << "Rectangle with length " << r.length;
    cout << " and width " << r.width;
    cout << " has an area of " << r.area << endl;
}
```

Notice in the final segment of this example that both versions of Display() are defined. One of the functions takes a Circle as the formal parameter, and the overloaded version takes a Rectangle as its formal parameter. Each function body accesses data members specific to each of its formal parameter types, yet the overall functionality of each function is similar in that, in each case, a specific shape (Circle or Rectangle) is displayed.

Let's take a look at the output for this full program example:

```
Circle with radius 5 has an area of 78.5397
Rectangle with length 2 and width 4 has an area of 8
Rectangle with length 4 and width 4 has an area of 16
```

Next, let's add to our discussion of function overloading by understanding how standard type conversion allows for one function to be used by multiple data types. This can allow function overloading to be used more selectively.

Eliminating excessive overloading with standard type conversion

Basic language types can be converted from one type to another automatically by the compiler. This allows the language to supply a smaller set of operators to manipulate standard types than would otherwise be necessary. Standard type conversion can also eliminate the need for function overloading when preserving the exact data type of the function parameters is not crucial. Promotion and demotion between standard types are often handled transparently, without explicit casting, in expressions including assignments and operations.

Here is an example illustrating simple standard type conversions. This example does not include function overloading:

```
https://github.com/PacktPublishing/Deciphering-Object-Orient-
ed-Programming-with-CPP/blob/main/Chapter02/Chp2-Ex7.cpp
```

```
#include <iostream>
```

```
using std::cout;      // preferred to: using namespace std;
using std::endl;

int Maximum(double, double);      // function prototype

int main()
{
    int result = 0;
    int m = 6, n = 10;
    float x = 5.7, y = 9.89;

    result =   Maximum(x, y);
    cout << "Result is: " << result << endl;
    cout << "The maximum is: " << Maximum(m, n) << endl;
    return 0;
}

int Maximum(double a, double b)   // function definition
{
    return (a > b)? a : b;
}
```

In this example, the Maximum() function takes two double precision floating-point numbers as parameters, and the function returns the result as an int. First, notice that int Maximum(double, double); is prototyped near the top of the program and is defined at the bottom of this same file.

Now, in the main() function, notice that we have three int variables defined: result, a, and x. The latter two are initialized with values of 6 and 10, respectively. We also have two floats defined and initialized: float x = 5.7, y = 9.89;. In the first call to function Maximum(), we use x and y as actual parameters. These two floating-point numbers are promoted to double precision floating-point numbers, and the function is called as expected.

This is an example of standard type conversion. Let's notice that the return value of int Maximum(double, double) is an integer – not a double. This means that the value returned from this function (either formal parameter a or b) will be a copy of a or b, first truncated to an integer before being used as a return value. This return value is neatly assigned to result, which has been declared an int in main(). These are all examples of standard type conversion.

Next, Maximum() is called with actual parameters m and n. Similar to the previous function call, the integers m and n are promoted to doubles and the function is called as expected. The return value will also be truncated back to an int, and this value will be passed to cout for printing as an integer.

The output for this example is as follows:

```
Result is: 9
The maximum is: 10
```

Now that we understand how function overloading and standard type conversions work, let's examine a situation where the two combined could create an ambiguous function call.

Ambiguities arising from function overloading and type conversion

When a function is invoked and the formal and actual parameters match exactly in type, no ambiguities arise with respect to which of a selection of overloaded functions should be called – the function with the exact match is the obvious choice. However, when a function is called and the formal and actual parameters differ in type, standard type conversion may be performed on the actual parameters, as necessary. There are situations, however, when the formal and actual parameter types do not match, and overloaded functions exist. In these cases, it may be difficult for the compiler to select which function should be selected as the best match. In these cases, a compiler error is generated indicating that the available choices paired with the function call itself are ambiguous. Explicit type casting or re-prototyping the desired choice in a more local scope can help correct these otherwise ambiguous situations.

Let's review a simple function illustrating the function overloading, standard type conversion, and potential ambiguity:

https://github.com/PacktPublishing/Deciphering-Object-Oriented-Programming-with-CPP/blob/main/Chapter02/Chp2-Ex8.cpp

```
#include <iostream>
using std::cout;      // preferred to: using namespace std;
using std::endl;

int Maximum (int, int);    // overloaded function prototypes
float Maximum (float, float);

int main()
{
```

```
    char a = 'A', b = 'B';
    float x = 5.7, y = 9.89;
    int m = 6, n = 10;
    cout << "The max is: " << Maximum(a, b) << endl;
    cout << "The max is: " << Maximum(x, y) << endl;
    cout << "The max is: " << Maximum(m, n) << endl;


    // The following (ambiguous) line generates a compiler
    // error - there are two equally good fn. candidates
    // cout << "The maximum is: " << Maximum(a, y) << endl;


    // We can force a choice by using an explicit typecast
    cout << "The max is: " <<
            Maximum(static_cast<float>(a), y) << endl;
    return 0;
}


int Maximum (int arg1, int arg2)     // function definition
{
    return (arg1 > arg2)? arg1 : arg2;
}


float Maximum (float arg1, float arg2)  // overloaded fn.
{
    return (arg1 > arg2)? arg1 : arg2;
}
```

In this preceding simple example, two versions of Maximum() are both prototyped and defined. These functions are overloaded; notice that their names are the same, but they differ in the types of arguments that they utilize. Also note that their return types differ; however, since return type is not part of a function's signature, the return types need not match.

Next, in `main()`, two variables each of type `char`, `int`, and `float` are declared and initialized. Next, `Maximum(a, b)` is called and the two `char` actual parameters are converted to integers (using their ASCII equivalents) to match the `Maximum(int, int)` version of this function. This is the match closest to the `char` argument types of a and b: `Maximum(int, int)` versus `Maximum(float, float)`. Then, `Maximum(x, y)` is called with two floats, and this call will exactly match the `Maximum(float, float)` version of this function. Similarly, `Maximum(m, n)` will be called and will perfectly match the `Maximum(int, int)` version of this function.

Now, notice the next function call (which, not coincidentally, is commented out): `Maximum(a, y)`. Here, the first actual parameter perfectly matches the first argument in `Maximum(int, int)`, yet the second actual parameter perfectly matches the second argument in `Maximum(float, float)`. And for the non-matching parameter, a type conversion *could* be applied – but it is not! Instead, this function call is flagged by the compiler as an ambiguous function call since either of the overloaded functions could be an appropriate match.

On the line of code `Maximum((float) a, y)`, notice that the function call to `Maximum((float) a, y)` forces an explicit typecast to the first actual parameter a, resolving any potential ambiguity of which overloaded function to call. With parameter a now cast to be a `float`, this function call easily matches `Maximum(float, float)`, and is no longer considered ambiguous. Type casting can be a tool to disambiguate crazy situations such as these.

Here is the output to accompany our example:

```
The maximum is: 66
The maximum is: 9.89
The maximum is: 10
The maximum is: 65
```

Summary

In this chapter, we learned about additional non-OO C++ features that are essential building blocks needed to base C++'s object-oriented features. These language necessities include using the `const` qualifier, understanding function prototypes, using default values in prototypes, function overloading, how standard type conversion affects overloaded function choices, and how possible ambiguities may arise (and be resolved).

Very importantly, you are now ready to move forward to the next chapter in which we will explore indirect addressing using pointers in reasonable detail. The matter-of-fact skills that you have accumulated in this chapter will help you more easily navigate each progressively more detailed chapter to ensure you are ready to easily tackle the OO concepts starting in *Chapter 5, Exploring Classes in Detail*.

Remember, C++ is a language filled with more gray areas than most other languages. The subtle nuances you are accumulating with your skill set will enhance your value as a C++ developer – one who can not only navigate and understand existing nuanced code but one who can create easily maintainable code.

Questions

1. What is the signature of a function and how is a function's signature related to name mangling in C++? How do you think this facilitates how overloaded functions are handled internally by the compiler?

2. Write a small C++ program to prompt a user to enter information regarding a Student, and print out the data. Use the following steps to write your code:

   a. Create a data type for Student using a class or struct. Student information should minimally include firstName, lastName, gpa, and the currentCourse in which the Student is registered. This information may be stored in a simple class. You may utilize either char arrays to represent the string fields since we have not yet covered pointers, or you may (preferably) utilize the string type. Also, you may read in this information in the main() function rather than creating a separate function to read in the data (since the latter will require knowledge of pointers or references). Please do not use global (that is, external variables).

   b. Create a function to print out all the data for the Student. Remember to prototype this function. Use a default value of 4.0 for gpa in the prototype of this function. Call this function two ways: once passing in each argument explicitly, and once using the default gpa.

   c. Now, overload the print function with one that either prints out selected data (for example, lastName and gpa) or with a version of this function that takes a Student as an argument (but not a pointer or reference to a Student – we'll do that later). Remember to prototype this function.

   d. Use iostreams for I/O.

3

Indirect Addressing – Pointers

This chapter will provide a thorough understanding of how to utilize pointers in C++. Though it is assumed that you have some prior experience with indirect addressing, we will start at the beginning. Pointers are a ground-level and pervasive feature of the language – one you must thoroughly understand and be able to utilize with ease. Many other languages use indirect addressing through references alone; however, in C++ you must roll up your sleeves and understand how to use and return heap memory correctly and effectively with pointers. You will see pointers heavily used throughout code from other programmers; there is no sensible way to ignore their use. Misusing pointers can create the most difficult errors to find in a program. A thorough understanding of indirect addressing using pointers is a necessity in C++ to create successful and maintainable code.

In this chapter, you will additionally preview the concept of a smart pointer, which can help alleviate the difficulty and potential pitfalls that may easily arise with native pointers. Nonetheless, you will need to have a facility with all types of pointers in order to successfully use existing class libraries or to integrate with or maintain existing code.

The goal of this chapter will be to build or enhance your understanding of indirect addressing using pointers so that you can easily understand and modify others' code, as well as write original, sophisticated, error-free C++ code yourself.

In this chapter, we will cover the following main topics:

- Pointer basics, including access, and memory allocation and release – for standard and user defined types
- Dynamically allocating arrays of 1, 2, and N dimensions, and managing their memory release
- Pointers as arguments to functions and as return values from functions
- Adding the const qualifier to pointer variables
- Using void pointers – pointers to objects of unspecified types
- Looking ahead to smart pointers to alleviate typical pointer usage errors

By the end of this chapter, you will understand how to allocate memory from the heap using `new()` for simple and complex data types, as well as mark the memory for return to the heap management facility using `delete()`. You will be able to dynamically allocate arrays of any data type and of any number of dimensions, as well as understand basic memory management for releasing memory when it is no longer needed in your applications, to avoid memory leakage. You will be able to pass pointers as arguments to functions with any level of indirection – that is, pointers to data, pointers to pointers to data, and so on. You will understand how and why to combine the const qualification with pointers – to the data, to the pointer itself, or to both. You will additionally understand how to declare and utilize generic pointers with no type – void pointers – and understand the situations in which they may prove useful. Lastly, you will preview the concept of a smart pointer to alleviate potential pointer conundrums and usage errors. These skills will be necessary in order to move forward with the next chapters in the book successfully.

Technical requirements

Online code for full program examples can be found in the following GitHub URL: `https://github.com/PacktPublishing/Deciphering-Object-Oriented-Programming-with-CPP/tree/main/Chapter03`. Each full program example can be found in the GitHub under the appropriate chapter heading (subdirectory) in a file that corresponds to the chapter number, followed by a dash, followed by the example number in the chapter at hand. For example, the first full program in this chapter can be found in the subdirectory `Chapter03` in a file named `Chp3-Ex1.cpp` under the aforementioned GitHub directory.

The CiA video for this chapter can be viewed at: `https://bit.ly/3AtBPlV`.

Understanding pointer basics and memory allocation

In this section, we will review pointer basics as well as introduce operators applicable to pointers, such as the address-of operator, the dereference operator, and operators `new()` and `delete()`. We will employ the address-of operator `&` to calculate the address of an existing variable, and conversely, we will apply the dereference operator `*` to a pointer variable to go to the address contained within the variable. We will see examples of memory allocation on the heap, as well as how to mark that same memory for potential reuse by returning it to the free list when we are done with it.

Using pointer variables allows our applications to have greater flexibility. At runtime, we can determine the quantity of a certain data type we may need (such as in a dynamically allocated array), organize data in data structures that facilitate sorting (such as in a linked list), or gain speed by passing an address of a large piece of data to a function (rather than passing a copy of the entire piece of data itself). Pointers have many uses, and we will see many examples throughout this chapter and throughout the course. Let's start at the beginning with pointer basics.

Revisiting pointer basics

First and foremost, let us review the meaning of a pointer variable. A pointer variable is one that may contain an address, and memory at that address may contain relevant data. It is typical to say that the pointer variable *points* to an address containing the relevant data. The value of the pointer variable itself is an address, not the data we are after. When we then go to that address, we find the data of interest. This is known as **indirect addressing**. To summarize, the content of a pointer variable is an address; if you then go to that address, you find the data. This is for a single level of indirection.

A pointer variable may point to the existing memory of a non-pointer variable, or it may point to memory that is dynamically allocated on the heap. The latter case is the most usual situation. Unless a pointer variable is properly initialized or assigned a value, the content of the pointer variable is meaningless and does not represent a usable address. A large mistake can be assuming that a pointer variable has been properly initialized when it may not have been. Let us look at some basic operators that are useful with pointers. We will start with the address-of & and the dereference operator *.

Using the address-of and dereference operators

The address-of operator & can be applied to a variable to determine its location in memory. The dereference operator * can be applied to a pointer variable to obtain the value of the data at the valid address contained within the pointer variable.

Let's see a simple example:

```
int x = 10;
int *pointerToX = nullptr; // pointer variable which may
                           // someday point to an integer
pointerToX = &x;   // assign memory loc. of x to pointerToX
cout << "x: " << x << " and *pointerToX: " << *pointerToX;
```

Notice in the previous segment of code that we first declare and initialize variable x to 10. Next, we declare int *pointerToX = nullptr; to state that variable pointerToX may someday point to an integer, yet it is initialized with a nullptr for safety. Had we not initialized this variable with a nullptr, it would have been uninitialized and, therefore, would not contain a valid memory address.

Moving forward in the code to the line pointerToX = &x;, we assign the memory location of x using the address-of operator (&) as the value of pointerToX, which is waiting to be filled with a valid address of some integer. On the last line of this code fragment, we print out both x and *pointerToX. Here, we are using the dereference operator * with the variable pointerToX. The dereference operator tells us to go to the address contained in the variable pointerToX. At that address, we find the data value of integer 10.

Here is the output this fragment would generate as a full program:

```
X: 10 and *pointerToX: 10
```

> **Important note**
>
> For efficiency, C++ does not neatly initialize all memory with zeros when an application starts, nor does C++ ensure that memory is conveniently empty, without values, when paired with a variable. The memory simply has in it what was previously stored there; C++ memory is not considered *clean*. Because memory is not given to a programmer *clean* in C++, the contents of a newly declared pointer variable, unless properly initialized or assigned a value, should not be construed to contain a valid address.

In the preceding example, we used the address-of operator & to calculate the address of an existing integer in memory, and we set our pointer variable to point to that memory. Instead, let us introduce operators new() and delete() to allow us to utilize dynamically allocated heap memory for use with pointer variables.

Using operators new() and delete()

Operator new() can be utilized to obtain dynamically allocated memory from the heap. A pointer variable may choose to point to memory that is dynamically allocated at runtime, rather than to point to another variable's existing memory. This gives us flexibility as to when we want to allocate the memory, and how many pieces of such memory we may choose to have. Operator delete() can then be applied to a pointer variable to mark memory we no longer require, returning the memory to the heap management facility for later reuse in the application. It is important to understand that once we delete() a pointer variable, we should no longer use the address contained within that variable as a valid address.

Let's take a look at simple memory allocation and release using a basic data type:

```
int *y = nullptr; // ptr y may someday point to an int
y = new int;    // y pts to uninit. memory allocated on heap
*y = 17;    // dereference y to load the newly allocated
         // memory with a value of 17
cout << "*y is: " << *y << endl;
delete y;  // relinquish the allocated memory

// alternative ptr declaration, mem alloc., initialization
int *z = new int(22);
cout <<  "*z is: " << *z << endl;
delete z;  // relinquish heap memory
```

In the previous program segment, we first declare pointer variable y with `int *y = nullptr;`. Here, y may someday contain the address of an integer, yet it is meanwhile safely initialized with a `nullptr`. On the next line, we allocate memory from the heap large enough to accommodate an integer with `y = new int;`, storing that address in pointer variable y. Next, with `*y = 17;`, we dereference y and store the value of `17` in the memory location pointed to by y. After printing out the value of `*y`, we then decide that we are done with the memory y points to and return it to the heap management facility by using operator `delete()`. It is important to note that variable y still contains the memory address it obtained with its call to `new()`; however, y should no longer use this relinquished memory.

Towards the end of the previous program segment, we alternatively declare pointer variable z, allocate heap memory for it to point to, and initialize that memory with `int *z = new int(22);`. Notice that we likewise deallocate the heap memory using `delete z;`.

> **Important note**
> It is the programmer's responsibility to remember that once memory has been deallocated, you should never again dereference that pointer variable; please understand that that address may have been reissued to another variable through another call to `new()` elsewhere in the program. A safeguard would be to reset a pointer to `nullptr` once its memory has been deallocated with `delete()`.

Now that we understand pointer basics with simple data types, let us move onward by allocating more complex data types, as well as understanding the notation necessary to utilize and access members of user defined data types.

Creating and using pointers to user defined types

Next, let us examine how to declare pointers to user defined types, and how to allocate their associated memory on the heap. To dynamically allocate a user defined type, the pointer will first be declared of that type. The pointer then must either be initialized or assigned a valid memory address – the memory can either be that of an existing variable or newly allocated heap memory. Once the address for the appropriate memory has been placed within the pointer variable, the `->` operator may be utilized to access struct or class members. Alternatively, the `(*ptr).member` notation may be used to access struct or class members.

Let's see a basic example:

https://github.com/PacktPublishing/Deciphering-Object-Oriented-Programming-with-CPP/blob/main/Chapter03/Chp3-Ex1.cpp

```
#include <iostream>
using std::cout;
```

```cpp
using std::endl;

struct collection
{
    int x;
    float y;
};

int main()
{
    collection *item = nullptr;   // pointer declaration
    item = new collection;    // memory allocation
    item->x = 9;           // use -> to access data member x
    (*item).y = 120.77; // alt. notation to access member y
    cout << (*item).x << " " << item->y << endl;
    delete item;             // relinquish memory
    return 0;
}
```

First, in the aforementioned program, we have declared a user defined type of collection, with data members x and y. Next, we declare item as a pointer to that type with collection *item = nullptr; while initializing the pointer with a nullptr for safety. Then, we allocate heap memory for item to point to, using operator new(). Now, we assign values to the x and y members of item, respectively, using either the -> operator or the (*). member access notation. In either case, the notation means to first dereference the pointer and then choose the appropriate data member. It's pretty straightforward with the (*). notation – the parentheses show us that the pointer dereference happens first, and then the choice of the member happens next with the . (member selection) operator. The -> shorthand notation indicates pointer dereference followed by member selection. After we use cout with the insertion operator << to print the appropriate values, we decide that we no longer need the memory associated with item and issue a delete item; to mark this segment of heap memory for return to the free list.

Let's take a look at this example's output:

```
9 120.77
```

Let us also take a look at the memory layout for this example. The memory address (9000) used is arbitrary – just an example address that may be generated by new().

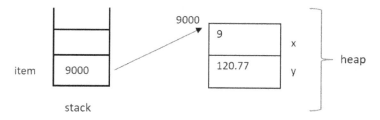

Figure 3.1 – Memory model for Chp3-Ex1.cpp

Now that we know how to allocate and release memory for user defined types, let's move forward and dynamically allocate arrays of any data type.

Allocating and deallocating arrays at runtime

Arrays may be dynamically allocated so that their size may be determined at runtime. Dynamically allocated arrays may be of any type, including user defined types. Determining the size of an array at runtime can be a space-saving advantage and gives us programming flexibility. Rather than allocating a fixed-sized array of the largest possible quantity needed (potentially wasting space), you can instead allocate the necessary size determined by various factors at runtime. You have the additional flexibility to delete and reallocate an array should the need arise to change an array's size. Arrays of any number of dimensions can be dynamically allocated.

In this section, we will examine how to dynamically allocate arrays of both basic and user defined data types, and of single and multiple dimensions. Let's get started.

Dynamically allocating single dimension arrays

Single-dimension arrays may be dynamically allocated so that their size may be determined at runtime. We will use a pointer to represent each array and will allocate the required memory with operator new(). Once the array is allocated, standard array notation can be used to access each array element.

Let's take a look at a simple example. We'll break it into two segments, however, the full program example can be found using the following link:

https://github.com/PacktPublishing/Deciphering-Object-Oriented-Programming-with-CPP/blob/main/Chapter03/Chp3-Ex2.cpp

```cpp
#include <iostream>
using std::cout;
using std::cin;
using std:::endl;
using std::flush;

struct collection
{
    int x;
    float y;
};

int main()
{
    int numElements = 0;
    int *intArray = nullptr;    // pointer declarations to
    collection *collectionArray = nullptr; // future arrays
    cout << "How many elements would you like? " << flush;
    cin >> numElements;

    intArray = new int[numElements]; // alloc. array bodies
    collectionArray = new collection[numElements];
    // continued …
```

In the first part of this program, we first declare a user defined type, collection, using a struct. Next, we declare an integer variable to hold the number of elements we would like to prompt the user to enter to select as the size for our two arrays. We also declare a pointer to an integer with int *intArray;, and a pointer to a collection using collection *collectionArray;. These declarations state that these pointers may one day each, respectively, point to one or more integers, or one or more objects of type collection. These variables, once allocated, will comprise our two arrays.

After prompting the user to enter the number of elements desired using `cin` and the extraction operator `>>`, we dynamically allocate both an array of integers of that size and an array of `collection` of that size. We use operator `new ()` in both cases: `intArray = new int [numElements];` and `collectionArray = new collection[numElements];`. The bracketed quantity of `numElements` indicates that the respective chunks of memory requested for each data type will be large enough to accommodate that many sequential elements of the relevant data type. That is, `intArray` will have memory allocated to accommodate `numElements` multiplied by the size needed for an integer. Note that an object's data type is known because the data type of what will be pointed to is included in the pointer declaration itself. The appropriate amount of memory for `collectionArray` will be similarly provided for with its respective call to operator `new ()`.

Let's continue by examining the remaining code in this example program:

```
// load each array with values
for (int i = 0; i < numElements; i++)
{
    intArray[i] = i; // load each array w values using
    collectionArray[i].x = i;   // array notation []
    collectionArray[i].y = i + .5;

    // alternatively use ptr notation to print values
    cout << * (intArray + i) << " ";
    cout << (* (collectionArray + i)).y << endl;
}
delete [] intArray;        // mark memory for deletion
delete [] collectionArray;
return 0;
}
```

Next, as we continue this example with the `for` loop, notice that we are using a typical array notation of `[]` to access each element of the two arrays, even though the arrays have been dynamically allocated. Because `collectionArray` is a dynamically allocated array of user defined types, we must also use `.` notation to access individual data members within each array element. Though using standard array notation makes accessing dynamically arrays quite simple, you may alternatively use pointer notation to access the memory.

Within the loop, notice that we incrementally print both the elements of `intArray` and the `y` member of `collectionArray` using pointer notation. In the expression `* (intArray +i)`, the identifier `intArray` represents the starting address of the array. By adding `i` offsets to this address, you are now at the address of the ith element in this array. By dereferencing this composite address with `*`, you will now go to the proper address to retrieve the relevant integer data, which is then printed using cout and the insertion operator <<. Likewise, with (*(collectionArray + i)).y, we first add i to the starting address of collectionArray, then using (), we dereference that address with *. Since this is a user defined type, we must then use . to select the appropriate data member y.

Lastly, in this example, we demonstrate how to deallocate memory that we no longer need using delete(). Though a simple statement of delete intArray; would suffice for the dynamically allocated array of standard types, we instead choose delete [] intArray; to be consistent with the required manner for deletion for dynamically allocated arrays of user defined types. That is, the more complex statement of delete [] collectionArray; is necessary for the proper deletion of an array of user defined types. In all cases, the memory associated with each dynamically allocated array will be returned to the free-list, and can then be reused when heap memory is again allocated with subsequent calls to operator new(). However, as we will later see, the [] used with delete() will allow for a special clean-up function to be applied to each array element of a user defined type before the memory is relinquished. Additionally, consistency is appreciated: if you allocate with new(), relinquish the memory with delete(); if you allocate with new [], then relinquish with delete []. This consistent pairing will also keep your program working as intended should any of these aforementioned operators be overloaded (that is, redefined) at a future date by the programmer.

It is crucial to remember not to dereference a pointer variable once its memory has been marked for deletion. Though that address will remain in the pointer variable until you assign the pointer a new address (or null pointer), once memory is marked for deletion, the memory in question might have been already reused by a subsequent call to new() elsewhere in the program. This is one of many ways you must be diligent when using pointers in C++.

The output to accompany the full program example is as follows:

```
How many elements would you like? 3
0  0.5
1  1.5
2  2.5
```

Let's additionally take a look at the memory layout for this example. The memory addresses (8500 and 9500) used are arbitrary – they are example addresses on the heap that may be generated by new().

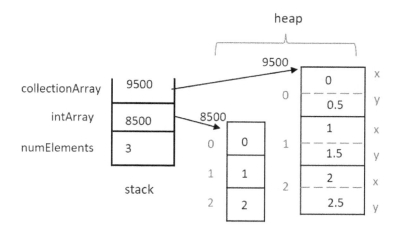

Figure 3.2 – Memory model for Chp3-Ex2.cpp

Next, let's move forward with our discussion on dynamically allocated arrays by allocating arrays of multiple dimensions.

Dynamically allocating 2-D arrays – an array of pointers

Two or more dimensioned arrays may also be dynamically allocated. For a 2-D array, the column dimension may be dynamically allocated and the row dimension may remain fixed, or both dimensions may be dynamically allocated. Allocating one or more dimensions dynamically allows the programmer to account for run time decisions regarding the array size.

Let's first consider the case where we have a fixed number of rows and a variable amount of entries in each of those rows (which would be the column dimension). For simplicity, we will assume that the number of entries in each row is the same from row to row, but it need not be. We can model a 2-D array with a fixed number of rows and a run-time-determined amount of entries in each of those rows (the column dimension) using an array of pointers.

Let's consider an example to illustrate a 2-D array where the column dimension is dynamically allocated:

https://github.com/PacktPublishing/Deciphering-Object-Oriented-Programming-with-CPP/blob/main/Chapter03/Chp3-Ex3.cpp

```
#include <iostream>
using std::cout;
using std::cin;
using std::endl;
using std::flush;
```

```cpp
constexpr int NUMROWS = 5; // convention to use uppercase
                   // since value is known at compile time
int main()
{
    float *TwoDimArray[NUMROWS] = { }; // init. to nullptrs
    int numColumns = 0;

    cout << "Enter number of columns: ";
    cin >> numColumns;

    for (int i = 0; i < NUMROWS; i++)
    {
        // allocate column quantity for each row
        TwoDimArray[i] = new float [numColumns];
        // load each column entry with data
        for (int j = 0; j < numColumns; j++)
        {
            TwoDimArray[i][j] = i + j + .05;
            cout << TwoDimArray[i][j] << " ";
        }
        cout << endl;  // print newline between rows
    }
    for (int i = 0; i < NUMROWS; i++)
        delete [] TwoDimArray[i];  // del col. for each row
    return 0;
}
```

In this example, notice that we initially declare an array of pointers to floats using `float *TwoDimArray[NUMROWS];`. For safety, we initialize each of these pointers to `nullptr`. Sometimes, it is helpful to read pointer declarations from right to left; that is, we have an array NUMROWS in size that contains pointers to floating-point numbers. More specifically, we have a fixed-sized array of pointers where each pointer entry can point to one or more contiguous floating-point numbers. The number of entries pointed to in each row comprises the column dimension.

Next, we prompt the user for the number of column entries. Here, we are assuming that each row will have the same number of entries in it (to make the column dimension); however, it is possible that each row could have a different total number of entries. By assuming each row will have a uniform number of entries, we have a straightforward loop using `i` to allocate the column quantity for each row using `TwoDimArray[i] = new float [numColumns];`.

In the nested loop that uses j as an index, we simply load values for each column entry of the row specified by i in the outer loop. The arbitrary assignment of TwoDimArray[i][j] = i + j + .05; loads an interesting value into each element. In the nested loop indexed on j, we also print out each column entry for row i.

Lastly, the program illustrates how to deallocate the dynamically allocated memory. Since the memory was allocated in a loop over a fixed number of rows – one memory allocation to gather memory to comprise each row's column entries – the deallocation will work similarly. For each of the rows, we utilize the statement: delete [] TwoDimArray[i];.

The output for the example is as follows:

```
Enter number of columns: 3
0.05 1.05 2.05
1.05 2.05 3.05
2.05 3.05 4.05
3.05 4.05 5.05
4.05 5.05 6.05
```

Next, let's take a look at the memory layout for this example. As in previous memory diagrams, the memory addresses used are arbitrary – they are example addresses on the heap as may be generated by new().

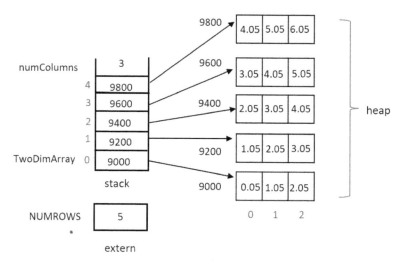

Figure 3.3 – Memory model for Chp3-Ex3.cpp

Now that we have seen how to utilize an array of pointers to model a 2-D array, let's move onward to see how we can model a 2-D array using a pointer to a pointer so that we may choose both dimensions at runtime.

Dynamically allocating 2-D arrays – pointers to pointers

Dynamically allocating both the row and column dimensions for an array can add necessary runtime flexibility to a program. To achieve this ultimate flexibility, a 2-D array can be modeled using a pointer to a pointer of the desired data type. Initially, the dimension representing the number of rows will be allocated. Next, for each row, the number of elements in each row will be allocated. As with the last example using an array of pointers, the number of elements in each row (the column entries) need not be uniform in size across rows. However, to accurately model the concept of a 2-D array, it is assumed that the column size will be allocated uniformly from row to row.

Let's consider an example to illustrate a 2-D array where both the row and column dimensions are dynamically allocated:

https://github.com/PacktPublishing/Deciphering-Object-Orient-ed-Programming-with-CPP/blob/main/Chapter03/Chp3-Ex4.cpp

```cpp
#include <iostream>
using std::cout;
using std::cin;
using std::endl;
using std::flush;

int main()
{
    int numRows = 0, numColumns = 0;
    float **TwoDimArray = nullptr;   // pointer to a pointer

    cout << "Enter number of rows: " << flush;
    cin >> numRows;
    TwoDimArray = new float * [numRows]; // alloc. row ptrs

    cout << "Enter number of Columns: ";
    cin >> numColumns;
```

```
    for (int i = 0; i < numRows; i++)
    {
        // allocate column quantity for each row
        TwoDimArray[i] = new float [numColumns];
        // load each column entry with data
        for (int j = 0; j < numColumns; j++)
        {
            TwoDimArray[i][j] = i + j + .05;
            cout << TwoDimArray[i][j] << " ";
        }
        cout << end;   // print newline between rows
    }
    for (i = 0; i < numRows; i++)
        delete [] TwoDimArray[i];   // del col. for each row
    delete [] TwoDimArray;   // delete allocated rows
    return 0;
}
```

In this example, notice that we initially declare a pointer to a pointer of type float using: float **TwoDimArray;. Reading this declaration from right to left, we see that TwoDimArray is a pointer to a pointer to float. More specifically, we understand that TwoDimArray will contain the address of one or more contiguous pointers, each of which may point to one or more contiguous floating-point numbers.

Now, we prompt the user for the number of row entries. We follow this input with the allocation to a set of float pointers, TwoDimArray = new float * [numRows];. This allocation creates a numRows quantity of contiguous float pointers.

Just as in the previous example, we prompt the user for how many columns in each row we would like to have. Just as before, in the outer loop indexed on i, we allocate the column entries for each row. In the nested loop indexed on j, we again assign values to our array entries and print them just as before.

Lastly, the program continues with the memory deallocation. Just as before, the column entries for each row are deallocated within a loop. Additionally, however, we need to deallocate the dynamically allocated number of row entries. We do this with delete [] TwoDimArray;.

The output for this program is slightly more flexible, as we can enter at runtime the number of both the desired rows and columns:

```
Enter number of rows: 3
Enter number of columns: 4
```

```
0.05 1.05 2.05 3.05
1.05 2.05 3.05 4.05
2.05 3.05 4.05 5.05
```

Let's again take a look at the memory model for this program. As a reminder, just as in previous memory diagrams, the memory addresses used are arbitrary – they are example addresses on the heap as may be generated by new().

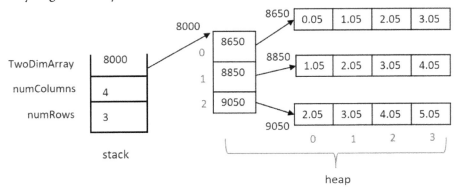

Figure 3.4 – Memory model for Chp3-Ex4.cpp

Now that we have seen how to utilize pointers to pointers to model a 2-D array, let's move onward to see how we may model arrays of any number of dimensions using pointers to pointers to pointers, and so on. In C++, you can model any dimensionality of a dynamically allocated array, so long as you can imagine it!

Dynamically allocating N-D arrays – pointers to pointers to pointers

In C++, you can model any dimensionality of a dynamically allocated array. You need only to be able to imagine it, declare the appropriate levels of pointers, and make the required levels of memory allocation (and eventual deallocation).

Let's take a look at the pattern you will need to follow:

https://github.com/PacktPublishing/Deciphering-Object-Orient-ed-Programming-with-CPP/blob/main/Chapter03/Chp3-Ex5.cpp

```cpp
#include <iostream>
using std::cout;
using std::cin;
using std::endl;
```

```cpp
using std::flush;

int main()
{
    int dim1 = 0, dim2 = 0, dim3 = 0;
    int ***ThreeDimArray = nullptr; // 3D dyn. alloc. array

    cout << "Enter dim 1, dim 2, dim 3: ";
    cin >> dim1 >> dim2 >> dim3;
    ThreeDimArray = new int ** [dim1]; // allocate dim 1

    for (int i = 0; i < dim1; i++)
    {
        ThreeDimArray[i] = new int * [dim2]; // alloc dim 2
        for (int j = 0; j < dim2; j++)
        {
            // allocate dim 3
            ThreeDimArray[i][j] = new int [dim3];
            for (int k = 0; k < dim3; k++)
            {
                ThreeDimArray[i][j][k] = i + j + k;
                cout << ThreeDimArray[i][j][k] << " ";
            }
            cout << endl;  // print '\n' between dimensions
        }
        cout << end;  // print '\n' between dimensions
    }
    for (int i = 0; i < dim1; i++)
    {
        for (int j = 0; j < dim2; j++)
            delete [] ThreeDimArray[i][j]; // release dim 3
        delete [] ThreeDimArray[i];  // release dim 2
    }
    delete [] ThreeDimArray;    // release dim 1
    return 0;
}
```

In this example, notice that we use three levels of indirection to specify the variable to represent the 3-D array int ***ThreeDimArray;. We subsequently allocate the required memory for each level of indirection. The first allocation is ThreeDimArray = new int ** [dim1];, which allocates dimension 1's set of pointers to pointers. Next, in a loop iterating over i, and for each element in dimension 1, we allocate ThreeDimArray[i] = new int * [dim2]; to allocate the pointers to integers for the second dimension of the array. And in a nested loop iterating over j, and for each element in dimension 2, we allocate ThreeDimArray[i][j] = new int [dim3]; to allocate the integers themselves in a quantity specified by dim3.

As in the last two examples, we initialize the array elements in the inner loops and print their values. At this point, you will undoubtedly notice the similarities between this program and its predecessor. A pattern for the allocation is emerging.

Lastly, we will deallocate the three levels of memory in a manner similar – yet in reverse – to the levels of allocation. We use a nested loop iterating over j to release the memory of the innermost level, followed by the memory release in the outer loop that iterates over i. Finally, we relinquish the memory for the initial dimension with a simple call to delete [] ThreeDimArray;.

The output for this example is as follows:

```
Enter dim1, dim2, dim3: 2 4 3
0 1 2
1 2 3
2 3 4
3 4 5

1 2 3
2 3 4
3 4 5
4 5 6
```

Now that we have seen how to model a 3-D array using pointers to pointers to pointers, a pattern has emerged to show us how to declare the required level and number of pointers to model an N-D array. We can also see the pattern for the necessary allocations. Multidimensional arrays can become quite large, especially if you were forced to model them with the largest potentially necessary fixed-sized array. The beauty of modeling with pointers to pointers (to pointers, and so on) for each level of a necessary multi-dimensional array is that you can allocate exactly a size that may be determined at runtime. To make usage easy, array notation using [] can be used as an alternative to pointer notation to access the elements in the dynamically allocated array. C++ has a lot of flexibility stemming from pointers. Dynamically allocated arrays demonstrate one such flexibility.

Let's now move forward with our understanding of pointers and consider their usage in functions.

Using pointers with functions

Functions in C++ will undoubtedly take arguments. We have seen many examples in the previous chapters illustrating function prototypes and function definitions. Now, let's augment our understanding of functions by passing pointers as arguments to functions, and using pointers as return values from a function.

Passing pointers as arguments to functions

Arguments passed from actual to formal parameters in a function call are by default copied on the stack. In order to modify the contents of a variable as an argument to a function, a pointer to that argument must instead be used as a function parameter.

Any time an actual parameter is passed to a function in C++, a copy of something is made and passed on the stack to that function. For example, if an integer is passed as an actual parameter to a function, a copy of that integer is made and then passed on the stack to the function to be received as the formal parameter. Changing the formal parameter in the scope of the function would only change the copy of the data that was passed into the function.

Should we instead require the ability to modify a function's parameters, it is then necessary that we pass a pointer to the desired data as a parameter to the function. In C++, passing a pointer as an actual parameter copies this address on the stack, and the copy of the address is received as the formal parameter in the function. However, using the copy of the address, we can still go to that address (by dereferencing that pointer) to access the desired data and make changes to the desired data.

To reiterate, something is always copied on the stack when you pass a parameter in C++. If you pass a non-pointer variable, you get a copy of that data passed on the stack to the function. Changes made to that data in the scope of that function are local changes only and do not persist when the function returns. The local copy is simply popped off the stack at the conclusion of the function. However, if you pass a pointer to a function, though the address stored in the pointer variable is still copied on the stack and passed to the function, you can still dereference the copy of the pointer to access the real data at the desired address.

You always need to be one step back from that which you want to modify. If you want to change a standard data type, pass a pointer to that type. If you want to change the value of the pointer itself (the address), you must pass a pointer to that pointer as a parameter to the function. Remember, a copy of something is passed to the function on the stack. You cannot change that copy beyond the scope of the function. Pass the address of that which you want to change – you are still passing a copy of that address, but using it will get you to the real data.

Let's take a few minutes to understand an example illustrating passing pointers as arguments to functions. Here, we will begin by examining two functions that contribute to the following full program example:

https://github.com/PacktPublishing/Deciphering-Object-Orient-ed-Programming-with-CPP/blob/main/Chapter03/Chp3-Ex6.cpp

```
void TryToAddOne(int arg)
{
    arg++;
}

void AddOne(int *arg)
{
    (*arg)++;
}
```

Examining the previous functions, notice that `TryToAddOne()` takes an `int` as a formal parameter, while `AddOne()` takes an `int *` as a formal parameter.

In `TryToAddOne()`, an integer passed to the function is merely a copy of the actual parameter sent to the function. This parameter is referred to as `arg` in the formal parameter list. Incrementing the value of `arg` by one in the body of the function is a local change only within `TryToAddOne()`. Once the function completes, the formal parameter, `arg`, is popped off the stack and the actual parameter in the call to this function will not have been modified.

However, notice that `AddOne()` takes an `int *` as a formal parameter. The address of the actual integer parameter will be copied on the stack and received as the formal parameter, `arg`. Using the copy of that address, we dereference the pointer `arg` using `*`, then increment the integer value at that address using `++` in the line of code: `(*arg)++;`. When this function completes, the actual parameter will have been modified because we have passed a copy of the pointer to that integer, rather than a copy of the integer itself.

Let's examine the remainder of this program:

```
#include <iostream>
using std::cout;
using std::endl;

void TryToAddOne(int); // function prototypes
void AddOne(int *);
```

```
int main()
{
    int x = 10, *y = nullptr;
    y = new int;      // allocate y's memory
    *y = 15;          // dereference y to assign a value
    cout << "x: " << x << " and *y: " << *y << endl;

    TryToAddOne(x);    // unsuccessful, call by value
    TryToAddOne(*y);   // still unsuccessful
    cout << "x: " << x << " and *y: " << *y << endl;

    AddOne(&x);     // successful, passing an address
    AddOne(y);      // also successful
    cout << "x: " << x << " and *y: " << *y << endl;
    delete y;       // relinquish heap memory
    return 0;
}
```

Notice the function prototypes at the top of this program segment. They will match the function definitions in the previous segment of code. Now, in the `main()` function, we declare and initialize `int x = 10;` and declare a pointer: `int *y;`. We allocate the memory for y using `new()` and then assign a value by dereferencing the pointer with `*y = 15;`. We print out the respective values of x and `*y` as a baseline.

Next, we call `TryToAddOne(x);` followed by `TryToAddOne(*y);`. In both cases, we are passing integers as actual parameters to the function. Variable x is declared to be an integer, and `*y` refers to the integer pointed to by y. Neither of these function calls will result in the actual parameter being changed, which we can verify when their respective values are next printed using `cout` and the insertion operator `<<`.

Finally, we call `AddOne(&x);` followed by `AddOne(y);`. In both cases, we are passing a copy of an address as the actual parameter to the function. Of course, `&x` is the address of variable x, so this works. Likewise, y itself is an address – it is declared as a pointer variable. Recall that inside the `AddOne()` function, the formal parameter is first dereferenced and then incremented in the body of the function: `(*arg)++;`. We can use a copy of a pointer to access actual data.

Here is the output for the full program example:

```
x: 10 and *y: 15
x: 10 and *y: 15
x: 11 and *y: 16
```

Next, let us add to our discussion of using pointers with functions by using pointers as return values from functions.

Using pointers as return values from functions

Functions may return pointers to data via their return statements. When returning a pointer via the return statement of a function, be sure that the memory that is pointed to will persist after the function call is completed. Do not return a pointer to stack memory that is local to the function. That is, do not return a pointer to local variables defined on the stack within the function. However, returning a pointer to memory allocated using new() within the function is acceptable. As the allocated memory will be on the heap, it will exist past the function call.

Let's see an example to illustrate these concepts:

https://github.com/PacktPublishing/Deciphering-Object-Orient-ed-Programming-with-CPP/blob/main/Chapter03/Chp3-Ex7.cpp

```cpp
#include <iostream>
#include <iomanip>
using std::cin;
using std::cout;
using std::endl;
using std::flush;
using std::setw;

constexpr int MAX = 20;
[[nodiscard]] char *createName();  // function prototype

int main()
{
    char *name = nullptr;   // pointer declaration and init.
    name = createName();    // function will allocate memory
    cout << "Name: " << name << endl;
    delete [] name;  // del alloc. memory (in a diff. scope
    return 0;   // than allocated); this can be error prone!
}

[[nodiscard]] char *createName()
{
```

```
    char *temp = new char[MAX];
    cout << "Enter name: " << flush;
    cin >> setw(MAX) >> temp; // ensure no overflow of temp
    return temp;
}
```

In this example, `constexpr int MAX = 20;` is defined and then `char *createName();` is prototyped, indicating that this function takes no arguments, yet returns a pointer to one or more characters.

In the `main()` function, a local variable: `char *name;` is defined, but not initialized. Next, `createName()` is called and its return value is used to assign a value to `name`. Notice that both `name` and the function's return type are of type `char *`.

In the call to `createName()`, notice that a local variable `char *temp = new char[MAX];` is both defined and allocated to point to a fixed amount of memory on the heap using operator `new()`. The user is then prompted to enter a name and that name is stored in `temp`. The local variable `temp` is then returned from `createName()`.

In `createName()`, it is important that the memory for `temp` be comprised of heap memory so that it will persist beyond the scope of this function. Here, a copy of the address stored in `temp` will be copied onto the stack in the area reserved for a return value from the function. Fortunately, that address refers to heap memory. The assignment `name = createName();` in `main()` will capture this address and copy it to be stored into the `name` variable, which is local to `main()`. Since the memory allocated in `createName()` is on the heap, this memory will exist once the function completes.

Just as important to note, had `temp` been defined as `char temp[MAX];` in `createName()`, the memory comprising `temp` would have existed on the stack and would have been local to `createName()`. Once `createName()` returns to `main()`, the memory for this variable would have been popped off the stack and been unavailable for proper use – even if that address had been captured in a pointer variable within `main()`. This is another potential pointer trap in C++. When returning a pointer from a function, always ensure that the memory to which the pointer points exists beyond the extent of the function.

The output for this example is:

```
Enter name: Gabrielle
Name: Gabrielle
```

Now that we understand how pointers can be used within parameters to functions and as return values from functions, let's move forward by examining further pointer nuances.

Using the const qualifier with pointers

The const qualifier can be used to qualify pointers in several different ways. The keyword const can be applied to the data pointed to, to the pointer itself, or both. By using the const qualifier in these ways, C++ offers means to protect values in a program that may be meant to be initialized but never again modified. Let's examine each of these various scenarios. We will also be combining const qualified pointers with return values from functions to understand which of these various scenarios are reasonable to implement.

Using pointers to constant objects

A pointer to a constant object may be specified so that the object that is pointed to may not be directly modified. A dereferenced pointer to this object may not be used as an l-value in any assignment. An **l-value** means a value that can be modified, and that occurs on the left-hand side of an assignment.

Let's introduce a simple example to understand the situation:

```cpp
// const qualified str; the data pointed to will be const
const char *constData = "constant";
const char *moreConstData = nullptr;


// regular strings, defined. One is loaded using strcpy()
char *regularString = nullptr;
char *anotherRegularString = new char[8];   // sized to fit
                                             // this string
strcpy(anotherRegularString, "regular");


// Trying to modify data marked as const will not work
// strcpy(constData, "Can I do this? ");   // NO!
// Trying to circumvent by having a char * point to
// a const char * also will not work
// regularString = constData; // NO!


// But we can treat a char * more strictly by assigning to
// const char *. It will be const from that viewpoint only
moreConstData = anotherRegularString; // Yes - can do this!
```

Here, we've introduced `const char *constData = "constant";`. The pointer points to data, which is initialized, and which may never again be modified through this identifier. For example, should we try to alter this value using a `strcpy`, where `constData` is the destination string, the compiler will issue an error.

Also, trying to circumvent the situation by trying to store `constData` into a pointer of the same (but not `const`) type, will generate a compiler error, such as in the line of code `regularString = constData;`. Of course, in C++, you can do anything if you try hard enough, so an explicit typecast here will work, but is purposely not shown. An explicit typecast will still generate a compiler warning to allow you to question whether this is truly something you intend to do. When we move forward with OO concepts, we will introduce ways to further protect data so that this type of circumvention can be eliminated.

On the last line of the previous code, notice that we store the address of a regular string into `const char *moreConstData`. This is allowed – you can always treat something with more respect than it was defined to have (just not less). This means that when using the identifier `moreConstData`, this string may not be modified. However, using its own identifier, which is defined as `char *anotherRegularString;`, this string may be changed. This seems inconsistent, but it is not. The `const char *` variable chose to point to a `char *` – elevating its protection for a particular situation. If the `const char *` truly wanted to point to an immutable object, it would have chosen to instead point to another `const char *` variable.

Next, let's see a variation on this theme.

Using constant pointers to objects

A constant pointer to an object is a pointer that is initialized to point to a specific object. This pointer may never be assigned to point to another object. This pointer itself may not be used as an l-value in an assignment.

Let's review a simple example:

```
// Define, allocate, load simple strings using strcpy()
char *regularString = new char[36]; // sized for str below
strcpy(regularString, "I am a modifiable string");
char *anotherRegularString = new char[21]; // sized for
                                            // string below
strcpy(anotherRegularString, "I am also modifiable");

// Define a const pointer to a string; must be initialized
char *const constPtrString = regularString; // Ok
```

```
// You may not modify a const pointer to point elsewhere
// constPtrString = anotherRegularString;   // No!

// But you may change the data which you point to
strcpy(constPtrString, "I can change the value"); // Yes
```

In this example, two regular char * variables (regularString and anotherRegularString) are defined and loaded with string literals. Next, char *const constPtrString = regularString; is defined and initialized to point to a modifiable string. Because the const qualification is on the pointer itself and not the data pointed to, the pointer itself must be initialized with a value at declaration. Notice that the line of code: constPtrString = anotherRegularString; would generate a compiler error because a const pointer cannot be on the left hand of an assignment. However, because the const qualification is not applicable to the data pointed to, a strcpy may be used to modify the value of the data as is seen in strcpy(constPtrString, "I can change the value");.

Next, let us combine the const qualifier on both the pointer and the data which is pointed to.

Using constant pointers to constant objects

A constant pointer to a constant object is a pointer that is established to point to a specific object and to unmodifiable data. The pointer itself must be initialized to a given object, which is (hopefully) initialized with appropriate values. Neither the object nor the pointer may be modified or used as l-values in assignments.

Here is an example:

```
// Define two regular strings and load using strcpy()
char *regularString = new char[36]; // sized for str below
strcpy(regularString, "I am a modifiable string");
char *anotherRegularString = new char[21]; // sized for
                                           // string below
strcpy(anotherRegularString, "I am also modifiable");

// Define const ptr to a const object; must be initialized
const char *const constStringandPtr = regularString; // Ok
```

```
// Trying to change the pointer or the data is illegal
constStringandPtr = anotherRegularString; // No! Can't
                                          // modify address
strcpy(constStringandPtr, "Nope"); // No! Can't modify data
```

In this example, two regular char * variables are declared, regularString and anotherRegularString. Each is initialized with a string literal. Next, we introduce const char *const constStringandPtr = regularString;, which is a const qualified pointer to data that is also treated as const. Notice that this variable must be initialized because the pointer itself cannot be an l-value in a later assignment. You will also want to ensure that this pointer is initialized with a meaningful value, as the data that is pointed to also cannot be changed (as illustrated by the strcpy statement, which would generate a compiler error). Combining const on the pointer as well as the data pointed to is a strict way to safeguard data.

> **Tip – deciphering pointer declarations**
>
> To read complex pointer declarations, it often helps to read the declaration backward – from right to left. For example, the pointer declaration const char *p1 = "hi!"; would be interpreted as p1 is a pointer to (one or more) characters that are constant. The declaration const char *const p2 = p1; would be read as p2 is a constant pointer to (one or more) characters that are constant.

Finally, let us move forward to understand the implications of const qualifying pointers, which serve as function parameters or as return values from functions.

Using pointers to constant objects as function arguments and as return types from functions

Copying arguments on the stack that are user defined types can be time-consuming. Passing a pointer as a function argument is speedier, yet permits the dereferenced object to possibly be modified in the scope of the function. Passing a pointer to a constant object as a function argument provides both speed and safety for the argument in question. The dereferenced pointer simply may not be an l-value in the scope of the function in question. The same principle holds true for the return value from a function. Constant qualifying the data pointed to insists that the caller of the function must also store the return value in a pointer to a constant object, ensuring the object's long-term immutability.

Let's take a look at an example to examine these ideas:

https://github.com/PacktPublishing/Deciphering-Object-Orient-ed-Programming-with-CPP/blob/main/Chapter03/Chp3-Ex8.cpp

```cpp
#include <iostream>
#include <iomanip>
#include <cstring>  // we'll generally prefer std::string,
         // however, let's understand ptr concept shown here
using std::cout;
using std::endl;
char suffix = 'A';
const char *GenId(const char *);  // function prototype

int main()
{
    const char *newId1, *newId2;   // pointer declarations
    newId1 = GenId("Group");  // func. will allocate memory
    newId2 = GenId("Group");
    cout << "New ids: " << newId1 << " " << newId2 << endl;
    delete [] newId1;  // delete allocated memory
    delete [] newId2;  // caution: deleting in different
                       // scope than allocation can
                       // lead to potential errors
    return 0;
}

const char *GenId(const char *base)
{
    char *temp = new char[strlen(base) + 2];
    strcpy(temp, base);  // use base to initialize string
    temp[strlen(base)] = suffix++; // Append suffix to base
    temp[strlen(base) + 1] = '\0'; // Add null character
    return temp; // temp will be upcast to a const char *
                 // to be treated more restrictively than
                 // it was defined
}
```

In this example, we begin with a global variable to store an initial suffix, `char *suffix = 'A';`, and the prototype for the function: `const char *GenId(const char *base);`. In `main()`, we declare, but do not initialize, `const char* newId1, *newId2;`, which will eventually hold the IDs generated by `GenId()`.

Next, we call `GenId()` twice, passing a string literal `"Group"` to this function as the actual parameter. This parameter is received as a formal parameter: `const char *base`. The return value of this function will be used to assign values to `newId1` and `newId2`, respectively.

Looking more closely, we see that the call to `GenId("Group")` passes the string literal `"Group"` as the actual parameter, which is received as `const char *base` in the formal parameter list of the function definition. This means that when using the identifier `base`, this string may not be modified.

Next, within `GenId()`, we declare local pointer variable `temp` on the stack and allocate enough heap memory for `temp` to point to, to accommodate the string pointed to by `base` plus an extra character for the suffix to be added, plus one for the null character to terminate the new string. Note that `strlen()` counts the number of characters in a string, excluding the null character. Now, by using `strcpy()`, `base` is copied into `temp`. Then, using the assignment `temp[strlen(base)] = suffix++;`, the letter stored in `suffix` is added to the string pointed to by `temp` (and `suffix` is incremented to the next letter for the next time we call this function). Remember that arrays are zero-based in C++ when adding characters to the end of a given string. For example, if `"Group"` comprises five characters in array `temp`'s positions 0 through 4, then the next character (from `suffix`) would be added at position 5 in `temp` (overwriting the current null character). In the next line of code, the null character is re-added to the end of the new string pointed to by `temp`, as all strings need to be null terminated. Note that, whereas `strcpy()` will automatically null-terminate a string, once you resort to a single-character replacement, such as by adding the suffix to the string, you then need to re-add the null character to the new overall string yourself.

Lastly, in this function, `temp` is returned. Notice that though `temp` is declared as a `char *`, it is returned as a `const char *`. This means that the string will be treated in a more restrictive fashion upon its return to `main()` than it was treated in the body of the function. In essence, it has been upcast to a `const char *`. The implication is that since the return value of this function is a `const char *`, only a pointer of type `const char *` can capture the return value of this function. This is required so that the string cannot be treated in a less restrictive fashion than intended by the creator of function `GenId()`. Had `newId1` and `newId2` been declared of type `char *` rather than `const char *`, they would not have been allowed to serve as l-values to capture the return value of `GenId()`.

At the end of `main()`, we delete the memory associated with `newId1` and `newId2`. Notice that the memory for these pointer variables was allocated and released in different scopes within the program. The programmer must always be diligent to keep track of memory allocation and release in C++. Forgetting to deallocate memory can lead to memory leakage within an application.

Here is the output to accompany our example:

```
New ids: GroupA GroupB
```

Now that we have an understanding of how and why to `const` qualify pointers, let's take a look at how and why we might choose a gener c pointer type by considering void pointers.

Using pointers to objects of unspecified types

Sometimes, programmers ask why they cannot simply have a generic pointer. That is, why must we always declare the type of data to which the pointer will eventually point, such as `int *ptr;`? C++ certainly does allow us to create pointers without associated types, but C++ then requires the programmer to keep track of things that would normally be done on their behalf. Nonetheless, we will see why void pointers are useful and what the programmer must undertake when using more generic void pointers in this section.

It is important to note that void pointers require careful handling, and their misuse can be extremely dangerous. We will, much later in the book, see a safer alternative to genericize types (including pointers) in *Chapter 13, Working with Templates*. Nonetheless, there are careful encapsulated techniques that use an underlying implementation of a `void *` for efficiency, paired with a safe wrapper of a template. We will see that templates are expanded for every type needed and can sometimes lead to *template bloat*. In these cases, a safe pairing of a template with an underlying `void *` implementation gives us both safety and efficiency.

To understand a void pointer, let us first consider why a type is typically associated with a pointer variable. Typically, declaring the type with the pointer gives C++ information about how to conduct pointer arithmetic or index into a dynamically allocated array of that pointer type. That is, if we have allocated `int *ptr = new int [10];`, we have 10 consecutive integers. Using either the array notation of `ptr[3] = 5;` or the pointer arithmetic of `*(ptr + 3) = 5;` to access one such element in this dynamically allocated set relies on the size of the data type `int` to internally allow C++ to understand how large each element is and how to move from one such item to the next. The data type also tells C++, once it has arrived at an appropriate memory address, how to interpret the memory. For example, an `int` and a `float` may have the same storage size on a given machine, however, the two's complement memory layout of an `int` versus the mantissa, exponent layout of a `float` is quite different. C++'s knowledge of how to interpret the given memory is crucial, and the data type of the pointer does just that.

However, the need still exists to have a more generic pointer. For example, you may want a pointer that might point to an integer in one situation, yet to a set of user defined types in another situation. Using a `void *` allows just this to happen. But what about type? What happens when you dereference a void pointer? If C++ does not know how many bytes to go from one element in a set to another, how can it index into a dynamically allocated array of void pointers? How will it interpret the bytes once at an address? What is the type?

The answer is that you, the programmer, must personally remember what you are pointing to at all times. Without the type associated with the pointer, the compiler cannot do this for you. And when it is time to dereference the void pointer, you will be in charge of correctly remembering the ultimate type involved and performing the appropriate type cast on that pointer.

Let's take a look at the mechanics and logistics of what is involved.

Creating void pointers

Pointers to objects of unspecified types may be specified by using void *. The void pointer may then point to an object of any type. Explicit casting must be used in order to dereference actual memory pointed to by the void *. Explicit casting must also be used in C++ to assign memory pointed to by a void * to a pointer variable of a known type. It is the programmer's responsibility to ensure that the dereferenced data types are the same before making the assignment. Should the programmer be incorrect, there will be an elusive pointer mistake to find elsewhere in the code.

Here is an example:

https://github.com/PacktPublishing/Deciphering-Object-Orient-ed-Programming-with-CPP/blob/main/Chapter03/Chp3-Ex9.cpp

```cpp
#include <iostream>
using std::cout;
using std::endl;

int main()
{
    void *unspecified = nullptr; // may point to any
                                 // data type
    int *x = nullptr;

    unspecified = new int; // void ptr now points to an int
    // void * must be cast to int * before dereferencing
    *(static_cast<int *>(unspecified)) = 89;

    // let x point to the memory that unspecified points to
    x = static_cast<int *>(unspecified);
    cout << *x << " " << *(static_cast<int *>(unspecified))
         << endl;
    delete static_cast<int *>(unspecified);
```

```
    return 0;
}
```

In this example, the declaration `void *unspecified;` creates a pointer that may, one day, point to memory that can be of any data type. The declaration `int *x;` declares a pointer that may someday point to one or more consecutive integers.

The assignment `*(static_cast<int *>(unspecified)) = 89;` first uses an explicit typecast to cast `unspecified` to an `(int *)` and then dereferences the `int *` to place the value of `89` in memory. It is important to note that this typecast must be done before `unspecified` may be dereferenced – otherwise, C++ does not understand how to interpret the memory that `unspecified` points to. Also note that if you accidentally typecast `unspecified` to the wrong type, the compiler would let you proceed, as typecasts are seen as a *"just do it"* command to the compiler. It is your job, as the programmer, to remember what type of data your `void *` points.

Lastly, we would like x to point to where `unspecified` points. Variable x is an integer and needs to point to one or more integers. Variable `unspecified` truly points to an integer, but since the data type of unspecified is `void *`, we must use an explicit typecast to make the following assignment work: `x = static_cast<int *>(unspecified) ;`. Also, programmatically, we hope that we are correct and that we have remembered that `unspecified` truly points to an `int`; knowing the correct memory layout is important should the `int *` ever be dereferenced. Otherwise, we have just forced an assignment between pointers of different types, leaving a lurking error in our program.

Here is the output to accompany our program:

```
89 89
```

There are many creative uses of void pointers in C++. Some techniques use `void *`'s for generic pointer manipulations and pair this inner processing with a thin layer on top to cast the data into a known data type. The thin top layers can be further genericized with the C++ feature of templates. Using templates, only one version of the explicit type casts is maintained by the programmer, yet many versions are truly made available on your behalf – one per actual concrete data type needed. These ideas encompass advanced techniques, but we will see several of them in the chapters ahead, starting with *Chapter 13*, *Working with Templates*.

Looking ahead to smart pointers for safety

We have seen many uses of pointers to add flexibility and efficiency to our programs. However, we have also seen that with the power that pointers can provide comes potential havoc! Dereferencing uninitialized pointers can take us to non-existent memory locations that will inevitably crash our programs. Accidentally dereferencing memory that we have marked for deletion is similarly destructive – the memory address may have already been reused by the heap management facility elsewhere in our program. Neglecting to delete dynamically allocated memory when we are done with it will cause memory leaks. Even more challenging is allocating memory in one scope and expecting to remember to delete that memory in another scope. Or, consider what happens when two or more pointers point to the same piece of heap memory. Which pointer is responsible for deleting the memory? This is an issue we will see several times throughout the book with various solutions. These issues are just a few of the potential landmines we may step on when we utilize pointers.

You may ask whether there is another way to have the benefits of dynamically allocated memory, and yet have a safety net to govern its use. Fortunately, the answer is yes. The concept is a **smart pointer**, and there are several types of smart pointers in C++, including `unique_ptr`, `shared_ptr`, and `weak_ptr`. The premise of a smart pointer is that it is a class to safely wrap the usage of a raw pointer, minimally handling the proper deallocation of heap memory when the outer smart pointer goes out of scope.

However, to best understand smart pointers, we will need to understand *Chapter 5, Exploring Classes in Detail, Chapter 12, Friends and Operator Overloading*, and *Chapter 13, Working with Templates*. After understanding these core C++ features, smart pointers will be a meaningful option for us to embrace for pointer safety in the new code that we create. Will you still need to understand how to use native pointers in C++? Yes. It is inevitable that you will utilize many class libraries in C++ that heavily use native pointers, so you will need to understand their usage as well. Additionally, you may be integrating with, or maintaining, existing C++ code that is heavily native pointer reliant. You may also look online at many C++ forums or tutorials, and native pointers will inevitably pop up there as well.

The bottom line is that as C++ programmers, we need to understand how to use native C++ pointers, yet also understand their dangers, potential misuse, and pitfalls. Then, once we have mastered classes, operator overloading, and templates, we can add smart pointers to our repertoire and wisely choose to use them in our wholly new code. Yet, we will be prepared for any C++ situation by also understanding native C++ pointers.

With that in mind, we will continue gaining facility with native C++ pointers until we have best laid the groundwork to add these useful smart pointer classes into our repertoire. Then, we will see each smart pointer type in full detail.

Summary

In this chapter, we have learned many aspects surrounding pointers in C++. We have seen how to allocate memory from the heap using `new()` and how to relinquish that memory to the heap management facility using `delete()`. We have seen examples using both standard and user defined types. We have also understood why we may want to dynamically allocate arrays and have seen how to do so for 1, 2, and N dimensions. We have seen how to release the corresponding memory using `delete[]`. We have reviewed functions by adding pointers as parameters to functions and as return values from functions. We have also learned how to `const` qualify pointers as well as the data to which they point (or both) and why you may want to do so. We have seen one way to genericize pointers by introducing void pointers. Lastly, we have looked ahead to the concept of smart pointers.

All of the skills using pointers from this chapter will be used freely in the upcoming chapters. C++ expects programmers to have great facility using pointers. Pointers allow the language great freedom and efficiency to utilize a vast number of data structures and to employ creative programming solutions. However, pointers can provide a massive way to introduce errors into a program with memory leakage, returning pointers to memory that no longer exists, dereferencing pointers that have been deleted, and so on. Not to worry; we will utilize many examples going forward using pointers so that you will be able to manipulate pointers with great facility. Additionally, we will later add specific types of smart pointers to our upcoming programming repertoire to allow us to use add pointer safety when constructing code from scratch.

Most importantly, you are now ready to move forward to *Chapter 4, Indirect Addressing – References*, in which we will explore indirect addressing using references. Once you have understood both types of indirect addressing – pointers and references – and can manipulate either with ease, we will take on the core object-oriented concepts in this book, starting in *Chapter 5, Exploring Classes in Detail*.

Questions

1. Modify and augment your C++ program from *Chapter 2, Adding Language Necessities, Question 2*, as follows:

    a. Create a function, `ReadData()`, which accepts a pointer to a Student as an argument to allow for `firstName`, `lastName`, and `gpa`, and the `currentCourseEnrolled` to be entered from the keyboard within the function and stored as the input parameter's data.

    b. Modify `firstName`, `lastName`, and `currentCourseEnrolled` to be modeled as `char *` (or `string`) in your `Student` class instead of using fixed-sized arrays (as they may have been modeled in *Chapter 2, Adding Language Necessities*). You may utilize a `temp` variable that is a fixed size to initially capture user input for these values, and then allocate the proper, respective sizes for each of these data members. Note that using a `string` will be the simplest and safest approach.

    c. Rewrite, if necessary, the `Print()` function from your solution in *Chapter 2, Adding Language Necessities*, to take a `Student` as a parameter for `Printd()`.

    d. Overload the `Print()` function with one that takes a `const Student *` as a parameter. Which one is more efficient? Why?

    e. In `main()`, create an array of pointers to `Student` to accommodate five students. Allocate each `Student`, call `ReadData()` for each `Student`, and then `Print()` each `Student` using a selection from your previous functions. When done, remember to `delete()` the memory for each `Student` allocated.

f.   Also in `main()`, create an array of void pointers that is the same size as the array of pointers to `Student`. Set each element in the array of `void` pointers to point to a corresponding `Student` from the array of `Student` pointers. Call the version of `Print()` that takes a `const Student *` as a parameter for each element in the `void *` array. Hint: you will need to cast `void *` elements to type `Student *` prior to making certain assignments and function calls.

2.   Write the following pointer declarations that include a `const` qualification:

a.   Write a declaration for a pointer to a constant object. Assume the object is of type `Student`. Hint: read your declaration from right to left to verify correctness.

b.   Write a declaration for a constant pointer to a non-constant object. Again, assume the object is of type `Student`.

c.   Write a declaration for a constant pointer to a constant object. The object will again be of type `Student`.

3.   Why does passing an argument of type `const Student *` to `Print()` in your preceding program make sense, yet passing a parameter of type `Student * const` does not make sense?

4.   Can you think of programming situations that may require a dynamically allocated 3-D array? What about a dynamically allocated array with more dimensions?

4

Indirect Addressing – References

This chapter will examine how to utilize references in C++. References can often, but not always, be used as an alternative to pointers for indirect addressing. Though you have prior experience with indirect addressing from our last chapter using pointers, we will start at the beginning to understand C++ references.

References, like pointers, are a language feature you must be able to utilize with ease. Many other languages use references for indirect addressing without requiring the thorough understanding that C++ imposes to correctly utilize both pointers and references. Just as with pointers, you will see references frequently used throughout code from other programmers. You may be pleased that using references will provide notational ease when writing applications compared to pointers.

Unfortunately, references cannot be used as a substitute for pointers in all situations requiring indirect addressing. Therefore, a thorough understanding of indirect addressing using both pointers and references is a necessity in C++ to create successful and maintainable code.

The goal of this chapter will be to complement your understanding of indirect addressing using pointers with knowing how to use C++ references as an alternative. Understanding both techniques of indirect addressing will enable you to be a better programmer, to easily understand and modify others' code, as well as to write original, mature, and competent C++ code yourself.

In this chapter, we will cover the following main topics:

- Reference basics – declaring, initializing, accessing, and referencing existing objects
- Using references with functions as arguments and as return values
- Using the const qualifier with references
- Understanding underlying implementation, and when references cannot be utilized

By the end of this chapter, you will understand how to declare, initialize, and access references; you will understand how to reference existing objects in memory. You will be able to use references as arguments to functions, and understand how they may be used as return values from functions.

You will also fathom how the `const` qualifier may apply to references as variables and be utilized with both a function's parameters and return type. You will be able to distinguish when references can be used in lieu of pointers, and in which situations they cannot provide a substitute for pointers. These skills will be necessary in order to move forward with the next chapters in the book successfully.

Technical requirements

Online code for full program examples can be found in the following GitHub URL: `https://github.com/PacktPublishing/Deciphering-Object-Oriented-Programming-with-CPP/tree/main/Chapter04`. Each full program example can be found in the GitHub under the appropriate chapter heading (subdirectory) in a file that corresponds to the chapter number, followed by a dash, followed by the example number in the chapter at hand. For example, the first full program in this chapter can be found in the subdirectory `Chapter04` in a file named `Chp4-Ex1.cpp` under the aforementioned GitHub directory.

The CiA video for this chapter can be viewed at: `https://bit.ly/3ptaMRK`.

Understanding reference basics

In this section, we will revisit reference basics as well as introduce operators applicable to references, such as the reference operator &. We will employ the reference operator (&) to establish a reference to the existing variable. Like pointer variables, reference variables refer to memory that is defined elsewhere.

Using reference variables allows us to use a more straightforward notation than the notation that pointers use when using indirectly accessed memory. Many programmers appreciate the clarity in the notation of a reference versus a pointer variable. But, behind the scenes, memory must always be properly allocated and released; some portion of memory that is referenced may come from the heap. The programmer will undoubtedly need to deal with pointers for some portion of their overall code.

We will discern when references and pointers are interchangeable, and when they are not. Let's get started with the basic notation for declaring and using reference variables.

Declaring, initializing, and accessing references

Let's begin with the meaning of a reference variable. A C++ **reference** is an alias or a means for referring to another variable. A reference is specified using the reference operator &. A reference must be initialized (at declaration) and may never be assigned to reference another object. The reference and the initializer must be of the same type. Since the reference and the object being referenced share the same memory, either variable may be used to modify the contents of the shared memory location.

A reference variable, behind the scenes, can be compared to a pointer variable in that it holds the address of the variable that it is referencing. Unlike a pointer variable, any usage of the reference variable automatically dereferences the variable to go to the address that it contains; the dereference operator * is simply not needed with references. Dereferencing is automatic and implied with each use of a reference variable.

Let's take a look at an example illustrating reference basics:

https://github.com/PacktPublishing/Deciphering-Object-Oriented-Programming-with-CPP/blob/main/Chapter04/Chp4-Ex1.cpp

```cpp
#include <iostream>
using std::cout;
using std::endl;

int main()
{
    int x = 10;
    int *p = new int;    // allocate memory for ptr variable
    *p = 20;             // dereference and assign value

    int &refInt1 = x;   // reference to an integer
    int &refInt2 = *p;  // also a reference to an integer
    cout << x << " " << *p << " ";
    cout << refInt1 << " " << refInt2 << endl;

    x++;        // updates x and refInt1
    (*p)++;     // updates *p and refInt2
    cout << x << " " << *p << " ";
    cout << refInt1 << " " << refInt2 << endl;

    refInt1++;      // updates refInt1 and x
    refInt2++;      // updates refInt2 and *p
    cout << x << " " << *p << " ";
    cout << refInt1 << " " << refInt2 << endl;
    delete p;       // relinquish p's memory
    return 0;
}
```

In the preceding example, we first declare and initialize `int x = 10;` and then declare and allocate `int *p = new int;`. We then assign the integer value `20` to `*p`.

Next, we declare and initialize two reference variables, `refInt1` and `refInt2`. In the first reference declaration and initialization, `int &refInt1 = x;`, we establish `refInt1` to refer to the variable x. It helps to read the reference declaration from right to left. Here, we are saying to use x to initialize `refInt1`, which is a reference (`&`) to an integer. Notice that both the initializer, x, is an integer and that `refInt1` is declared to be a reference to an integer; their types match. This is important. The code will not compile if the types differ. Likewise, the declaration and initialization `int &refInt2 = *p;` also establishes `refInt2` as a reference to an integer. Which one? The one pointed to by p. This is why p is dereferenced using `*` to go to the integer itself.

Now, we print out x, `*p`, `refInt1`, and `refInt2`; we can verify that x and `refInt1` have the same value of `10`, and `*p` and `refInt2` also have the same value of `20`.

Next, using the original variables, we increment both x and `*p` by one. Not only does this increment the values of x and `*p`, but the values of `refInt1` and `refInt2`. Repeating the printing of these four values, we again notice that x and `refInt1` have the value of `11`, while `*p` and `refInt2` have the value of `21`.

Finally, we use the reference variables to increment the shared memory. We increment both `refInt1` and `*refint2` by one and this also increments the values of the original variables x and `*p`. This is because the memory is one and the same between the original variable and the reference to that variable. That is, the reference can be thought of as an alias to the original variable. We conclude the program by again printing out the four variables.

Here is the output:

```
10 20 10 20
11 21 11 21
12 22 12 22
```

> **Important note**
>
> Remember, a reference variable must be initialized to the variable it will refer to. The reference may never be assigned to another variable. More precisely, we cannot rebind the reference to another entity. The reference and its initializer must be the same type.

Now that we have a handle on how to declare simple references, let's take a more complete look at referencing existing objects, such as those to user defined types.

Referencing existing objects of user defined types

Should a reference to an object of a `struct` or `class` type be defined, the object being referenced is simply accessed using the . (member selection) operator. Again, it is not necessary (such as it is with pointers) to first use the dereference operator to go to the object being referenced before choosing the desired member.

Let's take a look at an example in which we reference a user defined type:

https://github.com/PacktPublishing/Deciphering-Object-Oriented-Programming-with-CPP/blob/main/Chapter04/Chp4-Ex2.cpp

```cpp
#include <iostream>
using std::cout;
using std::endl;
using std::string;

class Student     // very simple class - we will add to it
{                 // in our next chapter
public:
    string name;
    float gpa;
};

int main()
{
    Student s1;
    Student &sRef = s1;  // establish a reference to s1
    s1.name = "Katje Katz";    // fill in the data
    s1.gpa = 3.75;
    cout << s1.name << " has GPA: " << s1.gpa << endl;
    cout << sRef.name << " has GPA: " << sRef.gpa << endl;

    sRef.name = "George Katz";  // change the data
    sRef.gpa = 3.25;
    cout << s1.name << " has GPA: " << s1.gpa << endl;
```

```
        cout << sRef.name << " has GPA: " << sRef.gpa << endl;
        return 0;
}
```

In the first part of this program, we define a user defined type, Student, using a class. Next, we declare a variable s1 of type Student using Student s1;. Now, we declare and initialize a reference to a Student using Student &sRef = s1;. Here, we declare sRef to reference a specific Student, namely s1. Notice that both s1 is of type Student and the reference type of sRef is also that of type Student.

Now, we load some initial data into s1.name and s1.gpa using two simple assignments. Consequently, this alters the value of sRef since s1 and sRef refer to the same memory. That is, sRef is an alias for s1.

We print out various data members for s1 and sRef and notice that they contain the same values.

Now, we load new values into sRef.name and sRef.gpa using assignments. Similarly, we print out various data members for s1 and sRef and notice that again, the values for both have changed. Again, we can see that they reference the same memory.

The output to accompany this program is as follows:

```
 Katje Katz has GPA: 3.75
 Katje Katz has GPA: 3.75
 George Katz has GPA: 3.25
 George Katz has GPA: 3.25
```

Let's now move forward with our understanding of references by considering their usage in functions.

Using references with functions

So far, we have minimally demonstrated references by using them to establish an alias for an existing variable. Instead, let's put forth a meaningful use of references, such as when they are used in function calls. We know most functions in C++ will take arguments, and we have seen many examples in the previous chapters illustrating function prototypes and function definitions. Now, let's augment our understanding of functions by passing references as arguments to functions, and using references as return values from functions.

Passing references as arguments to functions

References may be used as arguments to functions to achieve call-by-reference, rather than call-by-value, parameter passing. References can alleviate the need for pointer notation in the scope of the function in question as well as in the call to that function. Object or . (member selection) notation is used to access `struct` or `class` members for formal parameters that are references.

In order to modify the contents of a variable passed as an argument to a function, a reference (or pointer) to that argument must be used as a function parameter. Just as with a pointer, when a reference is passed to a function, a copy of the address representing the reference is passed to the function. However, within the function, any usage of a formal parameter that is a reference will automatically and implicitly be dereferenced, allowing the user to use object rather than pointer notation. As with passing a pointer variable, passing a reference variable to a function will allow the memory referenced by that parameter to be modified.

When examining a function call (apart from its prototype), it will not be obvious whether an object passed to that function is passed by value or by reference. That is, whether the entire object will be copied on the stack or whether a reference to that object will instead be passed on the stack. This is because object notation is used when manipulating references, and the function calls for these two scenarios will use the same syntax.

Diligent use of function prototypes will solve the mystery of what a function definition looks like and whether its arguments are objects or references to objects. Remember, a function definition may be defined in a separate file from any calls to that function, and not be easily available to view. Note that this ambiguity does not come up with pointers specified in a function call; it is immediately obvious that an address is being sent to a function based on how the variable is declared.

Let's take a few minutes to understand an example illustrating passing references as arguments to functions. Here, we will begin by examining three functions, which contribute to the following full program example:

https://github.com/PacktPublishing/Deciphering-Object-Oriented-Programming-with-CPP/blob/main/Chapter04/Chp4-Ex3.cpp

```
void AddOne(int &arg)    // These two fns. are overloaded
{
    arg++;
}


void AddOne(int *arg)    // Overloaded function definition
{
    (*arg)++;
}
```

```
void Display(int &arg)    // Function parameter establishes
                          // a reference to arg
{
    cout << arg << " " << flush;
}
```

Examining the previous functions, notice that AddOne(int &arg) takes a reference to an int as a formal parameter, while AddOne(int *arg) takes a pointer to an int as a formal parameter. These functions are overloaded. The types of their actual parameters will determine which version is called.

Now let's consider Display(int &arg). This function takes a reference to an integer. Notice that object (not pointer) notation is used to print arg within this function's definition.

Now, let's examine the remainder of this program:

```
#include <iostream>
using std::cout;
using std::flush;

void AddOne(int &);      // function prototypes
void AddOne(int *);
void Display(int &);

int main()
{
    int x = 10, *y = nullptr;
    y = new int;      // allocate y's memory
    *y = 15;          // dereference y to assign a value
    Display(x);
    Display(*y);

    AddOne(x);     // calls ref. version (with an object)
    AddOne(*y);    // also calls reference version
    Display(x);    // Based on prototype, we see we are
    Display(*y);   // passing by ref. Without prototype,
                   // we may have guessed it was by value.
    AddOne(&x);    // calls pointer version
    AddOne(y);     // also calls pointer version
```

```
    Display(x);
    Display(*y);
    delete y;      // relinquish y's memory
    return 0;
}
```

Notice the function prototypes at the top of this program segment. They will match the function definitions in the previous segment of code. Now, in the `main()` function, we declare and initialize `int x = 10;` and declare a pointer `int *y;`. We allocate the memory for y using `new()` and then assign a value by dereferencing the pointer with `*y = 15;`. We print out the respective values of x and `*y` as a baseline using successive calls to `Display()`.

Next, we call `AddOne(x)` followed by `AddOne(*y)`. Variable x is declared to be an integer and `*y` refers to the integer pointed to by y. In both cases, we are passing integers as actual parameters to the version of the overloaded function with the signature `void AddOne(int &);`. In both cases, the formal parameters will be changed in the function, as we are passing by reference. We can verify this when their respective values are next printed using successive calls to `Display()`. Note that in the function call `AddOne(x);`, the reference to the actual parameter x is established by the formal parameter `arg` (in the function's parameter list) at the time of the function call.

In comparison, we then call `AddOne(&x);` followed by `AddOne(y);`. In both cases, we are calling the overloaded version of this function with the signature `void AddOne(int *);`. In each case, we are passing a copy of an address as the actual parameter to the function. Naturally, `&x` is the address of variable x, so this works. Likewise, y itself is an address – it is declared as a pointer variable. We again verify that their respective values are again changed with two calls to `Display()`.

Notice, in each call to `Display()`, we pass an object of type `int`. Looking at the function call alone, we cannot determine whether this function will take an `int` as an actual parameter (which would imply the value could not be changed), or an `int &` as an actual parameter (which would imply that the value could be modified). Either of these is a possibility. However, by looking at the function prototype, we can clearly see that this function takes an `int &` as a parameter, and from this, we understand that the parameter may likely be modified. This is one of the many reasons function prototypes are helpful.

Here is the output for the full program example:

```
10 15 11 16 12 17
```

Now, let's add to our discussion of using references with functions by using references as return values from functions.

Using references as return values from functions

Functions may return references to data via their return statements. We will see a requirement to return data by reference when we overload operators for user defined types in *Chapter 12, Friends and Operator Overloading*. With operator overloading, returning a value from a function using a pointer will not be an option to preserve the operator's original syntax. We must return a reference (or a reference qualified with `const`); this will also allow overloaded operators to enjoy cascaded use. Additionally, understanding how to return objects by reference will be useful as we explore the C++ Standard Template Library in *Chapter 14, Understanding STL Basics*.

When returning a reference via the return statement of a function, be sure that the memory that is referred to will persist after the function call is completed. Do **not** return a reference to a local variable defined on the stack within the function; this memory will be popped off the stack the moment the function completes.

Since we cannot return a reference to a local variable within the function, and since returning a reference to an external variable is pointless, you may ask where the data that we return a reference to will reside. This data will inevitably be on the heap. Heap memory will exist past the extent of the function call. In most circumstances, the heap memory will have been allocated elsewhere; however, on rare occasions, the memory may have been allocated within this function. In this unusual situation, you must remember to relinquish the allocated heap memory when it is no longer required.

Deleting heap memory through a reference (versus pointer) variable will require you to use the address-of operator, &, to pass the required address to operator `delete()`. Even though reference variables contain the address of the object they are referencing, the use of a reference identifier is always in its dereferenced state. It is **rare** that the need may arise to delete memory using a reference variable; we will discuss a meaningful (yet rare) example in *Chapter 10, Implementing Association, Aggregation, and Composition*.

> Important note
>
> The following example illustrates syntactically how to return a reference from a function, which you will utilize when we overload operators to allow their cascaded use, for example. However, it is not recommended to use references to return newly allocated heap memory (in most cases, the heap memory will have been allocated elsewhere). It is a common convention to use references to signal to other programmers that there is no need for memory management for that variable. Nevertheless, rare scenarios for such deletions via references may be seen in existing code (as with the aforementioned rare usage with associations), so it is useful to see how such a rare deletion may be done.

Let's see an example to illustrate the mechanics of using a reference as a return value from a function:

https://github.com/PacktPublishing/Deciphering-Object-Orient-ed-Programming-with-CPP/blob/main/Chapter04/Chp4-Ex4.cpp

```cpp
#include <iostream>
using std::cout;
using std::endl;

int &CreateId();   // function prototype

int main()
{
    int &id1 = CreateId();   // reference established
    int &id2 = CreateId();
    cout << "Id1: " << id1 << " Id2: " << id2 << endl;
    delete &id1; // Here, '&' is address-of, not reference
    delete &id2; // to calculate address to pass delete()
    return 0;   // It is unusual to delete in fashion shown,
}             // using the addr. of a ref. Also, deleting in
              // a diff. scope than alloc. can be error prone

int &CreateId()    // Function returns a reference to an int
{
    static int count = 100;   // initialize with first id
    int *memory = new int;
    *memory = count++;   // use count as id, then increment
    return *memory;
}
```

In this example, we see int &CreateId(); prototyped towards the top of the program. This tells us that CreateId() will return a reference to an integer. The return value must be used to initialize a variable of type int &.

Toward the bottom of the program, we see the function definition for CreateId(). Notice that this function first declares a static counter, which is initialized exactly once to 100. Because this local variable is static, it will preserve its value from function call to function call. We then increment this counter by one a few lines later. The static variable, count, will be used as a basis to generate a unique ID.

Next, in `CreateId()`, we allocate space for an integer on the heap and point to it using the local variable `memory`. We then load `*memory` with the value of `count` and then increase `count` for the next time we enter this function. We then use `*memory` as the return value of this function. Notice that `*memory` is an integer (the one pointed to on the heap by the variable `memory`). When we return it from the function, it is returned as a reference to that integer. When returning a reference from a function, always ensure that the memory that is referenced exists beyond the extent of the function.

Now, let's look at our `main()` function. Here, we initialize a reference variable `id1` with the return value of our first call to `CreateId()` in the following function call and initialization: `int &id1 = CreateId();`. Note that the reference `id1` must be initialized when it is declared, and we have met that requirement with the aforementioned line of code.

We repeat this process with `id2`, initializing this reference with the return value of `CreateId()`. We then print both `id1` and `id2`. By printing both `id1` and `id2`, you can see that each ID variable has its own memory and maintains its own data values.

Next, we must remember to deallocate the memory that `CreateId()` allocated on our behalf. We must use operator `delete()`. Wait, operator `delete()` expects a pointer to the memory that will be deleted. Variables `id1` and `id2` are both references, not pointers. True, they each contain an address because each is inherently implemented as a pointer, but any use of their respective identifiers is always in a dereferenced state. To circumvent this dilemma, we simply take the address of reference variables `id1` and `id2` prior to calling `delete()`, such as `delete &id1;`. It is **rare** that you would need to delete memory via a reference variable, but now you know how to do so should the need arise.

The output for this example is as follows:

```
Id1: 100 Id2: 101
```

Now that we understand how references can be used within parameters to functions and as return values from functions, let's move forward by examining further reference nuances.

Using the const qualifier with references

The `const` qualifier can be used to qualify the data in which references are initialized or *refer to*. We can also use `const` qualified references as arguments to functions and as return values from functions.

It is important to understand that a reference is implemented as a constant pointer in C++. That is, the address contained within the reference variable is a fixed address. This explains why a reference variable must be initialized to the object to which it will refer, and may not later be updated using an assignment. This also explains why constant qualifying the reference itself (and not just the data that it refers to) does not make sense. This variety of `const` qualification is already implied with its underlying implementation.

Let's take a look at these various scenarios using `const` with references.

Using references to constant objects

The const qualifier can be used to indicate that the data to which references are initialized are unmodifiable. In this fashion, the alias always refers to a fixed piece of memory, and the value of that variable may not be changed using the alias itself. The reference, once specified as constant, implies that neither the reference nor its value may be changed. Again, the reference itself may not be changed due to its underlying implementation as a constant qualified pointer. A const qualified reference may not be used as an *l-value* in any assignment.

> **Note**
>
> Recall, an **l-value** is a value that can be modified and that occurs on the left-hand side of an assignment.

Let's introduce a simple example to understand the situation:

https://github.com/PacktPublishing/Deciphering-Object-Orient-ed-Programming-with-CPP/blob/main/Chapter04/Chp4-Ex5.cpp

```cpp
#include <iostream>
using std::cout;
using std::endl;

int main()
{
    int x = 5;
    const int &refInt = x;
    cout << x << " " << refInt << endl;
    // refInt = 6;   // Illegal -- refInt is const
    x = 7;    // we can inadvertently change refInt
    cout << x << " " << refInt << endl;
    return 0;
}
```

In the previous example, notice that we declare int x = 5; and then we establish a constant reference to that integer with the declaration: const int &refInt = x;. Next, we print out both values for a baseline and notice that they are identical. This makes sense; they reference the same integer memory.

Next, in the commented-out piece of code, `//refInt = 6;`, we try to modify the data that the reference refers to. Because `refInt` is qualified as `const`, this is illegal; this is the reason why we commented out this line of code.

However, on the following line of code, we assign `x` a value of 7. Since `refInt` refers to this same memory, its value will also be modified. Wait, isn't `refInt` constant? Yes, by qualifying `refInt` as `const`, we are indicating that its value will not be modified using the identifier `refInt`. This memory can still be modified using `x`.

But wait, isn't this a problem? No, if `refInt` truly wants to refer to something unmodifiable, it can instead initialize itself with a `const int`, not an `int`. This subtle point is something to remember in C++ so you can write code for exactly the scenario you intend to have, understanding the significance and consequences of each choice.

The output for this example is as follows:

```
5 5
7 7
```

Next, let's see a variation on the `const` qualification theme.

Using pointers to constant objects as function arguments and as return types from functions

Using `const` qualification with function parameters cannot just allow the speed of passing an argument by reference, but the safety of passing an argument by value. It is a useful feature in C++.

A function that takes a reference to an object as a parameter often has less overhead than a comparable version of the function that takes a copy of an object as a parameter. This most notably occurs when the object type that would be otherwise copied on the stack is large. Passing a reference as a formal parameter is speedier, yet permits the actual parameter to be potentially modified in the scope of the function. Passing a reference to a constant object as a function argument provides both speed and safety for the argument in question. The reference qualified as `const` in the parameter list simply may not be an *l-value* in the scope of the function in question.

The same benefit of `const` qualified references exists for the return value from a function. Constant qualifying the data referenced insists that the caller of the function must also store the return value in a reference to a constant object, ensuring the object may not be modified.

Let's take a look at an example:

https://github.com/PacktPublishing/Deciphering-Object-Orient-ed-Programming-with-CPP/blob/main/Chapter04/Chp4-Ex6.cpp

```cpp
#include <iostream>
using std::cout;
using std::cin;
using std::endl;

struct collection
{
    int x;
    float y;
};

void Update(collection &);    // function prototypes
void Print(const collection &);

int main()
{
    collection collect1, *collect2 = nullptr;
    collect2 = new collection;  // allocate mem. from heap
    Update(collect1);  // a ref to the object is passed
    Update(*collect2); // same here: *collect2 is an object
    Print(collect1);
    Print(*collect2);
    delete collect2;   // delete heap memory
    return 0;
}

void Update(collection &c)
{
    cout << "Enter <int> and <float> members: ";
    cin >> c.x >> c.y;
}
```

```
void Print(const collection &c)
{
    cout << "x member: " << c.x;
    cout << "   y member: " << c.y << endl;
}
```

In this example, we first define a simple `struct collection` with data members x and y. Next, we prototype `Update(collection &);` and `Print(const collection &);`. Notice that `Print()` constant qualifies the data being referenced as the input parameter. This means that this function will enjoy the speed of passing this parameter by reference, and the safety of passing the parameter by value.

Notice, towards the end of the program, we see the definitions for both `Update()` and `Print()`. Both take references as arguments, however, the parameter to `Print()` is constant qualified: `void Print(const collection &);`. Notice that both functions use the . (member selection) notation within each function body to access the relevant data members.

In `main()`, we declare two variables, `collect1` of type `collection`, and `collect2`, which is a pointer to a `collection` (and whose memory is subsequently allocated). We call `Update()` for both `collect1` and `*collect2`, and in each case, a reference to the applicable object is passed to the `Update()` function. In the case of `collect2`, which is a pointer variable, the actual parameter must first dereference `*collect2` to go to the object being referenced before calling this function.

Finally, in `main()`, we call `Print()` successively for both `collect1` and `*collect2`. Here, `Print()` will reference each object serving as a formal parameter as constant qualified referenced data, ensuring that no modifications of either input parameter are possible within the scope of the `Print()` function.

Here is the output to accompany our example:

```
Enter x and y members: 33 23.77
Enter x and y members: 10 12.11
x member: 33    y member: 23.77
x member: 10    y member: 12.11
```

Now that we have an understanding of when `const` qualified references are useful, let's take a look at when we can use references in lieu of pointers, and when we cannot.

Realizing underlying implementation and restrictions

References can ease the notation required for indirect referencing. However, there are situations in which references simply cannot take the place of pointers. To understand these situations, it is useful to review the underlying implementation of a reference in C++.

References are implemented as constant pointers, hence they must be initialized. Once initialized, references may not refer to a different object (though the value of the object being referenced can be changed).

To understand the implementation, let's consider a sample reference declaration: `int &intVar = x;`. From an implementation aspect, it is as though the former variable declaration is instead declared as `int *const intVar = &x;`. Note that the `&` symbol shown on the left-hand side of an initialization takes on the meaning of reference, whereas the `&` symbol shown on the right-hand side of an initialization or assignment implies address-of. These two declarations illustrate how a reference is defined versus its underlying implementation.

Even though a reference is implemented as a constant pointer, the usage of the reference variable is as if the underlying constant pointer has been dereferenced. For this reason, you cannot initialize a reference with a `nullptr` – not only can a `nullptr` not be dereferenced but since references can only be initialized and not reset, the opportunity would be lost to establish the reference variable to refer to a meaningful object. This also holds true for references to pointers.

Next, let's understand in which situations we cannot use references.

Understanding when we must use pointers instead of references

Based on the underlying implementation of references (as `const` pointers), most of the restrictions for reference usage make sense. For example, references to references are generally not allowed; each level of indirection would need to be initialized upfront and that often takes multiple steps, such as when using pointers. However, we will see **r-value references** (`&&`) in *Chapter 15, Testing Classes and Components*, where we will examine various *move* operations. Arrays of references are also not permitted (each element would need to be initialized immediately); nonetheless, arrays of pointers are always an option. Also, pointers to references are not permitted; however, references to pointers are permitted (as are pointers to pointers).

Let's take a look at the mechanics of an interesting allowable reference case that we have not yet explored:

https://github.com/PacktPublishing/Deciphering-Object-Orient-ed-Programming-with-CPP/blob/main/Chapter04/Chp4-Ex7.cpp

```cpp
#include <iostream>
using std::cout;
using std::endl;

int main()
{
    int *ptr = new int;
    *ptr = 20;
```

```
    int *&refPtr = ptr;   // establish a reference to a ptr
    cout << *ptr << " " << *refPtr << endl;
    delete ptr;
    return 0;
}
```

In this example, we declare int *ptr; and then allocate the memory for ptr (consolidated on one line). We then assign a value of 20 to *p.

Next, we declare int *&refPtr = ptr;, which is a reference to a pointer of type int. It helps to read the declaration from right to left. As such, we use ptr to initialize refPtr, which is a reference to a pointer to an int. In this case, the two types match; ptr is a pointer to an int, so refPtr must also then reference a pointer to an int. We then print out the value of both *ptr and *refPtr and can see that they are the same.

Here is the output to accompany our program:

```
20 20
```

With this example, we have seen yet another interesting use of references. We also understand the restrictions placed upon using references, all of which are driven by their underlying implementation.

Summary

In this chapter, we have learned numerous aspects of C++ references. We have taken the time to understand reference basics, such as declaring and initializing reference variables to existing objects, as well as how to access reference components for basic and user defined types.

We have seen how to utilize references in a meaningful fashion with functions, both as input parameters and as a return value. We have also seen when it is reasonable to apply the const qualifier to references, as well as seen how this concept can be combined with parameters and return values from functions. Lastly, we have seen the underlying implementation of references. This has helped explain some of the restrictions references encompass, as well as understand which cases of indirect addressing will require the use of pointers instead of references.

As with pointers, all of the skills using references from this chapter will be used freely in the upcoming chapters. C++ allows programmers to have a more convenient notation for indirect addressing using references; however, programmers are expected to utilize either for indirect addressing with relative ease.

Finally, you are now ready to move forward to *Chapter 5*, *Exploring Classes in Detail*, in which we begin the object-oriented features of C++. This is what we have been waiting for; let's get started!

Questions

1. Modify and augment your C++ program from *Chapter 3, Indirect Addressing – Pointers, Question 1* as follows:

   a. Overload your ReadData() function with a version that accepts a Student & parameter to allow firstName, lastName, currentCourseEnrolled, and gpa to be entered from the keyboard within the function.

   b. Replace the Print() function that takes a Student from your previous solution to instead take a const Student & as a parameter for Print().

   c. Create variables of type Student and of type Student * in main(). Now, call the various versions of ReadData(), and Print(). Do the pointer variables necessarily need to call the versions of these functions that accept pointers, and do the non-pointer variables necessarily need to call the versions of these functions that accept references? Why or why not?

Part 2: Implementing Object-Oriented Concepts in C++

The goal of this part is to understand how to implement OO designs using both C++ language features and proven programming techniques. C++ can be used for many paradigms of coding; programmers must strive to program in an OO fashion in C++ (it's not automatic). This is the largest section of the book, as understanding how to map language features and implementation techniques to OO concepts is paramount.

The initial chapter in this section explores classes in great detail, beginning by describing the OO concepts of encapsulation and information hiding. Language features such as member functions, the `this` pointer, access regions in detail, constructors in detail (including the copy constructor, the member initialization list, and in-class initialization), destructor, qualifiers on member functions (`const`, `static`, and `inline`), and qualifiers on data members (`const` and `static`) are examined in depth.

The next chapter in this section tackles single inheritance basics with the OO concepts of generalization and specialization, detailing inherited constructors through the member initialization list, the order of construction and destruction, and understanding inherited access regions. Final classes are explored. This chapter pushes deeper by exploring public versus protected and private base classes and how these language features change the OO meaning of inheritance.

The subsequent chapter delves into the OO concept of polymorphism with respect to understanding the concept as well as its implementation in C++ using virtual functions. The `virtual`, `override`, and `final` keywords are explored. Dynamic binding of an operation to a specific method is examined. The virtual function table is explored to explain runtime binding.

The next chapter explains abstract classes in detail, pairing the OO concept with its implementation using pure virtual functions. The OO concept of an interface (not explicitly in C++) is introduced and a method for implementation is reviewed. Casting up and down the inheritance hierarchy completes this chapter.

The next chapter explores multiple inheritance and the potential issues that may arise from using this feature. Virtual base classes are detailed as well as the OO concept of a discriminator to help determine whether multiple inheritance is the best design for a given scenario or if another may exist.

The final chapter in this section introduces the concepts of association, aggregation, and composition and how to implement these common object relationships using pointers or references, sets of pointers, or embedded objects.

This part comprises the following chapters:

- *Chapter 5, Exploring Classes in Detail*
- *Chapter 6, Implementing Hierarchies with Single Inheritance*
- *Chapter 7, Utilizing Dynamic Binding through Polymorphism*
- *Chapter 8, Mastering Abstract Classes*
- *Chapter 9, Exploring Multiple Inheritance*
- *Chapter 10, Implementing Association, Aggregation, and Composition*

5

Exploring Classes in Detail

This chapter will begin our pursuit of **object-oriented programming (OOP)** in C++. We will begin by introducing **object-oriented (OO)** concepts and then progress to understanding how these concepts can be implemented in C++. Many times, implementing OOP ideas will be through *direct language support*, such as the features in this chapter. Sometimes, however, we will utilize various programming techniques to implement object-oriented concepts. These techniques will be seen in later chapters. In all cases, it is important to understand the object-oriented concepts and how these concepts relate to well-thought-out designs, and then have a clear understanding of how to implement these designs with robust code.

This chapter will detail C++ class usage in extreme detail. Subtle features and nuances are detailed beyond the basics. The goal of this chapter will be to allow you to understand OO concepts, and for you to begin to think in terms of object-oriented programming. Embracing core OO ideals, such as encapsulation and information hiding, will allow you to write code that is easier to maintain, and will allow you to modify others' code more easily.

In this chapter, we will cover the following main topics:

- Defining object-oriented terminology and concepts – object, class, instance, encapsulation and information hiding

- Applying class and member function basics

- Examining member function internals; the `this` pointer

- Using access labels and access regions

- Understanding constructors – default, overloaded, copy, conversion constructors, and in-class initializers

- Understanding destructors and their proper usage

- Applying qualifiers to data members and member functions – `inline`, `const`, and `static`

By the end of this chapter, you will understand core object-oriented terminology applicable to classes, and how key OO ideas such as encapsulation and information hiding will lead to software that is easier to maintain.

You will also appreciate how C++ provides built-in language features to support object-oriented programming. You will become well versed in the use of member functions and will understand their underlying implementation through the `this` pointer. You will understand how to correctly use access labels and access regions to promote encapsulation and information hiding.

You will understand how constructors can be used to initialize objects, and the many varieties of constructors from basic to typical (overloaded) to the copy constructor, and even conversion constructors. Similarly, you will understand how to make proper use of the destructor prior to an object's end of existence.

You will also understand how qualifiers, such as `const`, `static`, and `inline`, may be applied to member functions to support either object-oriented concepts or efficiency. Likewise, you will understand how to apply qualifiers, such as `const` and `static`, to data members to additionally support OO ideals.

C++ can be used as an object-oriented programming language, but it is not automatic. To do so, you must understand OO concepts, ideology, and language features that will allow you to support this endeavor. Let us begin our pursuit of writing code that is easier to modify and maintain by understanding the core and essential building block found in object-oriented C++ programs, the C++ class.

Technical requirements

Online code for full program examples can be found in the following GitHub URL: `https://github.com/PacktPublishing/Deciphering-Object-Oriented-Programming-with-CPP/tree/main/Chapter05`. Each full program example can be found in the GitHub under the appropriate chapter heading (subdirectory) in a file that corresponds to the chapter number, followed by a dash, followed by the example number in the chapter at hand. For example, the first full program in this chapter can be found in the subdirectory `Chapter05` in a file named `Chp5-Ex1.cpp` under the aforementioned GitHub directory.

The CiA video for this chapter can be viewed at: `https://bit.ly/3KaiQ39`.

Introducing object-oriented terminology and concepts

In this section, we will introduce core object-oriented concepts as well as applicable terminology that will accompany these key ideas. Though new terms will come up throughout this chapter, we will begin with essential terms necessary to begin our journey in this section.

Let's get started with basic object-oriented terminology.

Understanding object-oriented terminology

We will begin with basic object-oriented terminology, and then as we introduce new concepts, we will extend the terminology to include C++ specific terminology.

The terms object, class, and instance are all important and related terms with which we can start our definitions. An **object** embodies a meaningful grouping of characteristics and behaviors. An object can be manipulated and can receive the action or consequences of a behavior. Objects may undergo transformations and can change repeatedly over time. Objects can interact with other objects.

The term object, at times, may be used to describe the blueprint for groupings of like items. The term **class** may be used interchangeably with this usage of an object. The term object may also (and more often) be used to describe a specific item in such a grouping. The term **instance** may be used interchangeably with this meaning of an object. The context of usage will often make clear which meaning of the term *object* is being applied. To avoid potential confusion, the terms *class* and *instance* can preferably be used.

Let's consider some examples, using the aforementioned terms:

Class	Instance
Student	A specific student at the University of Maryland
University	Temple University in Philadelphia, PA, USA
Bank	The First National Bank of Pittsburgh

Objects also have components. The characteristics of a class are referred to as **attributes**. Behaviors of a class are referred to as **operations**. The specific implementation of a behavior or operation is referred to as its **method**. In other words, the method is how an operation is implemented, or the body of code defining the function, whereas the operation is the function's prototype or protocol for usage.

Let's consider some high-level examples, using the aforementioned terms:

Class	Attributes	Operations	Methods
Bank	name	Set Interest Rate	Determine the prime rate and add 1%
	location	Invest IRA Accounts	Select between mutual funds, money markets, and bonds per customer's tolerance for risk
	federal reserve ID		

Each instance of a class will most likely have distinct values for its attributes. For example:

Class	Instance	Attributes	Attribute values
Bank	First National	name	The First National Bank of Pittsburgh
		location	23 South Street, Pittsburgh, PA, USA
		federal reserve ID	1834-PA-659

Now that we have the basic OO terms under our belt, let's move on to important object-oriented concepts that are relevant to this chapter.

Understanding object-oriented concepts

The key object-oriented concepts relating to this chapter are *encapsulation* and *information hiding*. Incorporating these interrelated ideals into your design will provide the basis for writing more easily modifiable and maintainable programs.

The grouping of meaningful characteristics (attributes) and behaviors (operations) that operate on those attributes, bundled together in a single unit, is known as **encapsulation**. In C++, we typically group these items together in a class. The interface to each class instance is made through operations that model the behaviors relevant to each class. These operations may additionally modify the internal state of the object by changing the values of its attributes. Concealing attributes within a class and providing an interface for operating on those details leads us to explore the supportive concept of information hiding.

Information hiding refers to the process of *abstracting* the details of performing an operation into a class method. That is, the user needs only to understand which operation to utilize and its overall purpose; the implementation details are hidden within the method (function's body). In this fashion, changing the underlying implementation (method) will not change the operation's interface. Information hiding can additionally refer to keeping the underlying implementation of a class' attributes hidden. We will explore this further when we introduce access regions. Information hiding is a means to achieve proper encapsulation of a class. A properly encapsulated class will enable proper class abstraction and thus the support of OO designs.

Object-oriented systems are inherently more easily maintained because classes allow upgrades and modifications to be made quickly and without impact to the entire system due to encapsulation and information hiding.

Understanding class and member function basics

A C++ **class** is a fundamental building block in C++ that allows a programmer to specify a user defined type, encapsulating related data and behaviors. A C++ class definition will contain attributes, operations, and sometimes methods. C++ classes support encapsulation.

Creating a variable of a class type is known as **instantiation**. The attributes in a class are known as **data members** in C++. Operations in a class are known as **member functions** in C++ and are used to model behaviors. In OO terms, an operation implies the signature of a function, or its prototype (declaration), and the method implies its underlying implementation or the body of the function (definition). In some OO languages, the term *method* is used more loosely to imply either the operation or its method, based on the context of usage. In C++, the terms *data member* and *member function* are most often used.

The prototype for a member function must be placed in a class definition. Most often, the member function definition is placed outside of the class definition. The scope resolution operator : : is then used to associate a given member function definition to the class in which it is a member. Dot . or arrow - > notation is used to access all class members, including member functions, depending on whether we are accessing members through an instance or through a pointer to an instance.

C++ structures may also be used to encapsulate data and their related behaviors. A C++ struct can do anything a C++ class can do; in fact, class is implemented in terms of struct in C++. Though structures and classes may behave identically (other than default visibility), classes are more often used to model objects, relationships between object types, and implement object-oriented systems.

Let's take a look at a simple example in which we instantiate a class and a struct, each with member functions, for comparison with one another. We will break this example into several segments. The full program example can be found in the GitHub repository:

https://github.com/PacktPublishing/Deciphering-Object-Oriented-Programming-with-CPP/blob/main/Chapter05/Chp5-Ex1.cpp

```cpp
#include <iostream>
using std::cout;    // preferred to: using namespace std;
using std::endl;
using std::string;

struct student
{
    string name;
    float gpa;
    void Initialize(string, float);  // fn. prototype
```

```
        void Print();
};


class University
{
public:
    string name;
    int numStudents;
    void Initialize(string, int);    // fn. prototype
    void Print();
};
```

In the preceding example, we first define a student type using a struct, and a University type using a class. Notice, by convention, that user defined types created using structures are not capitalized, yet user defined types created using classes begin with a capital letter. Also notice that the class definition requires the label public: toward the beginning of its definition. We will explore the use of this label later in this chapter; however, for now, the public label is present so that this class will have the same default visibility of its members as does the struct.

In both the class and struct definitions, notice the function prototypes for Initialize() and Print(). We will tie these prototypes to member function definitions in the next program segment using ::, the scope resolution operator.

Let's examine the various member function definitions:

```
void student::Initialize(string n, float avg)
{
    name = n;     // simple assignment
    gpa = avg;    // we'll see preferred init. shortly
}


void student::Print()
{
    cout << name << " GPA: " << gpa << endl;
}


void University::Initialize(string n, int num)
{
    name = n;                  // simple assignment; we will see
```

```
        numStudents = num;   // preferred initialization shortly
}

void University::Print()
{
    cout << name << " Enrollment: " << numStudents << endl;
}
```

Now, let's review the various member function definitions for each user defined type. The definitions for void student::Initialize(string, float), void student::Print(), void University::Initialize(string, int), and void University::Print() appear consecutively in the preceding fragment. Notice how the scope resolution operator :: allows us to tie the relevant function definition back to the class or struct in which it is a member.

Additionally, notice that in each Initialize() member function, the input parameters are used as values to load the relevant data members for a specific instance of a specific class or struct type. For example, in the function definition of void University::Initialize(string n, int num), the input parameter num is used to initialize numStudents for a particular University instance.

> **Note**
> The scope resolution operator :: associates member function definitions with the class (or struct) in which they are a member.

Let's see how member functions are called by considering main() in this example:

```
int main()
{
    student s1;   // instantiate a student (struct instance)
    s1.Initialize("Gabby Doone", 4.0);
    s1.Print();

    University u1;   // instantiate a University (class)
    u1.Initialize("GWU", 25600);
    u1.Print();

    University *u2;          // pointer declaration
    u2 = new University();   // instantiation with new()
    u2->Initialize("UMD", 40500);
```

```
    u2->Print();    // or alternatively: (*u2).Print();
    delete u2;
    return 0;
}
```

Here, in `main()`, we simply define a variable, `s1`, of type `student`, and a variable, `u1`, of type `University`. In object-oriented terms, it is preferable to say that `s1` is an instance of `student`, and `u1` is an instance of `University`. The instantiation occurs when the memory for an object is made available. For this reason, declaring pointer variable `u2` using: `University *u2;` does not instantiate a `University`; it merely declares a pointer to a possible future instance. Rather, on the following line, `u2 = new University();`, we instantiate a `University` when the memory is allocated.

For each of the instances, we initialize their data members by calling their respective `Initialize()` member functions, such as `s1.Initialize("Gabby Doone", 4.0);` or `u1.Initialize("UMD", 4500);`. We then call `Print()` through each respective instance, such as `u2->Print();`. Recall that `u2->Print();` may also be written as `(*u2).Print();`, which more easily allows us to remember that the instance here is `*u2`, whereas `u2` is a pointer to that instance.

Notice that when we call `Initialize()` through `s1`, we call `student::Initialize()` because `s1` is of type `student`, and we initialize `s1`'s data members in the body of this function. Similarly, when we call `Print()` through `u1` or `*u2`, we call `University::Print()` because `u1` and `*u2` are of type `University` and we subsequently print out a particular university's data members.

Since instance `u1` was dynamically allocated on the heap, we are responsible for releasing its memory using `delete()` toward the end of `main()`.

The output to accompany this program is as follows:

```
Gabby Doone GPA: 4.4
GWU Enrollment: 25600
UMD Enrollment: 40500
```

Now that we are creating class definitions with their associated member function definitions, it is important to know how developers typically organize their code in files. Most often, a class will be broken into a header (`.h`) file, which will contain the class definition, and a source code (`.cpp`) file, which will `#include` the header file, and then follow with the member function definitions themselves. For example, a class named `University` would have a `University.h` header file and a `University.cpp` source code file.

Now, let's move forward with our understanding of the details of how member functions work by examining the `this` pointer.

Examining member function internals; the "this" pointer

So far, we have noticed that member functions are invoked through objects. We have noticed that in the scope of a member function, it is the data members (and other member functions) of the particular object that invoked the function that may be utilized (in addition to any input parameters). Alas, how, and why does this work?

It turns out that most often, member functions are invoked through objects. Whenever a member function is invoked in this fashion, that member function receives a pointer to the instance that invoked the function. A pointer to the object calling the function is then passed as an implicit first argument to the function. The name of this pointer is **this**.

Though the this pointer may be referred to explicitly in the definition of each such member function, it usually is not. Even without its explicit use, the data members utilized in the scope of the function belong to this, a pointer to the object that invoked the function.

Let us take a look at a full program example. Though the example is broken into segments, the full program can be found in the following GitHub location:

https://github.com/PacktPublishing/Deciphering-Object-Oriented-Programming-with-CPP/blob/main/Chapter05/Chp5-Ex2.cpp

```cpp
#include <iostream>
#include <cstring>  // though we'll prefer std::string, one
                    // pointer data member will illustrate
                    // important concepts
using std::cout;    // preferred to: using namespace std;
using std::endl;
using std::string;


class Student
{
// for now, let's put everything public access region
public:
    string firstName;   // data members
    string lastName;
    char middleInitial;
    float gpa;
    char *currentCourse;   // ptr to illustrate key concepts
```

```
        // member function prototypes
        void Initialize(string, string, char, float,
                        const char *);
        void Print();
    };
```

In the first segment of the program, we define class `Student` with a variety of data members, and two member function prototypes. For now, we will place everything in the `public` access region.

Now, let's examine the member function definitions for `void Student::Initialize()` and `void Student::Print()`. We will also examine how each of these functions looks internally to C++:

```
    // Member function definition
    void Student::Initialize(string fn, string ln, char mi,
                             float gpa, const char *course)
    {
        firstName = fn;
        lastName = ln;
        this->middleInitial = mi;  // optional use of 'this'
        this->gpa = gpa;  // required, explicit use of 'this'
        // remember to allocate memory for ptr data members
        currentCourse = new char [strlen(course) + 1];
        strcpy(currentCourse, course);
    }
    // It is as if Student::Initialize() is written as:
    // void Student_Initialize_str_str_char_float_constchar*
    //     (Student *const this, string fn, string ln,
    //      char mi, float avg, const char *course)
    // {
    //     this->firstName = fn;
    //     this->lastName = ln;
    //     this->middleInitial = mi;
    //     this->gpa = avg;
    //     this->currentCourse = new char [strlen(course) + 1];
    //     strcpy(this->currentCourse, course);
    // }
```

```
// Member function definition
void Student::Print()
{
    cout << firstName << " ";
    cout << middleInitial << ". ";
    cout << lastName << " has a gpa of: ";
    cout << gpa << " and is enrolled in: ";
    cout << currentCourse << endl;
}
// It is as if Student::Print() is written as:
// void Student_Print(Student *const this)
// {
//     cout << this->firstName << " ";
//     cout << this->middleInitial << ". ";
//     cout << this->lastName << " has a gpa of: ";
//     cout << this->gpa << " and is enrolled in: ";
//     cout << this->currentCourse << endl;
// }
```

First, we see the member function definition for void Student::Initialize(), which takes a variety of parameters. Notice that in the body of this function, we first assign input parameter fn to data member firstName. We proceed similarly, using the various input parameters, to initialize the various data members for the specific object that will invoke this function. Also notice that we allocate memory for pointer data member currentCourse to be enough characters to hold what input parameter course requires (plus one for the terminating null character). We then strcpy() the string from the input parameter, course, to data member currentCourse.

Also, notice in void Student::Initialize(), the assignment this->middleInitial = mi;. Here, we have an optional, explicit use of the this pointer. It is not necessary or customary in this situation to qualify middleInitial with this, but we may choose to do so. However, in the assignment this->gpa = gpa;, the use of this is required. Why? Notice that the input parameter is named gpa and the data member is also gpa. Simply assigning gpa = gpa; would set the most local version of gpa (the input parameter) to itself and would not affect the data member. Here, disambiguating gpa with this on the left-hand side of the assignment indicates to set the data member, gpa, which is pointed to by this, to the value of the input parameter, gpa. Another solution is to use distinct names for data members versus input parameters, such as renaming gpa in the formal parameter list to avg (which we will do in later versions of this code).

Now, notice the commented-out version of void Student::Initialize(), which is below the utilized version of void Student::Initialize(). Here, we can see how most member functions are internally represented. First, notice that the name of the function is *name mangled* to include the data types of its parameters. This is internally how functions are represented, and consequentially, allows function overloading (that is, two functions with seemingly the same name; internally, each has a unique name). Next, notice that among the input parameters, there is an additional, first, input parameter. The name of this additional (hidden) input parameter is `this`, and it is defined as `Student *const this`.

Now, in the body of the internalized function view of `void Student::Initialize()`, notice that each data member's name is preceded with `this`. We are, in fact, accessing the data member of an object that is pointed to by `this`. Where is `this` defined? Recall that `this` is the implicit first input parameter to this function, and is a constant pointer to the object that invoked this function.

Similarly, we can review the member function definition for `void Student::Print()`. In this function, each data member is neatly printed out using `cout` and the insertion operator `<<`. However, notice below this function definition, the commented-out internal version of `void Student::Print()`. Again, `this` is actually an implicit input parameter of type `Student *const`. Also, each data member usage is preceded with access through the `this` pointer, such as `this->gpa`. Again, we can clearly see that a specific instance's members are accessed in the scope of a member function; these members are implicitly accessed through the `this` pointer.

Lastly, note that explicit use of `this` is permitted in the body of a member function. We can almost always precede usage of a data member or member function, accessed in the body of a member function, with explicit use of `this`. Later in this chapter, we will see the one contrary case (using a static method). Also, later in this book, we will see situations in which explicit usage of `this` will be necessary to implement more intermediate-level OO concepts.

Nonetheless, let's move forward by examining `main()` to complete this program example:

```
int main()
{
    Student s1;    // instance
    Student *s2 = new Student; // ptr to an instance
    s1.Initialize("Mary", "Jacobs", 'I', 3.9, "C++");
    s2->Initialize("Sam", "Nelson", 'B', 3.2, "C++");
    s1.Print();
    s2->Print(); // or use (*s2).Print();

    delete [] s1.currentCourse;    // delete dynamically
    delete [] s2->currentCourse; // allocated data members
    delete s2;    // delete dynamically allocated instance
    return 0;
}
```

In the last segment of this program, we instantiate `Student` twice in `main()`. `Student s1` is an instance, whereas `s2` is a pointer to a `Student`. Next, we utilize either `.` or `->` notation to invoke the various member functions through each relevant instance.

Note, when `s1` invokes `Initialize()`, the `this` pointer (in the scope of the member function) will point to `s1`. It will be as if `&s1` is passed as a first argument to this function. Likewise, when `*s2` invokes `Initialize()`, the `this` pointer will point to `s2`; it will be as if `s2` (which is already a pointer) is passed as an implicit first argument to this function.

After each instance invokes `Print()` to display the data members for each `Student`, notice that we release various levels of dynamically allocated memory. We start with the dynamically allocated data members for each instance, releasing each such member using `delete()`. Then, because `s2` is a pointer to an instance that we have dynamically allocated, we must also remember to release the heap memory comprising the instance itself. We again do so with `delete s2;`.

Here is the output for the full program example:

```
Mary I. Jacobs has a gpa of: 3.9 and is enrolled in: C++
Sam B. Nelson has a gpa of: 3.2 and is enrolled in: C++
```

Now, let's add to our understanding of classes and information hiding by examining access labels and regions.

Using access labels and access regions

Labels may be introduced into a class (or structure) definition to control the access or visibility of class (or structure) members. By controlling the direct access of members from various scopes in our application, we can support encapsulation and information hiding. That is, we can insist that users of our classes use the functions that we select, with the protocols we select, to manipulate data and other member functions within the class in ways we, the programmers, find reasonable and acceptable. Furthermore, we can hide the implementation details of the class by advertising to the user only the desired public interface for a given class.

Data members or member functions, collectively referred to as **members**, can be individually labeled, or grouped together into access regions. The three labels or **access regions** that may be specified are as follows:

- **private**: Data members and member functions in this access region are only accessible within the scope of the class. The scope of a class includes member functions of that class.

- **protected**: Behaves like `private` until we introduce inheritance. When inheritance is introduced, `protected` will provide a mechanism for allowing access within the derived class scope.

- **public**: Data members and member functions in this access region are accessible from any scope in the program.

> **Reminder**
>
> Data members and member functions are most always accessed via instances. You will ask, *in what scope is my instance?*, and *may I access a particular member from this particular scope?*

As many members as required by the programmer may be grouped under a given label or **access region**. Should access labels be omitted in a class definition, the default member access is `private`. If access labels are omitted in a structure definition, default member access is `public`. When access labels are explicitly introduced, rather than relying on default visibility, a `class` and a `struct` are identical. Nonetheless, in object-oriented programming, we tend to utilize classes for user defined types.

It is interesting to note that data members, when grouped into an access region collectively under the same access label, are guaranteed to be laid out in memory in the order specified. However, if multiple access regions exist containing data members within a given class, the compiler is free to reorder those respective groupings for efficient memory layout.

Let's examine an example to illustrate access regions. Though this example will be broken into several segments, the full example will be shown and can also be found in the GitHub repository:

https://github.com/PacktPublishing/Deciphering-Object-Oriented-Programming-with-CPP/blob/main/Chapter05/Chp5-Ex3.cpp

```
#include <iostream>
#include <cstring>    // though we'll prefer std::string,
// one ptr data member will illustrate important concepts
using std::cout;      // preferred to: using namespace std;
using std::endl;
using std::string;

class Student
{
// private members are accessible only within the scope of
// the class (that is, within member functions or friends)
private:
    string firstName;    // data members
    string lastName;
    char middleInitial;
    float gpa;
    char *currentCourse;  // ptr to illustrate key concepts
public:    // public members are accessible from any scope
```

```
    // member function prototypes
    void Initialize();
    void Initialize(string, string, char, float,
                    const char *);
    void CleanUp();
    void Print();
};
```

In this example, we first define the Student class. Notice that we add a private access region near the top of the class definition and place all of the data members within this region. This placement will ensure that these data members will only be able to be directly accessed and modified within the scope of this class, which means by member functions of this class (and friends, which we will much later see). By limiting the access of data members only to member functions of their own class, safe handling of those data members is ensured; only access through intended and safe functions that the class designer has introduced themself will be allowed.

Next, notice that the label public has been added in the class definition prior to the prototypes of the member functions. The implication is that these functions will be accessible in any scope of our program. Of course, we will generally need to access these functions each via an instance. But the instance can be in the scope of main() or any other function (even in the scope of another class' member functions) when the instance accesses these public member functions. This is known as a class' public interface.

Access regions support encapsulation and information hiding

A good rule of thumb is to place your data members in the private access region and then specify a safe, appropriate public interface to access them using public member functions. By doing so, the only access to data members is in manners that the class designer has intended, through member functions the class designer has written, which have been well tested. With this strategy, the underlying implementation of the class may also be changed without causing any calls to the public interface to change. This practice supports encapsulation and information hiding.

Let's continue by taking a look at the various member function definitions in our program:

```
void Student::Initialize()
{   // even though string data members are initialized with
    // empty strings, we are showing how to clear these
    // strings, should Initialize() be called more than 1x
    firstName.clear();
    lastName.clear();
```

```
        middleInitial = '\0';        // null character
        gpa = 0.0;
        currentCourse = nullptr;
}


// Overloaded member function definition
void Student::Initialize(string fn, string ln, char mi,
                         float avg, const char *course)
{
        firstName = fn;
        lastName = ln;
        middleInitial = mi;
        gpa = avg;
        // dynamically allocate memory for pointer data member
        currentCourse = new char [strlen(course) + 1];
        strcpy(currentCourse, course);
}


// Member function definition
void Student::CleanUp()
{    // deallocate previously allocated memory
        delete [] currentCourse;
}


// Member function definition
void Student::Print()
{
        cout << firstName << " " << middleInitial << ". ";
        cout << lastName << " has gpa: " << gpa;
        cout << " and enrolled in: " << currentCourse << endl;
}
```

Here, we have defined the various member functions that were prototyped in our class definition. Notice the use of the scope resolution operator : : to tie the class name to the member function name. Internally, these two identifiers are *name mangled* together to provide a unique, internal function name. Notice that the void Student::Initialize() function has been overloaded; one version simply initializes all data members to some form of null or zero, whereas the overloaded version uses input parameters to initialize the various data members.

Now, let's continue by examining our main() function in the following segment of code:

```
int main()
{
    Student s1;

    // Initialize() is public; accessible from any scope
    s1.Initialize("Ming", "Li", 'I', 3.9, "C++", "178GW");

    s1.Print(); // public Print() accessible from main()

    // Error! private firstName is not accessible in main()
    // cout << s1.firstName << endl;

    // CleanUp() is public, accessible from any scope
    s1.CleanUp();
    return 0;
}
```

In the aforementioned main() function, we first instantiate a Student with the declaration Student s1;. Next, s1 invokes the Initialize() function with the signature matching the parameters provided. Since this member function is in the public access region, it can be accessed in any scope of our program, including main(). Similarly, s1 invokes Print(), which is also public. These functions are in the Student class' public interface, and represent some of the core functionality for manipulating any given Student instance.

Next, in the commented-out line of code, notice that s1 tries to access firstName directly using s1.firstName. Because firstName is private, this data member can only be accessed in the scope of its own class, which means member functions (and later friends) of its class. The main() function is not a member function of Student, hence s1 may not access firstName in the scope of main(), that is, a scope outside its own class.

Lastly, we invoke s1.CleanUp();, which also works because CleanUp() is public and is hence accessible from any scope (including main()).

The output for this complete example is as follows:

```
Ming I. Li has gpa: 3.9 and enrolled in: C++
```

Now that we understand how access regions work, let's move forward by examining a concept known as a constructor, and the various types of constructors available within C++.

Understanding constructors

Did you notice how convenient it has been for the program examples in this chapter to have an `Initialize()` member function for each `class` or `struct`? Certainly, it is desirable to initialize all data members for a given instance. More so, it is crucial to ensure that data members for any instance have bonafide values, as we know that memory is not provided *clean* or *zeroed-out* by C++. Accessing an uninitialized data member, and utilizing its value as if it were bonafide, is a potential pitfall awaiting the careless programmer.

Initializing each data member individually each time a class is instantiated can be tedious work. What if we simply overlook setting a value? What if the values are `private`, and are therefore not directly accessible? We have seen that an `Initialize()` function is beneficial because once written, it provides a means to set all data members for a given instance. The only drawback is that the programmer must now remember to call `Initialize()` on each instance in the application. Instead, what if there is a way to ensure that an `Initialize()` function is called every time a class is instantiated? What if we could overload a variety of versions to initialize an instance, and the appropriate version could be called based on data available at the time? This premise is the basis for a constructor in C++. The language provides for an overloaded series of initialization functions, which will be automatically called once the memory for an instance becomes available.

Let's take a look at this family of initialization member functions by examining the C++ constructor.

Applying constructor basics and overloading constructors

A **constructor** is a member function that is automatically invoked after the memory for an instance is made available. Constructors are used to initialize the data members that comprise a newly instantiated object (except for static data members, which we'll examine later in this chapter). A constructor will have the same name as the class or struct in which it is a member. Constructors may be overloaded, which enables a `class` (or `struct`) to define multiple means by which to initialize an object. The return type of a constructor may not be specified.

Should your `class` or `struct` not contain a constructor, one will be made for you in the `public` access region, with no arguments. This is known as a default constructor. Behind the scenes, every time an object is instantiated, a constructor call is patched in by the compiler. When a class without a constructor is instantiated, the default constructor is patched in as a function call immediately following the instantiation. This system-supplied member function will have an empty body (method) and it will be linked into your program so that any compiler-added, implicit calls to this function upon instantiation can occur without a linker error. As needed per the design, a programmer may often write their own default (no-argument) constructor; that is, one that is used for the default means of instantiation with no arguments.

Most programmers provide at least one constructor, in addition to their own no-argument, default constructor. Recall that constructors can be overloaded. It is important to note that if you provide any constructor yourself, you will not then receive the system supplied no-argument default constructor, and that subsequently using such an interface for instantiation will cause a compiler error.

> **Reminder**
>
> Constructors have the same name as the class. You may not specify their return type. They can be overloaded. The compiler only creates a public, default (no-argument) constructor if you have not provided any constructors (that is, means for instantiation) in your class.

Let's introduce a simple example to understand constructor basics:

https://github.com/PacktPublishing/Deciphering-Object-Orient-ed-Programming-with-CPP/blob/main/Chapter05/Chp5-Ex4.cpp

```cpp
#include <iostream>
using std::cout;    // preferred to: using namespace std;
using std::endl;
using std::string;

class University
{
private:
    string name;
    int numStudents;
public:
    // constructor prototypes
    University(); // default constructor
    University(const string &, int);
    void Print();
    void CleanUp();
};

University::University()
{   // Because a string is a class type, all strings are
    // constructed with an empty value by default.
    // For that reason, we do not need to explicitly
    // initialize strings if an empty string is desired.
    // We'll see a preferred manner of initialization
    // for all data members shortly in this chapter.
    // Hence, name is constructed by default (empty string)
    numStudents = 0;
```

```
    }

University::University(const string &n, int num)
{    // any pointer data members should be allocated here
     name = n; // assignment between strings is deep assign.
     numStudents = num;
}

void University::Print()
{
     cout << "University: " << name;
     cout << " Enrollment: " << numStudents << endl;
}

void University::CleanUp()
{    // deallocate any previously allocated memory
}

int main()
{
     University u1; // Implicit call to default constructor
     // alternate constructor instantiation and invocation
     University u2("University of Delaware", 23800);
     University u3{"Temple University", 20500}; // note {}
     u1.Print();
     u2.Print();
     u3.Print();
     u1.CleanUp();
     u2.CleanUp();
     u3.CleanUp();
     return 0;
}
```

In the previous program segment, we first define class University; the data members are private, and the three member functions are public. Notice that the first two member functions prototyped are constructors. Both have the same name as the class; neither has its return type specified. The two constructors are overloaded, in that each has a different signature.

Next, notice that the three member functions are defined. Notice the use of the scope resolution operator `: :` preceding each member function name, in each of their definitions. Each constructor provides a different means for initializing an instance. The `void University::Print()` member function merely provides a means to provide simple output for our example.

Now, in `main()`, let's create three instances of `University`. The first line of code, `University u1;`, instantiates a `University` and then implicitly invokes the default constructor to initialize the data members. On the next line of code, `University u2("University of Delaware", 23800);`, we instantiate a second `University`. Once the memory for that instance has been made available on the stack in `main()`, the constructor matching the signature of the arguments provided, namely `University::University(const string &, int)`, will be implicitly invoked to initialize the instance.

Finally, we instantiate a third `University` using `University u3{"Temple University", 20500};`, which also makes use of the alternate constructor. Notice the use of `{ }`'s versus `()`'s in the instantiation and construction of `u3`. Either style may be utilized. The latter style was introduced in an effort to create uniformity; neither construct results in a performance advantage.

We can see that based upon how we instantiate an object, we can specify which constructor we would like to be called on our behalf to perform the initialization.

The output for this example is as follows:

```
University: Enrollment: 0
University: University of Delaware Enrollment: 23800
University: Temple Enrollment: 20000
```

Parameter comparison

Did you notice the signature to the alternate `University` constructor is `University(const string &, int);`? That is, the first parameter is a `const string &` rather than a `string`, as used in previous examples for our `Initialize()` member function? Both are acceptable. A `string` parameter will pass a copy of the formal parameter on the stack to the member function. If the formal parameter is a string literal in quotes (such as `"University of Delaware"`), a `string` instance will first be made to house this literal string of characters. In comparison, if the parameter to the constructor is a `const string &`, then a reference to the formal parameter will be passed to this function and the object referenced will be treated as `const`. In the body of the constructor, we use assignment to copy the value of the input parameter to the data member. Not to worry, the implementation of the assignment operator for the `string` class performs a deep assignment from the source to the destination string. The implication is that we do not have to worry about the data member sharing memory (that is, not having its own copy) with the initialization data (string). Therefore, either use of a `string` or `const string &` as a parameter for the constructor is acceptable.

Now, let's complement our use of constructors with in-class initializers.

Constructors and in-class initializers

In addition to initializing data members within a constructor, a class may optionally contain **in-class initializers**. That is, default values that can be specified in a class definition as a means to initialize data members in the absence of specific constructor initialization (or assignment) of those data members.

Let's consider a revision of our previous example:

```cpp
class University
{
private:
    string name {"None"}; // in-class initializer to be
    int numStudents {0};  // used when values not set in
                          // constructor
    // Above line same as: int numStudents = 0;
public:
    University(); // default constructor
    // assume remainder of class def is as previously shown
};


University::University()
{   // Because there are no initializations (or
    // assignments) of data members name, numStudents
    // in this constructor, the in-class initializer
    // values will persist.
    // This constructor, with its signature, is still
    // required for the instantiation below, in main()
}
// assume remaining member functions exist here

int main()
{
    University u1;  // in-class initializers are used
}
```

In the previous code fragment, notice that our class definition for University contains two in-class initializers for data members name and numStudents. These values will be used to initialize data members for a University instance when a University constructor does not otherwise set these values. More specifically, if a University constructor uses initialization to set these values, the in-class initializers will be ignored (we will see formal constructor initialization with the member initialization list shortly in this chapter).

Additionally, if a constructor sets these data members through assignment within the body of a constructor (as we have seen in the previous constructor example), the assignments will overwrite any in-class initialization that was otherwise done on our behalf. However, if we do not set data members in a constructor (as shown in the current code fragment), the in-class initializers will be utilized.

In-class initializers can be used to simplify default constructors or to alleviate default values specified within a constructor's prototype (a style that is becoming less popular).

As we have seen in this example, in-class initializers can lead to a default constructor having no work (that is, initialization) remaining to be conducted in the method body itself. Yet, we can see that in some cases, a default constructor is necessary if we would like to use the default interface for instantiation. In cases such as these, =default may be added to the prototype of the default constructor to indicate that the system-supplied default constructor (with an empty body) should be linked in on our behalf, alleviating our need to provide an empty, default constructor ourselves (as in our previous example).

With this improvement, our class definition will become the following:

```
class University
{
private:
    string name {"None"}; // in-class init. to be used when
    int numStudents {0};  // values not set in constructor
public:
    // request the default constructor be linked in
    University() = default;
    University(const string &, int);
    void Print();
    void CleanUp();
};
```

In the previous class definition, we have now requested the system-supplied default constructor (with an empty body) in a situation where we would not have otherwise gotten one automatically (because we have provided a constructor with another signature). We have saved specifying an empty-bodied default constructor ourselves, as in our original example.

Next, let's add to our knowledge of constructors by examining a copy constructor.

Creating copy constructors

A **copy constructor** is a specialized constructor that is invoked whenever a copy of an object may need to be made. Copy constructors may be invoked during the construction of another object. They may also be invoked when an object is passed by value to a function via an input parameter or returned by value from a function.

Often, it is easier to make a copy of an object and modify the copy slightly than to construct a new object with its individual attributes from scratch. This is especially true if a programmer requires a copy of an object that has undergone many changes during the life of the application. It may be impossible to recall the order of various transformations that may have been applied to the object in question in order to create a duplicate. Instead, having the means to copy an object is desirable, and possibly crucial.

The signature of a copy constructor is `ClassName::ClassName(const ClassName &);`. Notice that a single object is explicitly passed as a parameter, and that parameter will be a reference to a constant object. The copy constructor, as do most member functions, will receive an implicit argument to the function, the `this` pointer. The purpose of the copy constructor's definition will be to make a copy of the explicit parameter to initialize the object pointed to by `this`.

If no copy constructor is implemented by the `class` (or `struct`) designer, one will be provided for you (in the `public` access region) that performs a shallow, member-wise copy. This is unlikely not what you want if you have data members in your class that are pointers. Instead, the best thing to do is to write a copy constructor yourself, and write it to perform a deep copy (allocating memory as necessary) for data members that are pointers.

Should the programmer wish to disallow copying during construction, `=delete` can be used in the prototype of the copy constructor as follows:

```
// disallow copying during construction
Student(const Student &) = delete;    // prototype
```

Alternatively, if the programmer wishes to prohibit object copying, a copy constructor may be prototyped in the `private` access region. In this case, the compiler will link in the default copy constructor (which performs a shallow copy), but it will be considered private. Therefore, instantiations that would utilize the copy constructor outside the scope of the class will be prohibited. This technique is used less frequently since the advent of `=delete`; however, it may be seen in existing code, so it is useful to understand.

Let's examine a copy constructor, starting with the class definition. Though the program is presented in several fragments, the full program example may be found in the GitHub repository:

```
https://github.com/PacktPublishing/Deciphering-Object-Orient-
ed-Programming-with-CPP/blob/main/Chapter05/Chp5-Ex5.cpp
```

```
#include <iostream>
#include <cstring>    // though we'll prefer std::string,
// one ptr data member will illustrate important concepts
using std::cout;       // preferred to: using namespace std;
using std::endl;
```

```
using std::string;

class Student
{
private:
    // data members
    string firstName;
    string lastName;
    char middleInitial;
    float gpa;
    char *currentCourse;   // ptr to illustrate key concepts
public:
    // member function prototypes
    Student();   // default constructor
    Student(const string &, const string &, char, float,
            const char *);
    Student(const Student &);   // copy constructor proto.
    void CleanUp();
    void Print();
    void SetFirstName(const string &);
};
```

In this program segment, we start by defining class Student. Notice the usual assortment of private data members and public member function prototypes, including the default constructor and an overloaded constructor. Also notice the prototype for the copy constructor Student(const Student &);.

Next, let's take a look at the member function definitions with the following continuation of our program:

```
// default constructor
Student::Student()
{
    // Because firstName and lastName are member objects of
    // type string, they are default constructed and hence
    // 'empty' by default. They HAVE been initialized.
    middleInitial = '\0';   // with a relevant value
    gpa = 0.0;
    currentCourse = 0;
```

```
}

// Alternate constructor member function definition
Student::Student(const string &fn, const string &ln,
                char mi, float avg, const char *course)
{
    firstName = fn;  // not to worry, assignment for string
    lastName = ln;   // is a deep copy into destination str
    middleInitial = mi;
    gpa = avg;
    // dynamically allocate memory for pointer data member
    currentCourse = new char [strlen(course) + 1];
    strcpy(currentCourse, course);
}

// Copy constructor definition - implement a deep copy
Student::Student(const Student &s)
{   // assignment between strings will do a deep 'copy'
    firstName = s.firstName;
    lastName = s.lastName;
    middleInitial = s.middleInitial;
    gpa = s.gpa;
    // for ptr data members, ensure a deep copy
    // allocate memory for destination
    currentCourse = new char [strlen(s.currentCourse) + 1];
    // then copy contents from source to destination
    strcpy(currentCourse, s.currentCourse);
}

// Member function definition
void Student::CleanUp()
{   // deallocate any previously allocated memory
    delete [] currentCourse;
}
```

```
// Member function definitions
void Student::Print()
{
    cout << firstName << " " << middleInitial << ". ";
    cout << lastName << " has a gpa of: " << gpa;
    cout << " and is enrolled in: " << currentCourse;
    cout << endl;
}


void Student::SetFirstName(const string &fn)
{
    firstName = fn;
}
```

In the aforementioned code fragment, we have various member function definitions. Most notably, let's consider the copy constructor definition, which is the member function with the signature of Student::Student(const Student &s).

Notice that the input parameter, s, is a reference to a Student that is const. This means that the source object, which we will be copying from, may not be modified. The destination object, which we will be copying into, will be the object pointed to by the this pointer.

As we carefully navigate the copy constructor, notice that we successively allocate space, as necessary, for any pointer data members that belong to the object pointed to by this. The space allocated is the same size as required by the data members referred to by s. We then carefully copy from source data member to destination data member. We meticulously ensure that we make an exact copy in the destination object of the source object.

Notice that we are making a *deep copy* in the destination object. That is, rather than simply copying the pointers contained in s.currentCourse to this->currentCourse, for example, we instead allocate space for this->currentCourse and then copy over the source data. The result of a shallow copy would instead be that the pointer data members in each object would share the same dereferenced memory (that is, the memory to which each pointer points). This is most likely not what you would want in a copy. Also recall that the default behavior of a system-supplied copy constructor would be to provide a shallow copy from the source to the destination object. It is also worthy to note that the assignment between two strings such as firstName = s.firstName; in the copy constructor will perform a deep assignment from source to destination string because that is the behavior of the assignment operator defined by the string class.

Now, let's take a look at our `main()` function to see the various ways in which the copy constructor could be invoked:

```cpp
int main()
{
    // instantiate two Students
    Student s1("Zachary", "Moon", 'R', 3.7, "C++");
    Student s2("Gabrielle", "Doone", 'A', 3.7, "C++");

    // These inits implicitly invoke copy constructor
    Student s3(s1);
    Student s4 = s2;
    s3.SetFirstName("Zack");// alter each object slightly
    s4.SetFirstName("Gabby");

    // This sequence does not invoke copy constructor
    // This is instead an assignment.
    // Student s5("Giselle", "LeBrun", 'A', 3.1, "C++");
    // Student s6;
    // s6 = s5;   // this is assignment, not initialization

    s1.Print();   // print each instance
    s3.Print();
    s2.Print();
    s4.Print();

    s1.CleanUp(); // Since some data members are pointers,
    s2.CleanUp(); // let's call a function to delete() them
    s3.CleanUp();
    s4.CleanUp();
    return 0;
}
```

In `main()`, we declare two instances of `Student`, `s1` and `s2`, and each is initialized with the constructor that matches the signature of `Student::Student(const string &, const string &, char, float, const char *);`. Notice that the signature used in instantiation is how we select which constructor should be implicitly called.

Next, we instantiate s3 and pass as an argument to its constructor the object s1 with Student s3 (s1) ;. Here, s1 is of type Student, so this instantiation will match the constructor that accepts a reference to a Student, the copy constructor. Once in the copy constructor, we know that we will make a deep copy of s1 to initialize the newly instantiated object, s3, which will be pointed to by the this pointer in the scope of the copy constructor method.

Additionally, we instantiate s4 with the following line of code: Student s4 = s2;. Here, because this line of code is an initialization (that is, s4 is both declared and given a value in the same statement), the copy constructor will also be invoked. The source object of the copy will be s2 and the destination object will be s4. Notice that we then modify each of the copies (s3 and s4) slightly by modifying their firstName data members.

Next, in the commented-out section of code, we instantiate two objects of type Student, s5 and s6. We then try to assign one to the other with s5 = s6;. Though this looks similar to the initialization between s4 and s2, it is not. The line s5 = s6; is an assignment. Each of the objects existed previously. As such, the copy constructor is not called for this segment of code. Nonetheless, this code is legal and has similar implications as with the assignment operator. We will examine these details later in the book when we discuss operator overloading in *Chapter 12, Friends and Operator Overloading*.

We then print out objects s1, s2, s3, and s4. Then, we call Cleanup () on each of these four objects. Why? Each object contained data members that were pointers, so it is appropriate to delete the heap memory contained within each instance (that is, selected pointer data members) prior to these outer stack objects going out of scope.

Here is the output to accompany the full program example:

```
Zachary R. Moon has a gpa of: 3.7 and is enrolled in: C++
Zack R. Moon has a gpa of: 3.7 and is enrolled in: C++
Gabrielle A. Doone has a gpa of: 3.7 and is enrolled in: C++
Gabby A. Doone has a gpa of: 3.7 and is enrolled in: C++
```

The output for this example shows each original Student instance, paired with its copy. Notice that each copy has been modified slightly from the original (firstName differs).

> **Related topic**
>
> It is interesting to note that the assignment operator shares many similarities with the copy constructor, in that it can allow data to be copied from a source to destination instance. However, the copy constructor is implicitly invoked for the initialization of a new object, whereas the assignment operator will be invoked when performing an assignment between two existing objects. Nonetheless, the methods of each will look strikingly similar! We will examine overloading the assignment operator to customize its behavior to perform a deep assignment (much like a deep copy) in *Chapter 12, Friends and Operator Overloading*.

Now that we have a deep understanding of copy constructors, let's look at one last variety of constructor, the conversion constructor.

Creating conversion constructors

Type conversions can be performed from one user defined type to another, or from a standard type to a user defined type. A conversion constructor is a language mechanism that allows such conversions to occur.

A **conversion constructor** is a constructor that accepts one explicit argument of a standard or user defined type and applies a reasonable conversion or transformation on that object to initialize the object being instantiated.

Let's take a look at an example illustrating this idea. Though the example will be broken into several segments and also abbreviated, the full program can be found in the GitHub repository:

https://github.com/PacktPublishing/Deciphering-Object-Oriented-Programming-with-CPP/blob/main/Chapter05/Chp5-Ex6.cpp

```
#include <iostream>
#include <cstring>   // though we'll prefer std::string,
// one ptr data member will illustrate important concepts
using std::cout;     // preferred to: using namespace std;
using std::endl;
using std::string;


class Student;       // forward declaration of Student class

class Employee
{
private:
    string firstName;
    string lastName;
    float salary;
public:
    Employee();
    Employee(const string &, const string &, float);
    Employee(Student &);  // conversion constructor
    void Print();
};
```

```
class Student
{
private: // data members
    string firstName;
    string lastName;
    char middleInitial;
    float gpa;
    char *currentCourse;  // ptr to illustrate key concepts
public:
    // constructor prototypes
    Student();  // default constructor
    Student(const string &, const string &, char, float,
            const char *);
    Student(const Student &);  // copy constructor
    void Print();
    void CleanUp();
    float GetGpa(); // access function for private data mbr
    const string &GetFirstName();
    const string &GetLastName();
};
```

In the previous program segment, we first include a forward declaration to class Student; – this declaration allows us to refer to the Student type prior to its definition. We then define class Employee. Notice that this class includes several private data members and three constructor prototypes – a default, alternative, and conversion constructor. As a side note, notice that a copy constructor has not been programmer-specified. This means that a default (shallow) copy constructor will be provided by the compiler. In this case, since there are no pointer data members, the shallow copy is acceptable.

Nonetheless, let us continue by examining the Employee conversion constructor prototype. Notice that in the prototype, this constructor takes a single argument. The argument is a Student &, which is why we needed the forward declaration for Student. Preferably, we might use a const Student & as the parameter type, but we will need to understand const member functions (later in this chapter) in order to do so. The type conversion that will take place will be to convert a Student into a newly constructed Employee. It will be our job to provide a meaningful conversion to accomplish this in the definition for the conversion constructor, which we will see shortly.

Next, we define our Student class, which is much the same as we have seen in previous examples.

Now, let us continue with the example to see the member function definitions for `Employee` and `Student`, and our `main()` function, in the following code segment. To conserve space, selected member function definitions will be omitted, however, the online code will show the program in its entirety.

Moving onward, our member functions for `Employee` and `Student` are as follows:

```
Employee::Employee()   // default constructor
{
    // Remember, firstName, lastName are member objects of
    // type string; they are default constructed and hence
    // 'empty' by default. They HAVE been initialized.
    salary = 0.0;
}


// alternate constructor
Employee::Employee(const string &fn, const string &ln,
                   float money)
{
    firstName = fn;
    lastName = ln;
    salary = money;
}


// conversion constructor param. is a Student not Employee
// Eventually, we can properly const qualify parameter, but
// we'll need to learn about const member functions first…
Employee::Employee(Student &s)
{
    firstName = s.GetFirstName();
    lastName = s.GetLastName();
    if (s.GetGpa() >= 4.0)
        salary = 75000;
    else if (s.GetGpa() >= 3.0)
        salary = 60000;
    else
        salary = 50000;
}
```

```
void Employee::Print()
{
    cout << firstName << " " << lastName << " " << salary;
    cout << endl;
}


// Definitions for Student's default, alternate, copy
// constructors, Print()and CleanUp() have been omitted
// for space, but are same as the prior Student example.

float Student::GetGpa()
{
    return gpa;
}


const string &Student::GetFirstName()
{
    return firstName;
}


const string &Student::GetLastName()
{
    return lastName;
}
```

In the previous segment of code, we notice several constructor definitions for Employee. We have a default, alternate, and conversion constructor.

Examining the definition of the Employee conversion constructor, notice that the formal parameter for the source object is s, which is of type Student. The destination object will be the Employee that is being constructed, which will be pointed to by the this pointer. In the body of this function, we carefully copy the firstName and lastName from Student &s to the newly instantiated Employee. Note that we use access functions const string &Student::GetFirstName() and const string &Student::GetLastName() to do so (via an instance of Student), as these data members are private.

Let's continue with the conversion constructor. It is our job to provide a meaningful conversion from one type to another. In that endeavor, we try to establish an initial salary for the Employee based on the gpa of the source Student object. Because gpa is private, an access function, Student::GetGpa(), is used to retrieve this value (via the source Student). Notice that because Employee did not have any dynamically allocated data members, we did not need to allocate memory to assist in a deep copy in the body of this function.

To conserve space, the member function definitions for the Student default, alternate, and copy constructor have been omitted, as have the definition for the void Student::Print() and void Student::CleanUp() member functions. However, they are the same as in the previous full program example illustrating the Student class.

Notice that access functions for private data members in Student, such as float Student::GetGpa(), have been added to provide safe access to those data members. Note that the value returned from float Student::GetGpa() on the stack is a copy of the gpa data member. The original gpa is in no worry of being breached by the use of this function. The same applies for member functions const string &Student::GetFirstName() and const string &Student::GetLastName(), which each returns a const string &, ensuring that the data that will be returned will not be breached.

Let's complete our program by examining our main() function:

```
int main()
{
    Student s1("Giselle", "LeBrun", 'A', 3.5, "C++");
    Employee e1(s1);   // conversion constructor
    e1.Print();
    s1.CleanUp();   // CleanUp() will delete() s1's
    return 0;       // dynamically allocated data members
}
```

In our main() function, we instantiate a Student, namely s1, which is implicitly initialized with the matching constructor. Then we instantiate an Employee, e1, using the conversion constructor in the call Employee e1(s1);. At a quick glance, it may seem that we are utilizing the Employee copy constructor. But at a closer look, we notice that the actual parameter s1 is of type Student, not Employee. Hence, we are using Student s1 as a basis to initialize Employee e1. Note that in no manner is the Student, s1, harmed or altered in this conversion. For this reason, it would be preferable to define the source object as a const Student & in the formal parameter list; once we understand const member functions, which will then be required for usage in the body of the conversion constructor, we can do so.

To conclude this program, we print out the Employee using Employee::Print(), which enables us to visualize the conversion we applied for a Student to an Employee.

Here is the output to accompany our example:

```
Giselle LeBrun 60000
```

Before we move forward, there's one final, subtle detail about conversion constructors that is very important to understand.

> **Important note**
>
> Any constructor that takes a single argument is considered a conversion constructor, which can potentially be used to convert the parameter type to the object type of the class to which it belongs. For example, if you have a constructor in the Student class that takes only a float, this constructor could be employed not only in the manner shown in the preceding example but also in places where an argument of type Student is expected (such as a function call), when an argument of type float is instead supplied. This may not be what you intend, which is why this interesting feature is being called out. If you don't want implicit conversions to take place, you can disable this behavior by declaring the constructor with the explicit keyword at the beginning of its prototype.

Now that we understand basic, alternative, copy and conversion constructors in C++, let's move forward and explore the constructor's complementary member function, the C++ destructor.

Understanding destructors

Recall how conveniently a class constructor provides us with a way to initialize a newly instantiated object? Rather than having to remember to call an Initialize() method for each instance of a given type, the constructor allows initialization automatically. The signature used in construction helps specify which of a series of constructors should be used.

What about object clean-up? Many classes contain dynamically allocated data members, which are often allocated in a constructor. Shouldn't the memory comprising these data members be released when the programmer is done with an instance? Certainly. We have written a CleanUp() member function for several of our example programs. And we have remembered to call CleanUp(). Conveniently, similar to a constructor, C++ has an automatically built-in feature to serve as a clean-up function. This function is known as the destructor.

Let's look at the destructor to understand its proper usage.

Applying destructor basics and proper usage

A **destructor** is a member function whose purpose is to relinquish the resources an object may have acquired during its existence. A destructor is automatically invoked when a class or struct instance has either of the following occur:

- Goes out of scope (this applies to non-pointer variables)
- Is explicitly deallocated using `delete` (for pointers to objects)

A destructor should (most often) clean up any memory that may have been allocated by the constructor. The destructor's name is a ~ character followed by the `class` name. A destructor will have no arguments; therefore, it cannot be overloaded. Lastly, the return type for a destructor may not be specified. Both classes and structures may have destructors.

In addition to deallocating memory that a constructor may have allocated, a destructor may be used to perform other end-of-life tasks for an instance, such as logging a value to a database. More complex tasks may include informing objects pointed to by class data members (whose memory is not being released) that the object at hand will be concluding. This may be important if the linked object contains a pointer back to the terminating object. We will see examples of this later in the book, in *Chapter 10, Implementing Association, Aggregation, and Composition*.

If you have not provided a destructor, the compiler will create and link in a `public` destructor with an empty body. This is necessary because a destructor call is automatically patched in just prior to the point when local instances are popped off the stack, and with `delete()`, just prior to the memory release of dynamically allocated instances. It is easier for the compiler to always patch in this call, rather than constantly looking to see whether your class has a destructor or not. Be sure to provide a class destructor yourself when there are resources to clean up or dynamically allocated memory requiring release. If the destructor will be empty, consider using `=default` in its prototype to acknowledge its automatic inclusion (and to forego providing a definition yourself); this practice, however, adds unnecessary code and therefore is becoming less popular.

There are some potential pitfalls. For example, if you forget to delete a dynamically allocated instance, the destructor call will not be patched in for you. C++ is a language that gives you the flexibility and power to do (or not do) anything. If you do not delete memory using a given identifier (perhaps two pointers refer to the same memory), please remember to delete it through the other identifier at a later date.

There's one last item worth mentioning. Though you may call a destructor explicitly, you will rarely ever need to do so. Destructor calls are implicitly patched in by the compiler on your behalf in the aforementioned scenarios. Only in very few advanced programming situations will you need to explicitly call a destructor yourself.

Let's take a look at a simple example illustrating a class destructor, which will be broken into three segments. Its full example can be seen in the GitHub repository listed here:

https://github.com/PacktPublishing/Deciphering-Object-Orient-ed-Programming-with-CPP/blob/main/Chapter05/Chp5-Ex7.cpp

```cpp
#include <iostream>
#include <cstring>      // though we'll prefer std::string,
// one ptr data member will illustrate important concepts
using std::cout;        // preferred to: using namespace std;
using std::endl;
using std::string;

class University
{
private:
    char *name;     // ptr data member shows destructor
                    // purpose
    int numStudents;
public:
    // constructor prototypes
    University(); // default constructor
    University(const char *, int); // alternate constructor
    University(const University &);   // copy constructor
    ~University();   // destructor prototype
    void Print();
};
```

In the previous segment of code, we first define class University. Notice the private access region filled with data members, and the public interface, which includes prototypes for a default, alternate, and copy constructor, as well as for the destructor and a Print() method.

Next, let's take a look at the various member function definitions:

```cpp
University::University()  // default constructor
{
    name = nullptr;
    numStudents = 0;
}
```

```
University::University(const char *n, int num)
{   // allocate memory for pointer data member
    name = new char [strlen(n) + 1];
    strcpy(name, n);
    numStudents = num;
}


University::University(const University &u) // copy const
{
    name = new char [strlen(u.name) + 1];   // deep copy
    strcpy(name, u.name);
    numStudents = u.numStudents;
}


University::~University()   // destructor definition
{
    delete [] name;  // deallocate previously allocated mem
    cout << "Destructor called " << this << endl;
}

void University::Print()
{
    cout << "University: " << name;
    cout << " Enrollment: " << numStudents << endl;
}
```

In the aforementioned code fragment, we see the various overloaded constructors we are now accustomed to seeing, plus `void University::Print()`. The new addition is the destructor definition.

Notice the destructor `University::~University()` takes no arguments; it may not be overloaded. The destructor simply deallocates memory that may have been allocated in any of the constructors. Note that we simply `delete [] name;`, which will work whether `name` points to a valid address or contains a null pointer (yes, applying `delete` to a null pointer is OK). We additionally print the `this` pointer in the destructor, just for fun, so that we can see the address of the instance that is approaching non-existence.

Next, let's take a look at `main()` to see when the destructor may be called:

```
int main()
{
    University u1("Temple University", 39500);
    University *u2 = new University("Boston U", 32500);
    u1.Print();
    u2->Print();
    delete u2;  // destructor will be called before delete()
                // and destructor for u1 will be called
    return 0;   // before program completes
}
```

Here, we instantiate two `University` instances; `u1` is an instance, and `u2` points to an instance. We know that `u2` is instantiated when its memory becomes available with `new()` and that once the memory has become available, the applicable constructor is called. Next, we call `University::Print()` for both instances to have some output.

Finally, toward the end of `main()`, we delete `u2` to return this memory to the heap management facility. Just prior to memory release, with the call to `delete()`, C++ will patch in a call to the destructor for the object pointed to by `u2`. It is as if a secret function call `u2->~University();` has been patched in prior to `delete u2;` (note, this is done automatically, no need for you to do so as well). The implicit call to the destructor will delete the memory that may have been allocated for any data members within the class. The memory release is now complete for `u2`.

What about instance `u1`? Will its destructor be called? Yes; `u1` is a stack instance. Just prior to its memory being popped off the stack in `main()`, the compiler will have patched in a call to its destructor, as if the call `u1.~University();` was added on your behalf (again, no need to do so yourself). For the instance `u1`, the destructor will also deallocate any memory for data members that may have been allocated. Likewise, the memory release is now complete for `u1`.

Notice that in each destructor call, we have printed a message to illustrate when the destructor is called, and have also printed out the memory address for `this` to allow you to visualize each specific instance as it is destructed.

Here is the output to accompany our full program example:

```
University: Temple University Enrollment: 39500
University: Boston U Enrollment: 32500
Destructor called 0x10d1958
Destructor called 0x60fe74
```

With this example, we have now examined the destructor, the complement to the series of class constructors. Let us move on to another set of useful topics relating to classes: various keyword qualifications of data members and member functions.

Applying qualifiers to data members and member functions

In this section, we will investigate qualifiers that can be added to both data members and member functions. The various qualifiers – `inline`, `const`, and `static` – can support program efficiency, aid in keeping private data members safe, support encapsulation and information hiding, and additionally be used to implement various object-oriented concepts.

Let's get started with the various types of member qualifications.

Adding inline functions for potential efficiency

Imagine a set of short member functions in your program that are repeatedly called by various instances. As an object-oriented programmer, you appreciate using a `public` member function to provide safe and controlled access to `private` data. However, for very short functions, you worry about efficiency. That is, the overhead of calling a small function repeatedly. Certainly, it would be more efficient to just paste in the two or three lines of code comprising the function. Yet, you resist because that may mean providing `public` access to otherwise hidden class information, such as data members, which you are hesitant to do. An `inline` function can solve this dilemma, allowing you to have the safety of a member function to access and manipulate your private data, yet the efficiency of executing several lines of code without the overhead of a function call.

An **inline** function is a function whose invocation is substituted with the body of the function itself. Inline functions can help eliminate the overhead associated with calling very small functions.

Why would calling a function have overhead? When a function is called, input parameters (including `this`) are pushed onto the stack, space is reserved for a return value of the function (though sometimes registers are used), and moving to another section of code requires storing information in registers to jump to that section of code, and so on. Replacing very small function bodies with inline functions can add to program efficiency.

An inline function may be specified using either of the following mechanisms:

- Placing the function definition inside the class definition
- Placing the keyword `inline` prior to the return type in the (typical) function definition, found outside the class definition

Specifying a function as `inline` in one of the aforementioned two fashions is merely a request to the compiler to consider the substitution of the function body for its function call. This substitution is not guaranteed. When might the compiler not actually inline a given function? If a function is recursive, it cannot be made `inline`. Likewise, if a function is lengthy, the compiler will not inline the function. Also, if the function call is dynamically bound with the specific implementation determined at run time (virtual functions), it cannot be made `inline`.

An `inline` function definition should be declared in the header file with the corresponding class definition. This will allow for any revisions to the function to be re-expanded correctly should the need arise.

Let's see an example using `inline` functions. The program will be broken into two segments, with some well-known functions removed. However, the full program may be seen in the GitHub repository:

https://github.com/PacktPublishing/Deciphering-Object-Oriented-Programming-with-CPP/blob/main/Chapter05/Chp5-Ex8.cpp

```cpp
#include <iostream>
#include <cstring>    // though we'll prefer std::string,
                      // one ptr data member will illustrate
                      // important concepts
using std::cout;      // preferred to: using namespace std;
using std::endl;
using std::string;

class Student
{
private:
    // data members
    string firstName;
    string lastName;
    char middleInitial;
    float gpa;
    char *currentCourse;   // ptr to illustrate key concepts
public:
    // member function prototypes
    Student();  // default constructor
    Student(const student &, const student &, char, float,
            const char *);
```

```
    Student(const Student &);   // copy constructor
    ~Student();   // destructor
    void Print();
    // inline function definitions
    const string &GetFirstName() { return firstName; }
    const string &GetLastName() { return lastName; }
    char GetMiddleInitial() { return middleInitial; }
    float GetGpa() { return gpa; }
    const char *GetCurrentCourse()
        { return currentCourse; }
    // prototype only, see inline function definition below
    void SetCurrentCourse(const char *);
};


inline void Student::SetCurrentCourse(const char *c)
{   // notice the detailed work to reset ptr data member;
    // it's more involved than if currentCourse was a str
    delete [] currentCourse;
    currentCourse = new char [strlen(c) + 1];
    strcpy(currentCourse, c);

}
```

In the previous program fragment, let's start with the class definition. Notice that several access function definitions have been added in the class definition itself, namely, functions such as GetFirstName(), GetLastName(), and so on. Look closely; these functions are actually defined within the class definition. For example, float GetGpa() { return gpa; } is not just the prototype, but the full function definition. By virtue of the function placement within the class definition, functions such as these are considered inline.

These small functions provide safe access to private data members. Notice const char *GetCurrentCourse(), for example. This function returns a pointer to currentCourse, which is stored in the class as a char *. But because the return value of this function is a const char *, this means that anyone calling this function must treat the return value as a const char *, which means treating it as unmodifiable. Should this function's return value be stored in a variable, that variable must also be defined as const char *. By upcasting this pointer to an unmodifiable version of itself with the return value, we are adding the provision that no one can get their hands on a private data member (which is a pointer) and then change its value.

Now, notice toward the end of the class definition, we have a prototype for `void SetCurrentCourse(const char *);`. Then, outside of this class definition, we will see the definition for this member function. Notice the keyword `inline` prior to the `void` return type of this function definition. The keyword must be explicitly used here since the function is defined outside of the class definition. Remember, with either style of `inline` designation for a method, the `inline` specification is merely a request to the compiler to make the substitution of function body for function call. As with any function, if you provide a prototype (without `=default`), be sure to provide a function definition (or else the linker will definitely complain).

Let's continue this example by examining the remainder of our program:

```cpp
// Definitions for default, alternate, copy constructor,
// and Print() have been omitted for space,
// but are same as last example for class Student

// the destructor is shown because we have not yet seen
// an example destructor for the Student class
Student::~Student()
{   // deallocate previously allocated memory
    delete [] currentCourse;
}

int main()
{
    Student s1("Jo", "Muritz", 'Z', 4.0, "C++");
    cout << s1.GetFirstName() << " " << s1.GetLastName();
    cout << " Enrolled in: " << s1.GetCurrentCourse();
    cout << endl;
    s1.SetCurrentCourse("Advanced C++ Programming");
    cout << s1.GetFirstName() << " " << s1.GetLastName();
    cout << " New course: " << s1.GetCurrentCourse();
    cout << endl;
    return 0;
}
```

Notice that in the remainder of our program example, several member function definitions have been omitted. The bodies of these functions are identical to the previous example illustrating a `Student` class in full, and can also be viewed online.

Let's focus instead on our `main()` function. Here, we instantiate a `Student`, namely `s1`. We then invoke several `inline` function calls via `s1`, such as `s1.GetFirstName();`. Because `Student::GetFirstName()` is inline, it is as if we are accessing data member `firstName` directly, as the body of this function merely has a `return firstName;` statement. We have the safety of using a function to access a `private` data member (meaning that no one can modify this data member outside the scope of the class), but the speed of an inline function's code expansion to eliminate the overhead of a function call.

Throughout `main()`, we make several other calls to `inline` functions in this same manner, including `s1.SetCurrentCourse();`. We now have the safety of encapsulated access with the speed of direct access to data members using small `inline` functions.

Here is the output to accompany our full program example:

```
Jo Muritz Enrolled in: C++
Jo Muritz New course: Advanced C++ Programming
```

Let's now move onward by investigating another qualifier we can add to class members, the `const` qualifier.

Adding const data members and the member initialization list

We have already seen earlier in this book how to constant-qualify variables and the implications of doing so. To briefly recap, the implication of adding a `const` qualifier to a variable is that the variable must be initialized when it is declared and that its value may never again be modified. We previously also saw how to add `const` qualification to pointers, such that we could qualify the data being pointed to, the pointer itself, or both. Let us now examine what it means to add a `const` qualifier to data members within a class, and learn about specific language mechanisms that must be employed to initialize those data members.

Data members that should never be modified should be qualified as `const`. A **const data member** is one that may only be initialized, and never assigned a new value. Just as with `const` variables, *never modified* means that the data member may not be modified using its own identifier. It will then be our job to ensure that we do not initialize our data members that are pointers to `const` objects with objects that are not labeled as `const` (lest we provide a back door to change our private data).

Keep in mind that in C++, a programmer can always cast the const-ness away from a pointer variable. Not that they should. Nonetheless, we will employ safety measures to ensure that by using access regions and appropriate return values from access functions, we do not easily provide modifiable access to our `private` data members.

The **member initialization list** must be used in a constructor to initialize any data members that are constant, or that are references. A member initialization list offers a mechanism to initialize data members that may never be l-values in an assignment. A member initialization list may also be used to initialize non-const data members. For performance reasons, the member initialization list is most often the preferred way to initialize any data member (const or non-const). The member initialization list also provides a manner to specify preferred construction for any data members that are of class types themselves (that is, member objects).

A member initialization list may appear in any constructor, and to indicate this list, simply place a : after the formal parameter list, followed by a comma-separated list of data members, paired with the initial value for each data member in parentheses. For example, here we use the member initialization list to set two data members, gpa and middleInitial:

```
Student::Student(): gpa(0.0), middleInitial('\0')
{
    // Remember, firstName, lastName are member objects of
    // type string; they are default constructed and hence
    // 'empty' by default. They HAVE been initialized.
    currentCourse = nullptr; // don't worry - we'll change
}                            // currentCourse to a string next!
```

Though we have used the member initialization list to initialize two data members in the previous constructor, we could have used it to set all of the data members! We'll see this proposition (and preferred usage) momentarily.

Data members in the member initialization list are initialized in the order in which they appear (that is, declared) in the class definition (except static data members, which we will see shortly). Next, the body of the constructor is executed. It is a nice convention to order the data members in the member initialization list to appear in the same order as the class definition. But remember, the order of actual initialization matches the order that the data members are specified in the class definition, irrespective of member initialization list ordering.

It is interesting to note that a reference must use the member initialization list because references are implemented as constant pointers. That is, the pointer itself points to a specific other object and may not point elsewhere. The values of that object may change, but the reference always references a specific object, the one in which it was initialized.

Using const qualification with pointers can be tricky to determine which scenarios require initialization with this list, and which do not. For example, a pointer to a constant object does not need to be initialized with the member initialization list. The pointer could point to any object, but once it does, it may not change the dereferenced value. However, a constant pointer must be initialized with the member initialization list because the pointer itself is fixed to a specific address.

Let's take a look at a `const` data member and how to use the member initialization list to initialize its value in a full program example. We will also see how to use this list to initialize non-const data members. Though this example is segmented and not shown in its entirety, the full program can be found in the GitHub repository:

```
https://github.com/PacktPublishing/Deciphering-Object-Orient-
ed-Programming-with-CPP/blob/main/Chapter05/Chp5-Ex9.cpp
```

```cpp
#include <iostream>
using std::cout;      // preferred to: using namespace std;
using std::endl;
using std::string;

class Student
{
private:
    // data members
    string firstName;
    string lastName;
    char middleInitial;
    float gpa;
    string currentCourse; // let's finally change to string
    const int studentId;   // added, constant data member
public:
    // member function prototypes
    Student();  // default constructor
    Student(const string &, const string &, char, float,
            const string &, int);
    Student(const Student &);  // copy constructor
    ~Student();  // destructor
    void Print();
    const string &GetFirstName() { return firstName; }
    const string &GetLastName() { return lastName; }
    char GetMiddleInitial() { return middleInitial; }
    float GetGpa() { return gpa; }
    const string &GetCurrentCourse()
        { return currentCourse; }
```

```
     void SetCurrentCourse(const string &);   // proto. only
};
```

In the aforesaid `Student` class, notice that we have added a data member, `const int studentId;`, to the class definition. This data member will require the use of the member initialization list to initialize this constant data member in each of the constructors.

Let's take a look at how the use of the member initialization list will work with constructors:

```
// Definitions for the destructor, Print(), and
// SetCurrentCourse() have been omitted to save space.
// They are similar to what we have seen previously.


// Constructor w/ member init. list to set data mbrs
Student::Student(): firstName(), lastName(),
                    middleInitial('\0'), gpa(0.0),
                    currentCourse(), studentId(0)
{
    // You may still set data members here, but using above
    // initialization is more efficient than assignment
    // Note: firstName, lastName are shown in member init.
    // list selecting default constructor for init.
    // However, as this is the default action for member
    // objects (string), we don't need to explicitly incl.
    // these members in the member initialization list
    // (nor will we include them in future examples).
}


Student::Student(const string &fn, const string &ln,
        char mi, float avg, const string &course, int id):
        firstName(fn), lastName(ln), middleInitial(mi),
        gpa(avg), currentCourse(course), studentId (id)
{
    // For string data members, the above init. calls
    // the string constructor that matches the arg in ().
    // This is preferred to default constructing a string
    // and then resetting it via assignment in the
    // constructor body.
```

```
}

Student::Student(const Student &s): firstName(s.firstName),
      lastName(s.lastName), middleInitial(s.middleInitial),
      gpa(s.gpa), currentCourse(s.currentCourse),
      studentId(s.studentId)
{
    // remember to do a deep copy for any ptr data members
}

int main()
{

    Student s1("Renee", "Alexander", 'Z', 3.7,
              "C++", 1290);
    cout << s1.GetFirstName() << " " << s1.GetLastName();
    cout << " has gpa of: " << s1.GetGpa() << endl;
    return 0;
}
```

In the preceding code fragment, we see three Student constructors. Notice the various member initialization lists, designated by a : after the formal parameter list for each of the three constructors.

Of particular interest is the member initialization list usage for data members that are of type string (or as we'll later see, of any class type). In this usage, the string data members are constructed using the member initialization list using the specified constructor; that is, the one whose signature matches the argument in (). This is inevitably more efficient than default constructing each string (which is what happened previously behind the scenes) and then resetting its value via assignment within the constructor method body.

With this in mind, the default string constructor selection in the member initialization list of the Student default constructor – that is, :firstName(), lastName(), currentCourse() – is shown to emphasize that these data members are member objects (of type string) and will be constructed. In this case, they will each be default constructed, which will provide their contents with an empty string. However, member objects will always be default constructed unless otherwise directed using the member initialization list. For this reason, the :firstName(), lastName(), and currentCourse() specifications in the member initialization list are optional and will not be included in future examples.

Each constructor will make use of the member initialization list to set the values of data members that are `const`, such as `studentId`. Additionally, the member initialization list can be used as a simple (and more efficient) way to initialize any other data member. We can see examples of the member initialization list being used to simply set non-const data members by viewing the member initialization list in either the default or alternate constructor, for example, `Student::Student()` `: studentId(0), gpa(0.0)`. In this example, gpa is not `const`, so its use in the member initialization list is optional.

Here is the output to accompany our full program example:

```
Renee Alexander has gpa of: 3.7
```

> **Important note**
>
> Even though the constructor's member initialization list is the only mechanism that can be used to initialize const data members (or those that are references or member objects), it is also often the preferred mechanism to perform simple initialization for any data member for performance reasons. In many cases (such as member objects – for example, a string), this saves data members from being first initialized (constructed themselves) with a default state and then re-assigned a value in the body of the constructor.

It is interesting to note that programmers may choose to utilize either ( ) or { } in the member initialization list to initialize data members. Notice the use of { } in the following code:

```
Student::Student(const string &fn, const string &ln,
         char mi, float avg, const string &course, int id):
         firstName{fn}, lastName{ln}, middleInitial{mi},
         gpa{avg}, currentCourse{course}, studentId{id}
{
}
```

The { } as used here were originally added for instantiation in C++ (and hence with usage within member initialization lists to fully construct data members) in an effort to provide a uniform initialization syntax. The { } also potentially control the narrowing of data types. However, when `std::initializer_list` is used with templates (a feature we will see in *Chapter 13, Working with Templates*), the { } provides semantic confusion. Due to complexities such as these interfering with the goal of language uniformity, the next C++ standard may revert to preferring the use of ( ) and so shall we. It is interesting to note that neither ( ) nor { } has an advantage from the perspective of performance.

Next, let's now move forward by adding the `const` qualifier to member functions.

Using const member functions

We have seen the constant qualifier used quite exhaustively now with data. It can also be used in conjunction with member functions. C++ provides a language mechanism to ensure that selected functions may not modify data; this mechanism is the const qualifier as applied to member functions.

A **const member function** is a member function that specifies (and enforces) that the method can only perform read-only activities on the object invoking the function.

A constant member function means that no portion of this may be modified. However, because C++ allows typecasting, it is possible to cast this to its non-const counterpart and then change data members. However, if the class designer truly meant to be able to modify data members, they simply would not label a member function as const.

Constant instances declared in your program may only invoke const member functions. Otherwise, these objects could be directly modified.

To label a member function as const, the keyword const should be specified after the argument list in the function prototype and in the function definition.

Let's see an example. It will be divided into two sections with some portions omitted; however, the full example can be seen in the GitHub repository:

https://github.com/PacktPublishing/Deciphering-Object-Orient-ed-Programming-with-CPP/blob/master/Chapter05/Chp5-Ex10.cpp

```cpp
#include <iostream>
using std::cout;    // preferred to: using namespace std;
using std::endl;
using std::string;

class Student
{
private:
    // data members
    string firstName;
    string lastName;
    char middleInitial;
    float gpa;
    string currentCourse;
    const int studentId;    // constant data member
public:
```

```
    // member function prototypes
    Student();  // default constructor
    Student(const string &, const string &, char, float,
            const string &, int);
    Student(const Student &);  // copy constructor
    ~Student();  // destructor
    void Print() const;
    const string &GetFirstName() const
        { return firstName; }
    const string &GetLastName() const
        { return lastName; }
    char GetMiddleInitial() const { return middleInitial; }
    float GetGpa() const { return gpa; }
    const string &GetCurrentCourse() const
        { return currentCourse; }
    int GetStudentId() const { return studentId; }
    void SetCurrentCourse(const string &);  // proto. only
};
```

In the previous program fragment, we see a class definition for `Student`, which is becoming very familiar to us. Notice, however, that we have added the `const` qualifier to most of the access member functions, that is, to those methods that are only providing read-only access to data.

For example, let us consider `float GetGpa() const { return gpa; }`. The `const` keyword after the argument list indicates that this is a constant member function. Notice that this function does not modify any data member pointed to by `this`. It cannot do so, as it is marked as a `const` member function.

Now, let's move on to the remainder of this example:

```
// Definitions for the constructors, destructor, and
// SetCurrentCourse() have been omitted to save space.

// Student::Print() has been revised, so it is shown below:
void Student::Print() const
{
    cout << firstName << " " << middleInitial << ". ";
    cout << lastName << " with id: " << studentId;
    cout << " and gpa: " << gpa << " is enrolled in: ";
```

```
        cout << currentCourse << endl;
}

int main()
{
    Student s1("Zack", "Moon", 'R', 3.75, "C++", 1378);
    cout << s1.GetFirstName() << " " << s1.GetLastName();
    cout << " Enrolled in " << s1.GetCurrentCourse();
    cout << endl;
    s1.SetCurrentCourse("Advanced C++ Programming");
    cout << s1.GetFirstName() << " " << s1.GetLastName();
    cout << " New course: " << s1.GetCurrentCourse();
    cout << endl;

    const Student s2("Gabby", "Doone", 'A', 4.0,
                     "C+-", 2239);
    s2.Print();
    // Not allowed, s2 is const
    // s2.SetCurrentCourse("Advanced C++ Programming");
    return 0;
}
```

In the remainder of this program, notice that we have again chosen not to include the definitions for member functions with which we are already familiar, such as the constructors, the destructor, and void Student::SetCurrentCourse().

Instead, let's focus our attention on the member function with the signature: void Student::Print() const. Here, the const keyword after the argument list indicates that no data members pointed to by this can be altered in the scope of this function. And none are. Likewise, any member functions called on this within void Student::Print() must also be const member functions. Otherwise, they could modify this.

Moving forward to examine our main() function, we instantiate a Student, namely s1. This Student calls several member functions, including some that are const. Student s1 then changes their current course using Student::SetCurrentCourse(), and then the new value of this course is printed.

Next, we instantiate another Student, s2, which is qualified as const. Notice that once this student is instantiated, the only member functions that may be applied to s2 are those that are labeled as const. Otherwise, the instance may be modified. We then print out data for s2 using Student::Print();, which is a const member function.

Did you notice the commented-out line of code: `s2.SetCurrentCourse("Advanced C++ Programming");`? This line is illegal and would not compile, because `SetCurrentCourse()` is not a constant member function and is hence inappropriate to be called via a constant instance, such as `s2`.

Let's take a look at the output for the full program example:

```
Zack Moon Enrolled in C++
Zack Moon New course: Advanced C++ Programming
Gabby A. Doone with id: 2239 and gpa: 3.9 is enrolled in: C++
```

Now that we have fully explored `const` member functions, let's continue to the final section of this chapter to delve into `static` data members and `static` member functions.

Utilizing static data members and static member functions

Now that we have been using C++ classes to define and instantiate objects, let's add to our knowledge of object-oriented concepts by exploring the idea of a class attribute. A data member that is intended to be shared by all instances of a particular class is known as a **class attribute**.

Typically, each instance of a given class has distinct values for each of its data members. However, on occasion, it may be useful for all instances of a given class to share one data member containing a single value. The object-oriented concept of a class attribute can be modeled in C++ using a **static data member**.

Static data members themselves are implemented as external (global) variables whose scope is tied back to the class in question using *name mangling*. Hence, each static data member can have its scope limited to the class in question.

Static data members are designated in the class definition with the keyword `static` preceding the data type. To finish modeling a `static` data member, an external variable definition, outside the class, must additionally follow the `static` data member specification in the class definition. Storage for this *class member* is obtained by the external variable that comprises its underlying implementation.

A static member function is one that encapsulates access to `static` data members within a class or structure. A `static` member function does not receive a `this` pointer, hence it may only manipulate `static` data members and other external (global) variables.

To indicate a `static` member function, the keyword `static` must be specified in front of the function's return type in the member function prototype only. The keyword `static` must not appear in the member function definition. If the keyword `static` appears in the function definition, the function will additionally be `static` in the C programming sense; that is, the function will be limited in scope to the file in which it is defined.

Let's take a look at an example of static data member and member function usage. The following example will be broken into segments, however, it will appear without any functions omitted or abbreviated, as it is the final example in this chapter. It can also be found in full in the GitHub repository:

https://github.com/PacktPublishing/Deciphering-Object-Oriented-Programming-with-CPP/blob/main/Chapter05/Chp5-Ex11.cpp

```cpp
#include <iostream>
#include <cstring>    // though we'll prefer std::string,
// one pointer data member will illustrate one last concept
using std::cout;      // preferred to: using namespace std;
using std::endl;
using std::string;

class Student
{
private:
    // data members
    string firstName;
    string lastName;
    char middleInitial;
    float gpa;
    string currentCourse;
    const char *studentId;  // pointer to constant string
    static int numStudents; // static data member
public:
    // member function prototypes
    Student();  // default constructor
    Student(const string &, const string &, char, float,
            const string &, const char *);
    Student(const Student &);  // copy constructor
    ~Student();  // destructor
    void Print() const;
    const string &GetFirstName() const
        { return firstName; }
    const string &GetLastName() const { return lastName; }
    char GetMiddleInitial() const { return middleInitial; }
```

```
    float GetGpa() const { return gpa; }
    const string &GetCurrentCourse() const
        { return currentCourse; }
    const char *GetStudentId() const { return studentId; }
    void SetCurrentCourse(const string &);
    static int GetNumberStudents(); // static mbr function
};


// definition for static data member
// (which is implemented as an external variable)
int Student::numStudents = 0;  // notice initial value of 0
                    // which is default for integral values
// Definition for static member function
inline int Student::GetNumberStudents()
{
    return numStudents;
}


inline void Student::SetCurrentCourse(const char *c)
{
    // far easier implementation to reset using a string
    currentCourse = c;
}
```

In the first segment of code comprising our full example, we have our Student class definition. In the private access region, we have added a data member, static int numStudents;, to model the object-oriented concept of a class attribute, a data member that will be shared by all instances of this class.

Next, notice toward the end of this class definition that we have added a static member function, static int GetNumberStudents();, to provide encapsulated access to the private data member numStudents. Note the keyword static is added in the prototype only. If we glance outside of the class definition to find the member function definition of int Student::GetNumberStudents(), we notice that there is no usage of the static keyword within the definition of this function itself. The body of this member function simply returns the shared numStudents, the static data member.

Also notice that just below the class definition is the external variable definition to support the implementation of the static data member: int Student::numStudents = 0;. Notice with this declaration the use of :: (the scope resolution operator) to associate the class name to the identifier numStudents. Though this data member is implemented as an external variable, and because the data member is labeled as private, it may only be accessed by member functions within the Student class. The implementation of a static data member as an external variable helps us understand where the memory for this shared data comes from; it is not part of any instance of the class but stored as a separate entity in the global namespace. Also notice that the declaration int Student::numStudents = 0; initializes this shared variable to a value of zero.

As an interesting aside, notice that the data member studentId has been changed from a const int to const char *studentId; in this new version of our Student class. Keep in mind that this means studentId is a pointer to a constant string, not a constant pointer. Because the memory for the pointer itself is not const, this data member will not need to be initialized using the member initialization list, but it will require some special handling.

Let's continue onward to review additional member functions comprising this class:

```
// Default constructor (note member init. list usage)
// Note: firstName, lastName, currentCourse as member
// objects (type string), will be default constructed
// to empty strings
Student::Student(): middleInitial('\0'), gpa(0.0),
                    studentId(nullptr)
{
    numStudents++;          // increment static counter
}


// Alternate constructor member function definition
Student::Student(const char *fn, const char *ln, char mi,
        float avg, const char *course, const char *id):
        firstName(fn), lastName(ln), middleInitial(mi),
        gpa(avg), currentCourse(course)
{
    // Because studentId is a const char *, we can't change
    // value pointed to directly! We enlist temp for help.
    char *temp = new char [strlen(id) + 1];
    strcpy (temp, id);    // studentId can't be an l-value,
    studentId = temp;     // but temp can!
    numStudents++;        // increment static counter
```

```
}

// copy constructor
Student::Student(const Student &s): firstName(s.firstName),
        lastName(s.lastName),middleInitial(s.middleInitial),
        gpa(s.gpa), currentCourse(s.currentCourse)
{
    delete studentId;   // release prev. allocated studentId
    // Because studentId is a const char *, we can't change
    // value pointed to directly! Temp helps w deep copy.
    char *temp = new char [strlen(s.studentId) + 1];
    strcpy (temp, s.studentId); // studentId can't be an
    studentId = temp;           // l-value, but temp can!
    numStudents++;      // increment static counter
}


Student::~Student()      // destructor definition
{
    delete [] studentId;
    numStudents--;     // decrement static counter
}

void Student::Print() const
{
    cout << firstName << " " << middleInitial << ". ";
    cout << lastName << " with id: " << studentId;
    cout << " and gpa: " << gpa << " and is enrolled in: ";
    cout << currentCourse << endl;
}
```

In the previous program segment of member functions, most member functions look as we've grown accustomed to seeing, but there are some subtle differences.

One difference, which relates to our static data member, is that numStudents is incremented in each of the constructors and decremented in the destructor. Since this static data member is shared by all instances of class Student, each time a new Student is instantiated, the counter will increase, and when an instance of a Student ceases to exist and its destructor is implicitly called, the counter will be decremented to reflect the removal of such an instance. In this way, numStudents will accurately reflect how many Student instances exist in our application.

This section of code has a few other interesting details to notice, unrelated to `static` data members and member functions. For example, in our class definition, we changed `studentId` from a `const int` to a `const char *`. This means that the data pointed to is constant, not the pointer itself, so we are not required to use the member initialization list to initialize this data member.

Nonetheless, in the default constructor, we choose to use the member initialization list to initialize `studentId` to a null pointer, `nullptr`. Recall that we may use the member initialization list for any data member, but we must use them to initialize `const` data members. That is, if the `const` part equates to memory that is allocated with an instance. Since the memory allocated within the instance for data member `studentId` is a pointer and the pointer part of this data member is not `const` (just the data pointed to), we do not need to use the member initialization list for this data member. We just choose to.

However, because `studentId` is a `const char *`, this means that the identifier `studentId` may not serve as an l-value, or be on the left-hand side of an assignment. In the alternate and copy constructors, we wish to initialize `studentId` and need the ability to use `studentId` as an l-value. But we cannot. We circumvent this dilemma by instead declaring a helper variable, `char *temp;`, and allocating it to contain the amount of memory we need to load the desired data. Then, we load the desired data into `temp`, and finally, we have `studentId` point to `temp` to establish a value for `studentId`. When we leave each constructor, the local pointer `temp` is popped off the stack; however, the memory is now captured by `studentId` and treated as `const`.

Lastly, in the destructor, we delete the memory associated with `const char *studentid`, using `delete [] studentId;`. It is interesting to note that in less-recent compilers, we instead needed to typecast `studentId` to a non-constant `char *`; that is, `delete const_cast<char *> (studentId);`, as operator `delete` previously did not expect a constant qualified pointer.

Now that we have completed reviewing new details in the member functions, let us continue by examining the final portion of this program example:

```
int main()
{
    Student s1("Nick", "Cole , 'S', 3.65, "C++", "112HAV");
    Student s2("Alex", "Tost", 'A', 3.78, "C++", "674HOP");

    cout << s1.GetFirstName() << " " << s1.GetLastName();
    cout << " Enrolled in " << s1.GetCurrentCourse();
    cout << endl;
    cout << s2.GetFirstName() << " " << s2.GetLastName();
    cout << " Enrolled in " << s2.GetCurrentCourse();
    cout << endl;
```

```
        // call a static member function in the preferred manner
        cout << "There are " << Student::GetNumberStudents();
        cout << " students" << endl;

        // Though not preferable, we could also use:
        // cout << "There are " << s1.GetNumberStudents();
        // cout << " students" << endl;
        return 0;
    }
```

In the main() function of our program, we start by instantiating two Students, s1 and s2. As each instance is initialized with a constructor, the shared data member value of numStudents is incremented to reflect the number of students in our application. Note that the external variable Student::numStudents, which holds the memory for this shared data member, was initialized to 0 when the program started with the statement earlier in our code: int Student::numStudents = 0;.

After we print out some details for each Student, we then print out the static data member numStudents using a static access function Student::GetNumStudents(). The preferred way to call this function is Student::GetNumStudents();. Because numStudents is private, only a method of the Student class may access this data member. We have now provided safe, encapsulated access to a static data member using a static member function.

It is interesting to remember that static member functions do not receive a this pointer, therefore, the only data they may manipulate will be static data in the class (or other external variables). Likewise, the only other functions they may call will be other static member functions in the same class or external non-member functions.

It is also interesting to note that we can seemingly call Student::GetNumStudents() via any instance, such as s1.GetNumStudents();, as we see in the commented-out section of code. Though it seems as though we are calling the member function through an instance, the function will not receive a this pointer. Instead, the compiler reinterprets the call, which is seemingly through an instance, and replaces the invocation with a call to the internal, *name mangled* function. It is clearer from a programming point of view to call static member functions using the first calling method, and not seemingly through an instance that would never be passed along to the function itself.

Finally, here is the output for our full program example:

```
Nick Cole Enrolled in C++
Alex Tost Enrolled in C++
There are 2 students
```

Now that we have reviewed our final example of this chapter, it is time to recap everything that we have learned.

Summary

In this chapter, we have begun our journey with object-oriented programming. We have learned many object-oriented concepts and terms, and have seen how C++ has direct language support to implement these concepts. We have seen how C++ classes support encapsulation and information hiding, and how implementing designs supporting these ideals can lead to code that is easier to modify and maintain.

We have detailed class basics, including member functions. We've moved deeper into member functions by examining member function internals, including understanding what the `this` pointer is and how it works – including the underlying implementation of member functions that implicitly receive a `this` pointer.

We have explored access labels and access regions. By grouping our data members in the `private` access region and providing a suite of `public` member functions to manipulate these data members, we have found that we can provide a safe, well-controlled, and well-tested means to manipulate data from the confines of each class. We have seen that making changes to a class can be limited to the member functions themselves. The user of the class need not know the underlying representation of data members – these details are hidden and can be changed as needed without causing a wave of changes elsewhere in an application.

We have deeply explored the many facets of constructors, by examining default, typical (overloaded) constructors, copy constructors, and even conversion constructors. We have been introduced to the destructor, and understand its proper usage.

We've added additional flavor to our classes by using various qualifiers to both data members and member functions, such as `inline` for efficiency, `const` to safeguard data and to ensure functions will as well, `static` data members to model the OO concept of class attributes, and `static` methods to provide safe interfaces to these `static` data members.

By immersing ourselves in object-oriented programming, we have gained a comprehensive set of skills relating to classes in C++. With a well-rounded set of skills and experience using classes under our respective belts and an appreciation for object-oriented programming, we can now move forward with *Chapter 6, Implementing Hierarchies with Single Inheritance*, to learn how to grow a hierarchy of related classes. Let's move forward!

Questions

1. Create a C++ program to encapsulate a `Student`. You may use portions of your previous exercises. Try to do this yourself, rather than relying on any online code. You will need this class as a basis to move forward with future examples; now is a good time to try each feature on your own. Incorporate the following steps:

    a. Create, or modify your previous `Student` class to fully encapsulate a student. Be sure to include several data members that be dynamically allocated. Provide several overloaded constructors to provide the means to initialize your class. Be sure to include a copy constructor. Also, include a destructor to release any dynamically allocated data members.

    b. Add an assortment of access functions to your class to provide safe access to data members within your class. Decide for which data members you will offer a `GetDataMember()` interface, and whether any of these data members should have the ability to be reset after construction with a `SetDataMember()` interface. Apply the `const` and `inline` qualifiers to these methods as appropriate.

    c. Be sure to utilize appropriate access regions – `private` for data members, and possibly for some helper member functions to break up a larger task. Add `public` member functions as necessary above and beyond your previous access functions.

    d. Include at least one `const` data member in your class and utilize the member initialization list to set this member. Add at least one `static` data member and one `static` member function.

    e. Instantiate a `Student` using each constructor signature, including the copy constructor. Make several instances dynamically allocated using `new()`. Be sure to `delete()` each of these instances when you are done with them (so that their destructor will be called).

Implementing Hierarchies with Single Inheritance

This chapter will extend our pursuit of object-oriented programming in C++. We will begin by introducing additional OO concepts, such as **generalization** and **specialization**, and then understand how these concepts are implemented in C++ through *direct language support*. We will begin building hierarchies of related classes, and understand how each class can become an easier to maintain, potentially reusable building block in our applications. We will understand how the new OO concepts presented in this chapter will support well-planned designs, and we will have a clear understanding of how to implement these designs with robust code in C++.

In this chapter, we will cover the following main topics:

- Object-oriented concepts of generalization and specialization, and *Is-A* relationships

- Single inheritance basics – defining derived classes, accessing inherited members, understanding inherited access labels and regions, and `final` class specification

- Construction and destruction sequence in a single inheritance hierarchy; selecting base class constructors with the member initialization list

- Modifying access labels in the base class list – `public` versus `private` and `protected` base classes – to change the OO purpose of inheritance to *Implementation Inheritance*

By the end of this chapter, you will understand the object-oriented concepts of generalization and specialization and will know how to use inheritance in C++ as a mechanism to realize these ideals. You will understand terms such as base and derived classes, as well as OO motivations for building hierarchies, such as supporting Is-A relationships or supporting implementation inheritance.

Specifically, you will understand how to grow inheritance hierarchies using single inheritance, and how to access inherited data members and member functions. You will also understand which inherited members you may directly access, based on their defined access regions.

You will understand the order of constructor and destructor invocations when instances of derived class types are instantiated and destroyed. You will know how to make use of the member initialization list to select which, from a potential group, of inherited constructors a derived class object may need to utilize as part of its own construction.

You will also understand how changing access labels in a base class list changes the OO meaning for the type of inheritance hierarchy you are growing. By examining public versus private and protected base classes, you will understand different types of hierarchies, such as those built to support Is-A relationships versus those built to support implementation inheritance.

By understanding the direct language support of single inheritance in C++, you will be able to implement the OO concepts of generalization and specialization. Each class within your hierarchy will be a more easily maintained component and can serve as a potential building block for creating new, more specialized components. Let us further our understanding of C++ as an OOP language by detailing single inheritance.

Technical requirements

Online code for full program examples can be found in the following GitHub URL: `https://github.com/PacktPublishing/Deciphering-Object-Oriented-Programming-with-CPP/tree/main/Chapter06`. Each full program example can be found in the GitHub under the appropriate chapter heading (subdirectory) in a file that corresponds to the chapter number, followed by a dash, followed by the example number in the chapter at hand. For example, the first full program in this chapter can be found in the subdirectory `Chapter06` in a file named `Chp6-Ex1.cpp` under the aforementioned GitHub directory.

The CiA video for this chapter can be viewed at: `https://bit.ly/3R7uNci`.

Expanding object-oriented concepts and terminology

In this section, we will introduce essential object-oriented concepts, as well as applicable terminology that will accompany these key ideas.

From *Chapter 5, Exploring Classes in Detail*, you now understand the key OO ideas of encapsulation and information hiding, and how C++ supports these ideals through the C++ class. Now, we will look at how we can grow a hierarchy of related classes, using a very general class as a building block, and then extend that class by creating a more specific class. Through growing a hierarchy of related classes in this repeated fashion, OO systems provide building blocks of potential reuse. Each class within the hierarchy is encapsulated, so maintenance and upgrades to a specific class can be made more easily and without impact to the entire system. By incrementally refining each class with a more specific and more detailed class to build a hierarchy of related classes, specific maintenance for each component is in a focused area for maintenance and changes.

Let's start by extending our basic OO terminology.

Deciphering generalization and specialization

The main object-oriented concepts extending through this chapter are *generalization* and *specialization*. Incorporating these principles into your design will provide the basis for writing more easily modifiable and maintainable code, and for code that can potentially be reused in related applications.

Generalization describes the process of abstracting commonalities from a grouping of classes and creating a more generalized class for that grouping to house the common properties and behaviors. The more generalized class can be known as a **base** (or **parent**) class. Generalization can also be used to collect more general properties and behaviors of a single class into a base class with the expectation that the new, generalized class can later serve as a building block or basis for additional, more specific (derived) classes.

Specialization describes the process of deriving a new class from an existing, generalized base class, for the purpose of adding specific, distinguishable properties and behaviors to adequately represent the new class. The specialized class can also be referred to as a **derived** (or **child**) class. A hierarchy of classes can incrementally refine their respective properties and behaviors through specialization.

Though reuse is difficult to achieve, the OOP concepts such as generalization and specialization make reuse more easily obtainable. Reuse can potentially be realized in applications that are similar in nature or in the same project domain, in continuations of existing projects, or potentially in related domains where minimally the most generalized classes and associated components can be reused.

Building a hierarchy is a fundamental language feature of C++. Let's move forward by exploring single inheritance to put this idea into action.

Understanding single inheritance basics

Inheritance is the C++ language mechanism that allows the concepts of generalization and specialization to be realized. **Single inheritance** is when a given class has exactly one immediate base class. Both single inheritance and multiple inheritance are supported in C++; however, we will focus on single inheritance in this chapter and will cover multiple inheritance in a later chapter.

Inheritance hierarchies can be built using both classes and structures in C++. Classes, however, are most often utilized rather than structures to support inheritance and OOP.

Growing an inheritance hierarchy for the purpose of generalization and specialization supports an **Is-A** relationship. For example, given a base class of `Person` and a derived class of `Student`, we can say *a Student Is-A Person*. That is, a `Student` is a specialization of a `Person`, adding additional data members and member functions above and beyond those provided by its base class, `Person`. Specifying an Is-A relationship through generalization and specialization is the most typical reason inheritance is used to create base and derived classes. Later in this chapter, we will look at another reason to utilize inheritance.

Let's get started by looking at the language mechanics in C++ to specify base and derived classes and to define an inheritance hierarchy.

Defining base and derived classes and accessing inherited members

With single inheritance, the derived class specifies who its immediate ancestor or base class is. The base class does not specify that it has any derived classes.

The derived class simply creates a base class list by adding a : after its class name, followed by the keyword `public` (for now), and then the specific base class name. Whenever you see a `public` keyword in the base class list, it means that we are using inheritance to specify an Is-A relationship.

Here is a simple example to illustrate the basic syntax:

- `Student` *Is-A* derived class of `Person`:

```
class Person   // base class
{
private:
    string name;
    string title;
public:
    // constructors, destructor,
    // public access functions, public interface etc.
    const string &GetTitle() const { return title; }
};


class Student: public Person   // derived class
{
private:
    float gpa;
public:
    // constructors, destructor specific to Student,
    // public access functions, public interface, etc.
    float GetGpa() const { return gpa; }
};
```

Here, the base class is `Person`, and the derived class is `Student`. The derived class need only define additional data members and member functions that augment those specified in the base class.

Instances of a derived class may generally access `public` members specified by the derived class or by any ancestor of the derived class. Inherited members are accessed in the same fashion as those specified by the derived class. Recall, dot notation (.) is used to access members of objects, and arrow notation (->) is used to access members of pointers to objects.

Of course, to make this example complete, we will need to add the applicable constructors, which we currently assume exist. Naturally, there will be nuances with constructors relating to inheritance, which we will soon cover in this chapter.

- Simple access of inherited members can be seen using the aforementioned classes as follows:

```
int main()
{
    // Let's assume the applicable constructors exist
    Person p1("Cyrus Bond", "Mr.");
    Student *s1 = new Student("Anne Lin", "Ms.", 4.0);
    cout << p1.GetTitle() << " " << s1->GetTitle();
    cout << s1->GetGpa() << endl;
    delete s1; // remember to relinquish alloc. memory
    return 0;
}
```

In the previous code fragment, the derived class instance of `Student`, pointed to by `s1`, can access both base and derived class members, such as `Person::GetTitle()` and `Student::GetGpa()`. The base class instance of `Person`, `p1`, can only access its own members, such as `Person::GetTitle()`.

Looking at a memory model for the preceding example, we have the following:

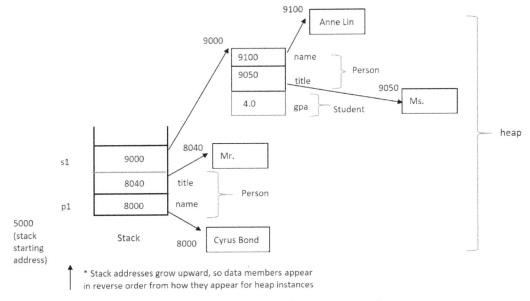

Figure 6.1 – Memory model for current example

Notice that in the preceding memory model, a `Student` instance is comprised of a `Person` subobject. That is, at the memory address indicating the beginning of `*s1`, a `Student`, we first see the memory layout of its `Person` data members. Then, we see the memory layout of its additional `Student` data members. Of course, `p1`, which is a `Person`, only contains `Person` data members.

Access to base and derived class members will be subject to the access regions specified by each class. Let's take a look to see how inherited access regions work.

Examining inherited access regions

Access regions, including inherited access regions, define from which scope members, including inherited members, are directly accessible.

The derived class inherits all members specified in its base class. However, direct access to those members is subject to the access regions specified in the base class.

Members (both data and function) inherited from the *base class* are accessible to the *derived class* as specified by the access regions that are imposed by the base class. The inherited access regions, and how they relate to derived class access, are as follows:

- **private** members defined in the base class are inaccessible outside the scope of the base class. The scope of a class includes member functions of that class.

- **protected** members defined in the base class are accessible in the scope of the base class and within the scope of the derived class, or its descendants. This means member functions of these classes.

- **public** members defined in the base class are accessible from any scope, including the scope of the derived class.

In the previous, simple example, we noticed that both a `Person` and a `Student` instance accessed the `public` member function `Person::GetTitle()` from the scope of `main()`. Also, we noticed that the `Student` instance accessed its `public` member `Student::GetGpa()` from `main()`. Generally, outside the scope of a given class, the only members that are accessible are those that are in the public interface, such as in this example.

We will soon see a larger, full program example in this chapter showcasing the `protected` access region. But first, let's discover an additional specifier that may be useful in determining the shape and extensibility of our inheritance hierarchy.

Specifying a class as final

In C++, we can indicate that a class may not be further extended within our inheritance hierarchy. This is known as a **leaf-node** and can support an OO design to enforce that a given class may not be further specialized. The keyword `final` is used in the base class list to designate a class as a **final** (unextendible) class or **leaf-node**.

Here is a simple example to illustrate the basic syntax:

- Given our previous base class `Person`, `Student` *Is-A* derived class of `Person`. Additionally, `GradStudent` *Is-A* final derived class of `Student`:

```
class GradStudent final: public Person // derived class
{
   // class definition
};
```

Here, `GradStudent` is specified as a final, unextendible class. Therefore, `GradStudent` may not appear in the base class list of a new derived class.

Next, let's review inherited constructors and destructors so that our upcoming full program example can provide greater overall utility.

Understanding inherited constructors and destructors

Through single inheritance, we can build a hierarchy of related classes. We have seen that when we instantiate a derived class object, memory for its base class data members is then followed by the additional memory required for the additional derived class data members. Each of these subobjects will need to be constructed. Luckily, each class will have defined a suite of constructors for just that purpose. We then need to understand how the language can be utilized to allow us to specify the appropriate base class constructor for the base class subobject when instantiating and constructing a derived class object.

Similarly, when an object of a derived class type is no longer needed and will be destructed, it is important to note that a destructor for each subobject comprising the derived class instance will be implicitly called on our behalf.

Let's take a look at the constructor and destructor sequence in a single inheritance hierarchy, and how we can make choices when more than one constructor is available for a base class subobject found in a derived class instance.

Implicit constructor and destructor invocations

Constructors and destructors are two types of member functions that are not explicitly inherited by the derived class. This means that the signature of a base class constructor may not be used to instantiate a derived class object. However, we will see that when a derived class object is instantiated, the memory for both the base and derived class portions of the overall object will be separately initialized using each class's respective constructors.

When an object of a derived class type is instantiated, not only will one of its constructors be invoked but so will one in each of its preceding base classes. The most generalized base class constructor will first be executed, followed by the constructors all the way down the hierarchy until we arrive at the derived class constructor that is the same type as the instance at hand.

Similarly, when a derived class instance goes out of scope (or is explicitly deallocated for pointers to instances), all the relevant destructors will be invoked, but in the opposite order of construction. First, the derived class destructor will be executed, then all the destructors in an upward fashion for each preceding base class will be invoked and executed until we reach the most generalized base class.

You may now ask, how may I choose from a set of potential base class constructors for my base class subobject when instantiating a derived class? Let's take a more detailed look at the member initialization list to discover the solution.

Usage of member initialization list to select a base class constructor

The member initialization list may be used to specify which base class constructor should be invoked when instantiating a derived class object. Each derived class constructor may specify that a different base class constructor should be used to initialize the given base class portion of the derived class object.

If the derived class constructor's member initialization list does not specify which base class constructor should be utilized, the default base class constructor will be invoked.

The member initialization list is specified using a : after the parameter list in the derived class constructor. To specify which base class constructor should be used, the name of the base class constructor, followed by parentheses including any values to be passed to that base class constructor, can be indicated. Based upon the signature of the parameters in the base class list following the base class name, the appropriate base class constructor will be selected to initialize the base class portion of the derived class object.

Here is a simple example to illustrate the basic syntax for base class constructor selection:

- Let's start with the basic class definitions (note that many member functions are omitted, as are some usual data members):

```
class Person
{
private:
    string name;
    string title;
public:
    Person() = default;  // various constructors
    Person(const string &, const string &);
    Person(const Person &);
    // Assume the public interface, access fns. exist
};

class Student: public Person
{
private:
    float gpa = 0.0;  // use in-class initializer
public:
    Student() = default;
    Student(const string &, const string &, float);
```

```
               // Assume the public interface, access fns. exist
      };
```

- The constructors for the previous class definitions would be as follows (notice two of the derived class constructors use the member initialization list):

```
// Base class constructors
// Note: default constructor is included by = default
// specification in Person constructor prototype

Person::Person(const string &n, const string &t):
               name(n), title(t)
{
}


Person::Person(const Person &p):
               name(p.name), title(p.title)
{
}


// Derived class constructors
// Note: default constructor is included by = default
// specification in Student constructor prototype and
// gpa is set with value of in-class initializer (0.0)

Student::Student(const char *n, const char *t,
                 float g): Person(n, t), gpa(g)
{
}


Student::Student(const Student &s): Person(s),
                                    gpa(s.gpa)
{
}
```

In the previous short segment of code, notice that the system-supplied default derived class constructor, `Student::Student()`, has been elected with `=default` added to the constructor prototype. With an alternate constructor in this class definition, this specification (or by writing the default constructor ourselves) is necessary if we would like to support this simple interface for class instantiation. Remember, we only get a system-supplied default constructor if there are no other constructors (that is, means for instantiation) in our class definition.

Next, notice in the alternate derived class constructor, `Student::Student(const string &, const string &, float)`, the use of the member initialization list for base class construction specification. Here, the `Person` constructor matching the signature of `Person::Person(const string &, const string &)` is selected to initialize the `Person` subobject at hand. Also, notice that parameters from the `Student` constructor, n and t, are passed up to the aforementioned `Person` constructor to help complete the `Person` subobject initialization. Had we not specified which `Person` base class constructor should be used in the member initialization list, the default `Person` constructor will be used to initialize the `Person` base class subobject of `Student`. The member initialization list is additionally used in this constructor to initialize data members introduced within the `Student` class definition (such as gpa).

Now, notice in the copy constructor for the derived class, `Student::Student(const Student &)`, the member initialization list is used to select the `Person` copy constructor, passing s as a parameter to the `Person` copy constructor. Here, the object referenced by s is a `Student`, however, the top part of `Student` memory contains `Person` data members. Hence, it is acceptable to implicitly upcast the `Student` to a `Person` to allow the `Person` copy constructor to initialize the `Person` subobject. Also, in the member initialization list of the `Student` copy constructor, the additional data member added by the `Student` class definition is initialized, namely, by initializing gpa (`s.gpa`). These additional data members could have also been set in the body of this constructor.

Now that we understand how to utilize the member initialization list to specify a base class constructor, let's move forward with a complete program example.

Putting all the pieces together

So far in this chapter, we have seen many pieces contributing to a full program example. It is important to see our code in action, with all its various components. We need to see the basic mechanics of inheritance, how the member initialization list is used to specify which base class constructor should implicitly be invoked, and the significance of the `protected` access region.

Let's take a look at a more complex, full program example to fully illustrate single inheritance. This example will be broken into segments; the full program can be found in the following GitHub location:

https://github.com/PacktPublishing/Deciphering-Object-Orient-ed-Programming-with-CPP/blob/main/Chapter06/Chp6-Ex1.cpp

```cpp
#include <iostream>
#include <iomanip>
using std::cout;  // preferred to: using namespace std;
using std::endl;
using std::setprecision;
using std::string;
using std::to_string;

class Person
{
private:
    // data members
    string firstName; // str mbrs are default constructed,
    string lastName;  // so don't need in-class initializers
    char middleInitial = '\0';  // in-class initialization
    string title;  // Mr., Ms., Mrs., Miss, Dr., etc.
protected: // make avail. to derived classes in their scope
    void ModifyTitle(const string &);
public:
    Person() = default;   // default constructor
    Person(const string &, const string &, char,
           const string &);
    // We get default copy constructor and destructor even
    // without the below protypes; hence, commented out
    // Person(const Person &) = default;  // def. copy ctor
    // ~Person() = default;  // use default destructor

    // inline function definitions
    const string &GetFirstName() const { return firstName; }
    const string &GetLastName() const { return lastName; }
    const string &GetTitle() const { return title; }
```

```
     char GetMiddleInitial() const { return middleInitial; }
};
```

In the previous class definition, we now have a fully-fledged class definition for `Person`, with many more details than our simple syntax examples formerly used in this section. Notice that we have introduced a `protected` access region and placed member function `void ModifyTitle(const string &);` in this access region.

Moving onward, let's examine the non-line member function definitions for `Person`:

```
// Default constructor included with = default in prototype
// With in-class initialization, it is often not necessary
// to write the default constructor yourself.

// alternate constructor
Person::Person(const string &fn, const string &ln, char mi,
               const string &t): firstName(fn),
               lastName(ln), middleInitial(mi), title(t)
{
    // dynamically allocate memory for any ptr data members
}


// We are using default copy constructor; let's see what
// it would look like if we prototyped/defined it ourselves
// (so we may better understand an upcoming discussion with
// the upcoming derived class copy constructor). Also,
// this is what the system-supplied version may look like.
// Person::Person(const Person &p): firstName(p.firstName),
//     lastName(p.lastName), middleInitial(p.middleInitial),
//     title(p.title)
// {
        // deep copy any pointer data members here
// }


// Using default destructor - no need to write it ourselves

void Person::ModifyTitle(const string &newTitle)
{
```

```
      title = newTitle;
}
```

The implementation for the aforesaid `Person` member functions is as expected. Now, let's add the class definition for the derived class, `Student`, along with its inline function definitions:

```
class Student: public Person
{
private:
   // data members
   float gpa = 0.0;    // in-class initialization
   string currentCourse;
   const string studentId;  // studentId is not modifiable
   static int numStudents; // static data mbr. init. occurs
public:                     // outside of the class definition
   // member function prototypes
   Student();   // we will provide default constructor
   Student(const string &, const string &, char,
          const string &, float, const string &,
          const string &);
   Student(const Student &);  // copy constructor
   ~Student();  // we will provide destructor
   void Print() const;
   void EarnPhD();  // public interface to inherited
                    // protected member
   // inline function definitions
   float GetGpa() const { return gpa; }
   const string &GetCurrentCourse() const
      { return currentCourse; }
   const string &GetStudentId() const { return studentId; }
   // prototype only, see inline function definition below
   void SetCurrentCourse(const string &);
   static int GetNumberStudents(); // static mbr function
};

// definition for static data mbr. (implemented as extern)
int Student::numStudents = 0;  // notice initial value of 0
```

```
inline void Student::SetCurrentCourse(const string &c)
{
    currentCourse = c;
}


// Definition for static member function (it's also inline)
inline int Student::GetNumberStudents()
{
    return numStudents;
}
```

In the preceding definition of `Student`, `class Student` is derived from `Person` using `public` inheritance (that is, a public base class), which supports an Is-A relationship. Notice the `public` access label after the base class list following the `:` in the derived class definition (that is, `class Student: public Person`). Notice that our `Student` class has added data members and member functions above and beyond those that it automatically inherits from `Person`.

Next, adding in the non-inline `Student` member functions, we continue growing our code:

```
// Default constructor uses in-class init. for gpa, while
// currentCourse (string mbr object) is default constructed
Student::Student(): studentId(to_string(numStudents + 100)
                                  + "Id")
{
    // Since studentId is const, we need to initialize it
    // during construction using member init list (above)
    // Also, remember to dynamically allocate memory for any
    // pointer data mbrs. here (not needed in this example)
    numStudents++;    // increment static counter
}


// alternate constructor
Student::Student(const string &fn, const string &ln,
                 char mi, const string &t, float avg,
                 const string &course, const string &id):
            Person(fn, ln, mi, t),
            gpa(avg), currentCourse(course), studentId(id)
```

```
{
    // Remember to dynamically allocate memory for any
    // pointer data members (none in this example)
    numStudents++;    // increment static counter
}


// copy constructor
Student::Student(const Student &s): Person(s), gpa(s.gpa),
                currentCourse(s.currentCourse),
                studentId(s.studentId)
{
    // deep copy any ptr data mbrs (none in this example)
    numStudents++;    // increment static counter
}

// destructor definition
Student::~Student()
{
    // Remember to release memory for any dynamically
    // allocated data members (none in this example)
    numStudents--;  // decrement static counter
}


void Student::Print() const
{
    // Private members of Person are not directly accessible
    // within the scope of Student, so we use access fns.
    cout << GetTitle() << " " << GetFirstName() << " ";
    cout << GetMiddleInitial() << ". " << GetLastName();
    cout << " with id: " << studentId << " gpa: ";
    cout << setprecision(2) << gpa;
    cout << " course: " << currentCourse << endl;
}


void Student::EarnPhD()
{
```

```
    // Protected members defined by the base class are
    // accessible within the scope of the derived class.
    // EarnPhd() provides a public interface to this
    // functionality for derived class instances.
    ModifyTitle("Dr.");
}
```

In the aforementioned segment of code, we define the non-inline member functions of `Student`. Notice that the default constructor merely uses the member initialization list to initialize a data member, as we did in the last chapter. Since no `Person` constructor has been specified in the member initialization list of the default `Student` constructor, the default `Person` constructor will be used to initialize the `Person` subobject when instantiating a `Student` with its default constructor.

Next, the alternate constructor for `Student` uses the member initialization list to specify that the alternate constructor of `Person` should be utilized to construct the `Person` subobject contained within a given `Student` instance. Notice that the selected constructor will match the signature `Person::Person(const string &, const string &, char, const string &)`, and that selected input parameters from the `Student` constructor (namely `fn`, `ln`, `mi`, and `t`) will be passed as parameters to the `Person` alternate constructor. The `Student` constructor's member initialization list is then used to initialize additional data members introduced by the `Student` class.

In the copy constructor for `Student`, the member initialization list is used to specify that the `Person` copy constructor should be called to initialize the `Person` subobject of the `Student` instance that is being constructed. The `Student` `&` will be implicitly upcast to a `Person` `&` as the `Person` copy constructor is called. Recall that the top part of a `Student` object *Is-A* `Person`, so this is fine. Next, in the remainder of the copy constructor's member initialization list for `Student`, we initialize any remaining data members the `Student` class has defined. Any data members requiring a deep copy (such as those that are pointers) may be handled in the body of the copy constructor.

Moving onward, we see a comment indicating the `Student` destructor. Implicitly, as the *last* line of code in this method (whether the destructor is system-supplied or user-written), a call to the `Person` destructor is patched in for us by the compiler. This is how the destructor sequence is automated for us. Consequently, the most specialized portion of the object, the `Student` pieces, will first be destructed, followed by the implicit call to the `Person` destructor to destruct the base class subobject.

Next, in the `Print()` method for `Student`, notice that we would like to print out various data members that are inherited from `Person`. Alas, these data members are `private`. We may not access them outside the scope of the `Person` class. Nevertheless, the `Person` class has left us with a public interface, such as `Person::GetTitle()` and `Person::GetFirstName()`, so that we may access these data members from any scope of our application, including from `Student::Print()`.

Finally, we come to the `Student::EarnPhD()` method. Notice that all this method does is invoke the `protected` member function `Person::ModifyTitle("Dr.");`. Recall that `protected` members defined by the base class are accessible within the scope of the derived class. `Student::EarnPhD()` is a member function of the derived class. `EarnPhD()` provides a public interface to modify the title of a `Person`, perhaps after checking whether the student has met graduation requirements. Because `Person::ModifyTitle()` is not `public`, instances of `Person` or `Student` must go through a controlled `public` interface to change their respective titles. Such interfaces might include methods such as `Student::EarnPhD()` or `Person::GetMarried()`, and so on.

Nonetheless, let's complete our full program example by examining `main()`:

```
int main()
{
    Student s1("Jo", "Li", 'U', "Ms.", 3.8,
               "C++", "178PSU");
    // Public members of Person and Student are accessible
    // outside the scope of their respective classes....
    s1.Print();
    s1.SetCurrentCourse("Doctoral Thesis");
    s1.EarnPhD();
    s1.Print();
    cout << "Total number of students: " <<
            Student::GetNumberStudents() << endl;
    return 0;
}
```

In the last segment of this program, in `main()`, we simply instantiate a `Student`, namely `s1`. The `Student` utilizes `Student::Print()` to print its current data. The `Student` then sets her current course set to `"Doctoral Thesis"` and then invokes `Student::EarnPhD();`. Note that any `public` members of `Student` or `Person` are available for `s1` to utilize outside the scope of their class, such as in `main()`. To complete the example, `s1` reprints her details using `Student::Print()`.

Here is the output for the full program example:

```
Ms. Jo U. Li with id: 178PSU gpa: 3.9 course: C++
Dr. Jo U. Li with id: 178PSU gpa: 3.9 course: Doctoral Thesis
Total number of students: 1
```

Now that we have competency with the basic mechanics of single inheritance, and have used single inheritance to model an Is-A relationship, let's move onward to see how inheritance can be used to model a different concept by exploring `protected` and `private` base classes.

Implementation inheritance – changing the purpose of inheritance

So far we demonstrated using a public base class, known also as *public inheritance*. Public base classes are used to model Is-A relationships and provide the primary motivation behind building an inheritance hierarchy. This usage supports the concepts of generalization and specialization.

Occasionally, inheritance may be used as a tool to implement one class in terms of another, that is, by one class using another as its underlying implementation. This is known as **implementation inheritance** and it does not support the ideals of generalization and specialization. Yet, implementation inheritance can provide a quick and easily reusable way to implement one class based upon another. It is fast and relatively error-free. Many class libraries use this tool without the knowledge of their class users. It is important to distinguish implementation inheritance from traditional hierarchy building for the motivation of specifying Is-A relationships.

Implementation inheritance, supported in C++ with private and protected base classes, is exclusive to C++. Other OOP languages choose to only embrace inheritance for the purpose of modeling Is-A relationships, which is supported in C++ through public base classes. An OO purist would endeavor to use inheritance only to support generalization and specialization (*Is-A*). However, using C++, we will understand appropriate uses of implementation inheritance so that we may use this language feature wisely.

Let's move onward to understand how and why we might utilize this type of inheritance.

Modifying access labels in the base class list by using protected or private base classes

To reiterate, the usual type of inheritance is `public` inheritance. The `public` label is used in the base class list for a given derived class. However, in the base class list, the keywords `protected` and `private` are also options.

That is, in addition to labeling access regions within a class or structure definition, an access label can be used in the base class list of a derived class definition to designate how members defined in a base class are inherited by derived classes.

Inherited members can only be made more restrictive than they were designated to be in the base class. When the derived class specifies that inherited members should be treated in a more restrictive fashion, any descendants of that derived class will also be subject to these specifications.

Let's see a quick example of the *base class list*:

- Recall that most often, a `public` access label will be specified in the base class list.

- In this example, a `public` access label is used to specify that a `Person` is a `public` base class of `Student`. That is, a `Student` *Is-A* `Person`:

```
class Student: public Person
{
    // usual class definition
};
```

Access labels specified in the *base class list* modify inherited access regions as follows:

- **public**: Public members in the base class are accessible from any scope; protected members in the base class are accessible from the scope of the base and derived classes. We are familiar with using a public base class.

- **protected**: Public and protected members in the base class act as though they are defined as protected by the derived class (that is, accessible from the scope of the base and derived classes and any descendants of the derived class).

- **private**: Public and protected members in the base class act as though they are defined as private by the derived class, allowing these members to be accessible within the scope of the derived class, but not within the scope of any of the derived class descendants.

> **Note**
> In all cases, class members labeled as private within a class definition, are accessible only within the scope of the defining class. Modifying the access labels in the base class list can only treat inherited members more restrictively, never less restrictively.

In the absence of an access label specified in conjunction with the base class, `private` will be assumed if the user defined type is a `class`, and `public` will be the default if the user defined type is a `struct`. A good rule of thumb is to always include the access label in the base class list for a derived class (or structure) definition.

Creating a base class to illustrate implementation inheritance

To understand implementation inheritance, let's review a base class that may serve as a basis to implement other classes. We will examine a typical pair of classes to implement an encapsulated `LinkList`. Though this example will be broken into several segments, the full example will be shown, and can also be found in the GitHub:

https://github.com/PacktPublishing/Deciphering-Object-Oriented-Programming-with-CPP/blob/main/Chapter06/Chp6-Ex2.cpp

```cpp
#include <iostream>
using std::cout;      // preferred to: using namespace std;
using std::endl;


using Item = int;


class LinkListElement   // a 'node' or element of a LinkList
{
private:
    void *data = nullptr;      // in-class initialization
    LinkListElement *next = nullptr;
public:
    LinkListElement() = default;
    LinkListElement(Item *i) : data(i), next(nullptr) { }
    ~LinkListElement()
        { delete static_cast<Item *>(data);
          next = nullptr; }
    void *GetData() const { return data; }
    LinkListElement *GetNext() const { return next; }
    void SetNext(LinkListElement *e) { next = e; }
};


class LinkList    // an encapsulated LinkList
{
private:
    LinkListElement *head = nullptr;   // in-class init.
    LinkListElement *tail = nullptr;
    LinkListElement *current = nullptr;
```

```
public:
    LinkList() = default; // required to keep default
                          // interface
    LinkList(LinkListElement *);
    ~LinkList();
    void InsertAtFront(Item *);
    LinkListElement *RemoveAtFront();
    void DeleteAtFront();
    int IsEmpty() const { return head == nullptr; }
    void Print() const;
};
```

We begin the previous segment of code with class definitions for both `LinkListElement` and `LinkList`. The `LinkList` class will contain data members that are pointers to the head, `tail`, and `current` element in the `LinkList`. Each of these pointers is of type `LinkListElement`. A variety of typical `LinkList` processing methods are included, such as `InsertAtFront()`, `RemoveAtFront()`, `DeleteAtFront()`, `IsEmpty()`, and `Print()`. Let's take a quick peek at the implementation of these methods with the next segment of code:

```
// default constructor - not necessary to write it
// ourselves with in-class initialization above

LinkList::LinkList(LinkListElement *element)
{
    head = tail = current = element;
}

void LinkList::InsertAtFront(Item *theItem)
{
    LinkListElement *newHead = new
                                LinkListElement(theItem);

    newHead->SetNext(head);  // newHead->next = head;
    head = newHead;
}

LinkListElement *LinkList::RemoveAtFront()
{
```

```
    LinkListElement *remove = head;
    head = head->GetNext();   // head = head->next;
    current = head;     // reset current for usage elsewhere
    return remove;
}


void LinkList::DeleteAtFront()
{
    LinkListElement *deallocate;
    deallocate = RemoveAtFront();
    delete deallocate;   // destructor will both delete data
}                        // and will set next to nullptr


void LinkList::Print() const
{
    if (!head)
       cout << "<EMPTY>";
    LinkListElement *traverse = head;
    while (traverse)
    {
        Item output = *(static_cast<Item *>
                        (traverse->GetData()));
        cout << output << " ";
        traverse = traverse->GetNext();
    }
    cout << endl;
}


LinkList::~LinkList()
{
    while (!IsEmpty())
        DeleteAtFront();
}
```

In the previously mentioned member function definitions, we note that a LinkList can be constructed either empty or with one element (note the two available constructors). LinkList::InsertAtFront() adds an item to the front of the list for efficiency. LinkList::RemoveAtFront() removes an item and returns it to the user, whereas LinkList::DeleteAtFront() removes and deletes the front item. The LinkList::Print() function allows us to view the LinkList whenever necessary.

Next, let's see a typical main() function to illustrate how a LinkList can be instantiated and manipulated:

```
int main()
{
    // Create a few items, to be data for LinkListElements
    Item *item1 = new Item;
    *item1 = 100;
    Item *item2 = new Item(200);

    // create an element for the Linked List
    LinkListElement *element1 = new LinkListElement(item1);
    // create a linked list and initialize with one element
    LinkList list1(element1);

    // Add some new items to the list and print
    list1.InsertAtFront(item2);
    list1.InsertAtFront(new Item(50));  // add nameless item
    cout << "List 1: ";
    list1.Print();            // print out contents of list

    // delete elements from list, one by one
    while (!(list1.IsEmpty()))
    {
        list1.DeleteAtFront();
        cout << "List 1 after removing an item: ";
        list1.Print();
    }

    // create a second linked list, add some items, print
    LinkList list2;
    list2.InsertAtFront(new Item (3000));
```

```
list2.InsertAtFront(new Item (600));
list2.InsertAtFront(new Item (475));
cout << "List 2: ";
list2.Print();

// delete elements from list, one by one
while (!(list2.IsEmpty()))
{
    list2.DeleteAtFront();
    cout << "List 2 after removing an item: ";
    list2.Print();
}
return 0;
}
```

In `main()`, we create a few items, of type `Item`, which will later be data for `LinkListElement`. We then instantiate a `LinkListElement`, namely `element1`, and add it to a newly constructed `LinkList`, using `LinkList list1(element1);`. We then add several items to the list using `LinkList::InsertAtFront()`, and call `LinkList::Print()` to print out `list1` for a baseline. Next, we delete elements from `list1`, one by one, printing as we go, using `LinkList::DeleteAtFront()` and `LinkList::Print()`, respectively.

Now, we instantiate a second `LinkList`, namely `list2`, which starts out empty. We gradually insert several items using `LinkList::InsertAtFront()`, then print the list, and then delete each element, one by one, using `LinkList::DeleteAtFront()`, printing the revised list with each step.

The point of this example is not to exhaustively review the inner workings of this code. You are undoubtedly familiar with the concept of a `LinkList`. More so, the point is to establish this set of classes, `LinkListElement` and `LinkList`, as a set of building blocks in which several *Abstract Data Types* can be built.

Nonetheless, the output for the preceding example is as follows:

```
List 1: 50 200 100
List 1 after removing an item: 200 100
List 1 after removing an item: 100
List 1 after removing an item: <EMPTY>
List 2: 475 600 3000
List 2 after removing an item: 600 3000
```

```
List 2 after removing an item: 3000
List 2 after removing an item: <EMPTY>
```

Next, let's see how `LinkList` can be used as a private base class.

Using a private base class to implement one class in terms of another

We have just created a `LinkList` class to support the basic handling of an encapsulated linked list data structure. Now, let's imagine that we would like to implement an **Abstract Data Type** (**ADT**), such as a stack. A **stack** is an ADT in that it has a set of expected operations to define its interface, such as `Push()`, `Pop()`, `IsEmpty()`, and perhaps `Print()`.

You may ask how a stack is implemented. The answer is that the implementation does not matter, so long as it supports the expected interface of the ADT being modeled. Perhaps a stack is implemented using an array, or perhaps it is implemented in a file. Perhaps it is implemented using a `LinkedList`. Each implementation has pros and cons. In fact, the underlying implementation of the ADT might change, yet users of the ADT should not be affected by such a change. This is the basis of *implementation inheritance*. A derived class is implemented in terms of a base class, yet the underlying details of the base class from which the new class is derived are effectively hidden. These details cannot be directly used by instances of the derived class (in this case, the ADT). Nonetheless, the base class silently provides the implementation for the derived class.

We will use this approach to implement a `Stack` using a `LinkedList` as its underlying implementation. To do this, we will have `class Stack` extend `LinkedList` using a `private` base class. `Stack` will define a public interface for its users to establish the interface for this ADT, such as `Push()`, `Pop()`, `IsEmpty()`, and `Print()`. The implementation of these member functions will make use of selected `LinkedList` member functions, but `Stack` users will not see this, nor will `Stack` instances be able to use any `LinkList` members directly themselves.

Here, we are not saying a `Stack` *Is-A* `LinkList`, but rather, a `Stack` is implemented in terms of a `LinkedList` at the moment—and that underlying implementation could change!

The code to implement `Stack` is simple. Assume we are using the `LinkList` and `LinkListElement` classes from the previous example. Let's add the `Stack` class here. The full program example can be found in our GitHub:

https://github.com/PacktPublishing/Deciphering-Object-Orient-ed-Programming-with-CPP/blob/main/Chapter06/Chp6-Ex3.cpp

```cpp
class Stack: private LinkList
{
private:
```

```cpp
    // no new data members are necessary
public:
    // Constructor / destructor prototypes shown below are
    // not needed; we get both without these prototypes!
    // Commented to remind what's automatically provided
    // Stack() = default; // will call :LinkList() def ctor
    // ~Stack() = default;
    // the public interface for Stack
    void Push(Item *i) { InsertAtFront(i); }
    Item *Pop();
    // It is necessary to redefine these operations because
    // LinkList is a private base class of Stack
    int IsEmpty() const { return LinkList::IsEmpty(); }
    void Print() { LinkList::Print(); }
};


Item *Stack::Pop()
{
    LinkListElement *top;
    top = RemoveAtFront();
    // copy top's data
    Item *item = new Item(*(static_cast<Item *>
                        (top->GetData())));
    delete top;
    return item;
}


int main()
{
    Stack stack1;     // create a Stack
    // Add some items to the stack, using public interface
    stack1.Push(new Item (3000));
    stack1.Push(new Item (600));
    stack1.Push(new Item (475));
    cout << "Stack 1: ";
    stack1.Print();
```

```
    // Pop elements from stack, one by one
    while (!(stack1.IsEmpty()))
    {
        stack1.Pop();
        cout << "Stack 1 after popping an item: ";
        stack1.Print();
    }
    return 0;
}
```

Notice how compact the aforementioned code is for our Stack class! We begin by specifying that Stack has a private base class of LinkList. Recall that a private base class means that the protected and public members inherited from LinkList act as though they were defined by Stack as private (and are only accessible within the scope of Stack, that is, member functions of Stack). This means that instances of Stack may not use the *former* public interface of LinkList. This also means that the underlying implementation of Stack as a LinkList is effectively hidden. Of course, LinkList instances are not affected in any way and may use their public interface as usual.

We notice that =default has been added to both the Stack constructor and destructor prototypes. Neither of these methods has work to do because we are not adding any data members to this class; therefore, the default system-supplied versions are acceptable. Note that if we omitted both the default constructor and destructor prototypes, we get both system-supplied versions linked in.

We easily define Stack::Push() to simply call LinkList::InsertAtFront(), just as Stack::Pop() does little more than call LinkList::RemoveAtFront(). Even though Stack would love to simply use the inherited implementations of LinkList::IsEmpty() and LinkList::Print(), due to LinkList being a private base class, these functions are not part of the public interface of Stack. Accordingly, Stack adds an IsEmpty() method that simply calls LinkList::IsEmpty();. Notice the use of the scope resolution operator to specify the LinkList::IsEmpty() method; without the base class qualification, we would be adding a recursive function call! This call to the base class method is allowed because Stack member functions can call the *once public* methods of LinkList (they are now treated as private within Stack). Similarly, Stack::Print() merely calls LinkList::Print().

In the scope of main(), we instantiate a Stack, namely stack1. Using the public interface of Stack, we easily manipulate stack1 using Stack::Push(), Stack::Pop(), Stack::IsEmpty(), and Stack::Print().

The output for this example is as follows:

```
Stack 1: 475 600 3000
Stack 1 after popping an item: 600 3000
Stack 1 after popping an item: 3000
Stack 1 after popping an item: <EMPTY>
```

It is important to note that a pointer to a Stack instance cannot be upcast to be stored as a pointer to a LinkList. Upcasting is not allowed across a private base class boundary. This would allow a Stack to reveal its underlying implementation; C++ does not allow this to happen. Here, we see that a Stack is merely implemented in terms of a LinkList; we are not saying that a Stack *Is-A* LinkedList. This is the concept of implementation inheritance in its best light; this example illustrates implementation inheritance favorably.

Next, let's move forward to see how we can use a protected base class, and how that differs from a private base class using implementation inheritance.

Using a protected base class to implement one class in terms of another

We have just implemented a Stack in terms of a LinkList using a private base class. Now, let's implement a Queue and a PriorityQueue. We will implement a Queue using LinkList as a protected base class, and a PriorityQueue using Queue as a public base class.

Again, Queue and PriorityQueue are Abstract Data Types. It is (relatively) unimportant how a Queue is implemented. The underlying implementation may change. Implementation inheritance allows us to implement our Queue using a LinkedList without revealing the underlying implementation to users of the Queue class.

Now, our class Queue will use LinkedList as a protected base class. Queue will define a public interface for its users to establish the expected interface for this ADT, such as Enqueue(), Dequeue(), IsEmpty(), and Print(). The implementation of these member functions will make use of selected LinkedList member functions, but Queue users will not see this, nor will Queue instances be able to use any LinkList members directly themselves.

Furthermore, our class PriorityQueue will extend Queue using public inheritance. That's right, we're back to Is-A. We are saying that a PriorityQueue *Is-A* Queue, and a Queue is implemented using a LinkedList.

We will just add a priority enqueuing method to our PriorityQueue class; this class will be glad to inherit the public interface from Queue (but obviously not from LinkList, which luckily is hidden behind a protected base class at its parent's level).

The code to implement Queue and PriorityQueue is again straightforward. The LinkList base class needs to be augmented to be more fully functional in order to proceed. The LinkListElement class can remain the same. We will show the basics of the revised LinkList class with only its class definition. The full code for both Queue and PriorityQueue will be shown in a separate segment. The full program example can be found in our GitHub:

https://github.com/PacktPublishing/Deciphering-Object-Orient-ed-Programming-with-CPP/blob/main/Chapter06/Chp6-Ex4.cpp

```cpp
// class LinkListElement is as shown previously
// The enhanced class definition of LinkList is:
class LinkList
{
private:
    LinkListElement *head = nullptr;
    LinkListElement *tail = nullptr;
    LinkListElement *current = nullptr;
public:
    LinkList() = default;
    LinkList(LinkListElement *);
    ~LinkList();
    void InsertAtFront(Item *);
    LinkListElement *RemoveAtFront();
    void DeleteAtFront();

    // Notice additional member functions added
    void InsertBeforeItem(Item *, Item *);
    LinkListElement *RemoveSpecificItem(Item *);
    void DeleteSpecificItem(Item *);
    void InsertAtEnd(Item *);
    LinkListElement *RemoveAtEnd();
    void DeleteAtEnd();

    int IsEmpty() const { return head == nullptr; }
    void Print() const;
};
// Assume we have the implementation for the methods here…
```

Notice that `LinkList` has been expanded to have a fuller set of features, such as being able to add, remove, and delete elements at various positions within the `LinkList`. To keep our examined code together brief, we will not show the implementation of these methods.

Now, let's add the class definitions for `Queue` and `PriorityQueue` in the next code segment:

```
class Queue: protected LinkList
{
private:
    // no new data members are necessary
public:
    // Constructor prototype shown below is not needed;
    // we get default w/o prototype (since no other ctor)
    // Commented to remind what's automatically provided
    // Queue() = default;  // calls :LinkList() def. ctor
    // Destructor prototype is needed (per virtual keyword)
    virtual ~Queue() = default; // we'll see virtual Chp. 7
    // public interface of Queue
    void Enqueue(Item *i) { InsertAtEnd(i); }
    Item *Dequeue();
    // redefine these methods, LinkList is prot. base class
    int IsEmpty() const { return LinkList::IsEmpty(); }
    void Print() { LinkList::Print(); }
};


Item *Queue::Dequeue()
{
    LinkListElement *front;
    front = RemoveAtFront();
    // make copy of front's data
    Item *item = new Item(*(static_cast<Item *>
                             (front->GetData())));
    delete front;
    return item;
}


class PriorityQueue: public Queue
{
```

```
private:
    // no new data members are necessary
public:
    // Constructor prototype shown below is not needed;
    // we get default w/o protoype (since no other ctor)
    // Commented to remind what's automatically provided
    // PriorityQueue() = default; // calls :Queue()
                                  // default constructor
    // destructor proto. is not needed for overriden dtor
    // ~PriorityQueue() override = default; // see Chp 7
    void PriorityEnqueue(Item *i1, Item *i2)
    { InsertBeforeItem(i1, i2); } // accessible in this
};                                // scope
```

In the previous segment of code, we define the Queue and PriorityQue classes. Notice that Queue has a protected base class of LinkList. With a protected base class, the protected and public members inherited from LinkList act as though they are defined by Queue as protected, which means that these inherited members are not only accessible within the scope of Queue, but also within any potential descendants of Queue. As before, these restrictions only apply to the Queue class, its descendants, and their instances; the LinkList class and its instances are unaffected.

In the Queue class, no new data members are necessary. The internal implementation is handled by LinkList. With a protected base class, we are saying that the Queue is implemented using a LinkList. Nonetheless, we must provide the public interface for Queue and we do so by adding methods such as Queue::Enqueue(), Queue::Dequeue(), Queue::IsEmpty() and Queue::Print(). Notice that in their implementations, these methods merely call LinkList methods to perform the necessary operations. Users of Queue must use Queue's public interface; the *once public* LinkList interface is hidden to Queue instances.

Next, we define PriorityQueue, another ADT. Notice that PriorityQueue defines Queue as a public base class. We are back to inheritance to support an Is-A relationship. A PriorityQueue *Is-A* Queue and can do everything a Queue can do, just a little more. As such, PriorityQueue inherits as usual from Queue, including Queue's public interface. PriorityQueue needs only to add an additional method for priority enqueuing, namely PriorityQueue::PriorityEnqueue().

Since Queue has a protected base class of LinkList, the public interface from LinkList is considered protected to Queue and its descendants, including PriorityQueue, so that LinkList's *once public* methods are considered protected to both Queue and PriorityQueue. Notice that PriorityQueue::PriorityEnqueue() makes use of LinkList::InsertBeforeItem(). This would not be possible if LinkList were a private, versus a protected, base class of Queue.

With the class definitions and implementation in place, let's continue with our `main()` function:

```
int main()
{
    Queue q1;      // Queue instance
    q1.Enqueue(new Item(50));
    q1.Enqueue(new Item(67));
    q1.Enqueue(new Item(80));
    q1.Print();
    while (!(q1.IsEmpty()))
    {
        q1.Dequeue();
        q1.Print();
    }

    PriorityQueue q2;      // PriorityQueue instance
    Item *item = new Item(167); // save a handle to item
    q2.Enqueue(new Item(67));    // first item added
    q2.Enqueue(item);            // second item
    q2.Enqueue(new Item(180));   // third item
    // add new item before an existing item
    q2.PriorityEnqueue(new Item(100), item); // 4th item
    q2.Print();
    while (!(q2.IsEmpty()))
    {
        q2.Dequeue();
        q2.Print();
    }
    return 0;
}
```

Now, in `main()`, we instantiate a `Queue`, namely `q1`, which utilizes the public interface of `Queue`. Note that `q1` may not use the *once public* interface of `LinkList`. The `Queue` may only behave like a `Queue`, not a `LinkList`. The ADT of `Queue` is preserved.

Finally, we instantiate a `PriorityQueue`, namely `q2`, which utilizes the public interface of both `Queue` and `PriorityQueue`, such as `Queue::Enqueue()` and `PriorityQueue::PriorityEnqueue()`, respectively. Because a `Queue` *Is-A* `PriorityQueue` (`Queue` is the `public` base class), the typical mechanics of inheritance are in place, allowing `PriorityQueue` to utilize the public interface of its ancestors.

The output for this example is as follows:

```
50 67 80
67 80
80
<EMPTY>
67 100 167 180
100 167 180
167 180
180
<EMPTY>
```

Finally, we have seen two examples of using implementation inheritance; it is not an often-used feature of C++. However, you now understand `protected` or `private` base classes should you run across them in library code, application code that you are maintaining, or the rare opportunity in which this technique may prove useful for a programming task you may encounter.

Optional uses for =default

We have seen `=default` used in constructor and destructor prototypes to alleviate the user need to supply such method definitions. However, let's recall some guidelines for when a constructor (or destructor) is provided for us automatically. In such cases, using `=default` with a constructor or destructor prototype will be more documentative in nature than a requirement; we will get the same system-supplied method in the absence of the `=default` prototype.

Using an `=default` prototype is not necessary if the default constructor is the only constructor in a class; recall that you will get a system-supplied default constructor if a class has no constructors (to provide an interface to instantiate the class). Using `=default` with a default constructor prototype is crucial, however, if there are other constructors in the class (not including the copy constructor) and you want to maintain the default object creation (construction) interface. For the copy constructor, if the default system-supplied version is adequate, you will get this method regardless of whether you use an `=default` prototype or omit the prototype entirely. Likewise, with the destructor, if the system-supplied version is adequate, you will get this version linked in regardless of whether you use an `=default` prototype or omit the prototype altogether; the latter style is becoming more prevalent.

We have now covered the basic features of single inheritance in C++. Let's quickly review what we've covered before moving to the next chapter.

Summary

In this chapter, we have moved further along our journey with object-oriented programming. We have added additional OO concepts and terms, and have seen how C++ has direct language support for these concepts. We have seen how inheritance in C++ supports generalization and specialization. We have seen how to incrementally build a hierarchy of related classes.

We have seen how to grow inheritance hierarchies using single inheritance, and how to access inherited data members and member functions. We have reviewed access regions to understand which inherited members may be directly accessed, based upon the access regions in which the members are defined in the base class. We know that having a `public` base class equates to defining an Is-A relationship, which supports the ideals of generalization and specialization, which is the most commonly used reason for inheritance.

We have detailed the order of constructor and destructor invocations when instances of derived class types are instantiated and destroyed. We have seen the member initialization list to select which inherited constructor a derived class object may choose to utilize as part of its own construction (for its base class subobject).

We have seen how changing access labels in a base class list changes the OO meaning for the type of inheritance being used. By comparing `public` versus `private` and `protected` base classes, we now understand different types of hierarchies, such as those built to support Is-A relationships versus those built to support implementation inheritance.

We have seen that base classes in our hierarchies may serve as potential building blocks for more specialized components, leading to potential reuse. Any potential reuse of existing code saves development time and cuts down on maintenance of otherwise duplicated code.

Through extending our OOP knowledge, we have gained a preliminary set of skills relating to inheritance and hierarchy building in C++. With the basic mechanics of single inheritance under our belts, we can now move forward to learn about many more interesting object-oriented concepts and details relating to inheritance. Continuing to *Chapter 7, Utilizing Dynamic Binding through Polymorphism*, we will next learn how to dynamically bind methods to their respective operations in a hierarchy of related classes.

Questions

1. Using your *Chapter 5, Exploring Classes in Detail*, solution, create a C++ program to build an inheritance hierarchy, generalizing `Person` as a base class from the derived class of `Student`.

   a. Decide which data members and member functions of your `Student` class are more generic and would be better positioned in a `Person` class. Build your `Person` class with these members, including appropriate constructors (default, alternate, and copy), a destructor, access member functions, and a suitable public interface. Be sure to place your data members in the private access region.

   b. Using a `public` base class, derive `Student` from `Person`. Remove members from `Student` that are now represented in `Person`. Adjust constructors and the destructor accordingly. Use the member initialization list to specify base class constructors as needed.

   c. Instantiate both `Student` and `Person` several times and utilize the appropriate `public` interfaces on each. Be sure to dynamically allocate several instances.

   d. Add a message using `cout` as the first line in each of your constructors and as the first line in your destructors so that you can see the construction and destruction order of each instance.

2. (Optional) Complete the class hierarchy, which includes `LinkList`, `Queue`, and `PriorityQueue`, using the online code as a basis. Complete the remaining operations in the `LinkList` class, and call them as appropriate in the public interface of `Queue` and `PriorityQueue`.

   a. Be sure to add copy constructors for each class (or prototype them in the private access region or use `=delete` in the prototype to suppress copying if you truly do not want to allow copies).

   b. Instantiate `LinkList` using either constructor, then demonstrate how each of your operations works. Be sure to invoke `Print()` after adding or deleting an element.

   c. Instantiate `Queue` and `PriorityQueue`, and demonstrate that each of the operations in their `public` interfaces works correctly. Remember to demonstrate the inherited operations in the `public` interface of `Queue` for instances of `PriorityQueue`.

7
Utilizing Dynamic Binding through Polymorphism

This chapter will further extend our knowledge of object-oriented programming in C++. We will begin by introducing a powerful OO concept, **polymorphism**, and then understand how this idea is implemented in C++ through *direct language support*. We will implement polymorphism using virtual functions in hierarchies of related classes, and understand how we can achieve runtime binding of a specific derived class method to a more generic, base class operation. We will understand how the OO concept of polymorphism presented in this chapter will support versatile and robust designs and easily extensible code in C++.

In this chapter, we will cover the following main topics:

- Understanding the OO concept of polymorphism and why it is important to OOP
- Defining virtual functions, understanding how virtual functions override base class methods (or halt the overriding process with the `final` specifier), generalizing derived class objects, the need for virtual destructors, as well as understanding function hiding
- Understand dynamic (runtime) binding of methods to operations
- Detailed understanding of the **virtual function table** (v-table)

By the end of this chapter, you will understand the OO concept of polymorphism, and how to implement this idea in C++ through virtual functions. You will understand how virtual functions enable the runtime binding of methods to operations in C++. You will see how an operation can be specified in a base class and overridden with a preferred implementation in a derived class. You will understand when and why it is important to utilize a virtual destructor.

You will see how instances of derived classes are often stored using base class pointers and why this is significant. We will discover that, regardless of how an instance is stored (as its own type or as that of a base class), the correct version of a virtual function will always be applied through dynamic binding. Specifically, you will see how runtime binding works under the hood as we examine virtual function pointers and virtual function tables in C++.

By understanding the direct language support of polymorphism in C++ using virtual functions, you will be on your way to creating an extensible hierarchy of related classes, featuring dynamic binding of methods to operations. Let us augment our understanding of C++ as an OOP language by detailing these ideals.

Technical requirements

Online code for full program examples can be found in the following GitHub URL: `https://github.com/PacktPublishing/Deciphering-Object-Oriented-Programming-with-CPP/tree/main/Chapter07`. Each full program example can be found in the GitHub under the appropriate chapter heading (subdirectory) in a file that corresponds to the chapter number, followed by a dash, followed by the example number in the chapter at hand. For example, the first full program in this chapter can be found in the subdirectory `Chapter07` in a file named `Chp7-Ex1.cpp` under the aforementioned GitHub directory.

The CiA video for this chapter can be viewed at: `https://bit.ly/3QQUGxg`.

Understanding the OO concept of polymorphism

In this section, we will introduce an essential object-oriented concept, polymorphism.

From *Chapter 5*, *Exploring Classes in Detail*, and *Chapter 6*, *Implementing Hierarchies with Single Inheritance*, you now understand the key OO ideas of encapsulation, information hiding, generalization, and specialization. You know how to encapsulate a class, how to build inheritance hierarchies using single inheritance, and the various reasons to build hierarchies (such as supporting Is-A relationships or for the lesser-used reason of supporting implementation inheritance). Let's begin by extending our basic OO terminology by exploring **polymorphism**.

When a base class specifies an operation such that a derived class may redefine the operation in its class with a more suitable method, the operation is said to be **polymorphic**. Let's revisit our definitions of **operation** and **method**, as well as their implications, to understand how these concepts lay the groundwork for polymorphism:

- In C++, an **operation** maps to the complete signature of the member function (name plus type and number of arguments – no return type).

- Additionally, in C++, a **method** maps to the definition or body of the operation (that is, the implementation or body of the member function).

- Recall that in OO terms, an **operation** implements a behavior of a class. The implementation of a base class operation may be via several distinct derived class **methods**.

Polymorphism gives an object the ability to take on *many forms*, yet have its most relevant behaviors applied, even when the object may be represented in a more genericized (base class) state than it is originally defined. This happens in C++ with public inheritance. A derived class object may be more generically pointed to by a base class pointer through upcasting. Yet, if an operation is defined to be polymorphic, the specific method applied when the operation is invoked will be the derived class version, as that is the most appropriate method for the object (irrespective of how it may currently be genericized as a base class object). Here, the derived class object fulfills the Is-A relationship of the base class. For example, a Student *Is-A* Person. Yet, a polymorphic operation will allow Student behaviors to be revealed on Student objects, even when they have *taken on the form* of a Person.

As we progress through this chapter, we will see derived class objects taking on the form of their public base classes, that is, taking on *many forms* (**polymorphism**). We will see how a polymorphic operation can be specified in a base class and overridden with a preferred implementation in a derived class.

Let's start by looking at the C++ language feature that allows us to implement polymorphism, namely, virtual functions.

Implementing polymorphism with virtual functions

Polymorphism allows dynamic binding of a method to an operation. Dynamic, or runtime, binding of a method to an operation is important because derived class instances may be pointed to by base class objects (that is, by pointers of a base class type). In these situations, the pointer type does not provide adequate information regarding the correct method that should be applied to the referenced instance. We need another way – one done at runtime – to determine which method applies to each instance.

Often, it is the case that a pointer to an instance of a derived class type will be generalized as a pointer to the base class type. When an operation is applied to the pointer, the correct method for what the object truly is should be applied, rather than the method that *seems* appropriate for the generalized pointer type.

Let's begin with the relevant keywords and logistics necessary to define virtual functions so that we may implement polymorphism.

Defining virtual functions and overriding base class methods

Virtual functions in C++ directly support polymorphism. A **virtual function** is as follows:

- A member function that correctly allows methods for a given operation to be overridden successively in a hierarchy to provide more suitable definitions

- A member function that allows dynamic, rather than the usual static, binding for methods

A virtual function is specified using the keyword **virtual** with the following nuances:

- The keyword `virtual` should precede the return type of the function in its prototype.

- Functions in the derived class with the same name and signature of a virtual function in any ancestor class redefine the virtual function in those base classes. Here, the keyword `virtual` is optional in the derived class prototype.

- Optionally and preferred, the keyword `override` can be added as part of the extended signature in the derived class prototype. This recommended practice will allow the compiler to flag an error if the signature of the intended overridden method does not match the signature as specified in the base class. The `override` keyword can eliminate unintended function hiding.

- Functions with the same name, yet a different signature in a derived class, do not redefine a virtual function in their base class; rather, they hide the methods found in their base classes.

- Additionally, the keyword `final` can be added as part of the extended signature of a virtual function prototype if the virtual function in question is not intended to be further overridden in a derived class.

The derived class need not redefine virtual functions specified in its base class if the inherited methods are suitable. However, should a derived class redefine an operation with a new method, the same signature (as specified by the base class) must be used for the overridden method. Furthermore, derived classes should only redefine virtual functions.

Here is a simple example to illustrate the basic syntax:

- `Print()` is a virtual function defined in the base class `Person`. It will be overridden with a more appropriate implementation in the `Student` class:

```
class Person   // base class
{
private:
    string name;
    string title;
public:
    // constructors/destructor (will soon be virtual),
    // public access functions, public interface etc.
    virtual void Print() const
    {
        cout << title << " " << name << endl;
    }
};
```

Here, the base class `Person`, introduces a virtual function, `Print()`. By labeling this function as `virtual`, the `Person` class is inviting any future descendants to redefine this function with a more suitable implementation or method, should they be so motivated.

- The virtual function defined in the base class `Person` is, in fact, overridden with a more appropriate implementation in the `Student` class:

```
class Student: public Person  // derived class
{
private:
    float gpa = 0.0;  // in-class initialization
public:
    // constructors, destructor specific to Student,
    // public access functions, public interface, etc.
    void Print() const override
    {
        Person::Print(); // call base class fn to help
        cout << " is a student. GPA: " << gpa << endl;
    }
};
```

Notice here that the derived class `Student` introduces a new implementation of `Print()` that will override (that is, replace), the definition in `Person`. Note that if the implementation of `Person::Print()` were acceptable to `Student`, `Student` would not be obligated to override this function, even if it is marked as `virtual` in the base class. The mechanics of public inheritance would simply allow the derived class to inherit this method.

But because this function is `virtual` in `Person`, `Student` may opt to redefine this operation with a more suitable method. Here, it does. In the `Student::Print()` implementation, `Student` first calls `Person::Print()` to take advantage of the aforementioned base class function, then prints additional information itself. `Student::Print()` is choosing to call a base class function for help; it is not required to do so if the desired functionality can be implemented fully within its own class scope.

Notice that when `Student::Print()` is defined to override `Person::Print()`, the same signature as specified by the base class is used. This is important. Should a new signature have been used, we would get into a potential function hiding scenario, which we will soon discuss in our *Considering function hiding* subsection within this chapter.

Note that though the virtual functions in `Person` and `Student` are written inline, a virtual function will almost never be expanded as inline code by the compiler since the specific method for the operation must be determined at runtime. A very few cases exist for compiler devirtualization, involving final methods or knowing an instance's dynamic type; such rare cases would allow a virtual function to be inlined.

Remember, polymorphic functions are meant to have the ability to override or replace base class versions of a given function. Function overriding differs from function overloading.

> **Important distinction**
> **Function overriding** is defined by introducing the same function name with the same signature in a hierarchy of related classes (via virtual functions), whereas the derived class version is meant to replace the base class version. In contrast, **function overloading** is defined when two or more functions with the same name, but with different signatures, exist in the same scope of the program (such as in the same class).

Additionally, operations not initially specified as virtual when introduced in a base class definition are not polymorphic and, therefore, should not be overridden in any derived class. This means that if a base class does not use the keyword `virtual` when defining an operation, the base class does not intend for the derived class to redefine this operation with a more suitable derived class method. The base class instead is insisting that the implementation it has provided is suitable for *any* of its descendants. Should the derived class attempt to redefine a non-virtual base class operation, a subtle bug will be introduced into the application. The error will be that derived class instances stored using derived class pointers will use the derived class method, yet derived class instances stored using base class pointers will use the base class definition. Instances should always use their own behavior irrespective of how they are stored – this is the point of polymorphism. Never redefine a non-virtual function.

> **Important note**
> Operations not specified in a base class as virtual in C++ are not polymorphic, and should never be overridden by a derived class.

Let's move forward and discover scenarios when we may want to collect derived class objects by a base class type, and when we may then need to qualify our destructors as virtual.

Generalizing derived class objects

When we view an inheritance hierarchy, it is typically one that employs public base classes; that is, it is a hierarchy that utilizes public inheritance to express Is-A relationships. When using inheritance in this manner, we may be motivated to collect groups of related instances together. For example, a hierarchy of `Student` specializations might include `GraduateStudent`, `UnderGraduateStudent`, and `NonDegreeStudent`. Assuming each of these derived classes has a public base class of `Student`, it would be appropriate to say a `GraduateStudent` *Is-A* `Student`, and so on.

We may find a reason in our application to group these *somewhat-like* instances together into one common set. For example, imagine that we are implementing a billing system for a university. The university may wish for us to collect all students, regardless of their derived class types, into one set to process them uniformly, so as to calculate their semester bills.

The `Student` class may have a polymorphic operation to `CalculateSemesterBill()`, which is implemented as a virtual function in `Student` with a default method. However, selected derived classes, such as `GraduateStudent`, may have preferred implementations that they wish to provide by overriding the operation in their own class with a more appropriate method. A `GraduateStudent`, for example, may have a different method to compute their total bill versus a `NonDegreeStudent`. Hence, each derived class may override the default implementation of `CalculateSemesterBill()` in each of their classes.

Nonetheless, in our bursar application, we can create a set of pointers of type `Student`, though each pointer will inevitably point to instances of the derived class types, such as `GraduateStudent`, `UnderGraduateStudent`, and `NonDegreeStudent`. When instances of derived class types have been generalized in this fashion, it is appropriate to apply functions (often virtual) to the set as defined in the base class level corresponding to the pointer type of the collection. Virtual functions allow these generalized instances to invoke a polymorphic operation to yield their individual derived class methods or implementations of these functions. This is exactly what we want. But, there are still more details to understand.

This basic premise of generalizing derived class instances will allow us to understand why we may need virtual destructors within many of our class definitions. Let's take a look.

Utilizing virtual destructors

We now can conceptualize situations when grouping derived class instances into a *somewhat-like* set stored by their common base class type may be useful. It is actually very powerful to collect sibling type derived class instances by their base class type and employ virtual functions to allow their distinct behaviors to shine through.

But, let's consider what happens when the memory for a derived class instance stored by a base class pointer goes away. We know its destructor is called, but which one? We actually know that a chain of destructors is called, starting with the destructor of the object type in question. But how do we know the actual derived class object type if the instance has been genericized by being stored using a base class pointer? A **virtual destructor** solves this issue.

By labeling a destructor as `virtual`, we are allowing it to be overridden as the *starting point* in the destruction sequence for a class and any of its descendants. The choice as to which destructor to use as the entry point of destruction will be deferred to runtime using dynamic binding, based on the object's actual type, not what the pointer type may be that references it. We will soon see how this process is automated by examining C++'s underlying virtual function table.

A virtual destructor, unlike all other virtual functions, actually specifies the starting point for a full sequence of functions to be executed. Recall that as the last line of code in a destructor, the compiler automatically patches in a call to call the immediate base class destructor, and so on, until we reach the initial base class in the hierarchy. The destruction chain exists to provide a forum to release dynamically allocated data members in all subobjects of a given instance. Contrasting this behavior to other virtual functions, those merely allow the single, correct version of the function to be executed (unless the programmer chooses to call a base class version of the same function as a helper function during the derived method implementation).

You may ask why it is important to start the destruction sequence at the proper level. That is, starting at the level that matches the object's actual type (versus a generalized pointer type that may point to the object). Recall that each class may have dynamically allocated data members. The destructor will deallocate these data members. Starting with the correct level destructor will ensure that you do not introduce any memory leaks into your application by forgoing appropriate destructors and their corresponding memory deallocations.

Are virtual destructors always necessary? That is a good question! Virtual destructors are always necessary when using a public base class hierarchy, that is, when using public inheritance. Recall that public base classes support Is-A relationships, which easily lead to allowing a derived class instance to be stored using a pointer of its base class type. For example, a `GraduateStudent` Is-A `Student`, so we can store a `GraduateStudent` as a `Student` in times when we require more generic processing along with its sibling types. We can always upcast in this fashion across a public inheritance boundary. However, when we use implementation inheritance (that is, private or protected base classes), upcasting is not allowed. So, for hierarchies employing private or protected inheritance, virtual destructors are not necessary because upcasting is simply disallowed; hence, it would never be ambiguous as to which destructor should be the entry point for classes in private and protected base class hierarchies. As a second example, we did not include a virtual destructor in our `LinkedList` class in *Chapter 6*, *Implementing Hierarchies with Single Inheritance*; therefore, `LinkedList` should only be extended as a protected or private base class. We did, however, include a virtual destructor in our `Queue` and `PriorityQueue` classes because `PriorityQueue` uses `Queue` as a public base class. A `PriorityQueue` may be upcast to a `Queue` (but not to a `LinkedList`), necessitating the virtual destructor introduction at the `Queue` and its descendent levels in the hierarchy.

Are the optional keywords `virtual` and `override` recommended when overriding a virtual destructor? Those are also good questions. We know that an overridden destructor is only the starting point in the destruction sequence. We also know that, unlike other virtual functions, the derived class destructor will have a unique name from the base class destructor. Even though a derived class destructor automatically overrides a base class destructor that has been declared as `virtual`, the usage of the *optional* keyword `override` is recommended in the derived class destructor prototype for documentation. However, the usage of the *optional* keyword `virtual` in the derived class destructor is generally no longer used. The reasoning is that the `override` keyword is meant to provide a safety net to catch spelling mistakes between originally defined and overridden functions. With destructors, the function names are not the same, hence this safety net is not an error-checking advantage, but more documentative.

Let's continue by putting all the necessary pieces together so we can see virtual functions of all varieties, including destructors, in action.

Putting all the pieces together

So far in this chapter, we have understood the nuances of virtual functions, including virtual destructors. It is important to see our code in action with all its various components and details. We need to see in one cohesive program the basic syntax to specify virtual functions, including how we may collect derived class instances by base class types, and see how virtual destructors play a role.

Let's take a look at a more complex, full program example to fully illustrate polymorphism, implemented using virtual functions in C++. This example will be broken into many segments; the full program can be found in the following GitHub location:

https://github.com/PacktPublishing/Deciphering-Object-Orient-ed-Programming-with-CPP/blob/main/Chapter07/Chp7-Ex1.cpp

```cpp
#include <iostream>
#include <iomanip>
using std::cout;      //preferred to: using namespace std;
using std::endl;
using std::setprecision;
using std::string;
using std::to_string;


constexpr int MAX = 5;


class Person
{
private:
    string firstName;
    string lastName;
    char middleInitial = '\0';  // in-class initialization
    string title;   // Mr., Ms., Mrs., Miss, Dr., etc.
protected:
    void ModifyTitle(const string &);
public:
    Person() = default;   // default constructor
    Person(const string &, const string &, char,
           const string &);
```

```
    // copy constructor =default prototype not needed; we
    // get the default version w/o the =default prototype
    // Person(const Person &) = default;  // copy const.
    virtual ~Person();  // virtual destructor
    const string &GetFirstName() const
        { return firstName; }
    const string &GetLastName() const { return lastName; }
    const string &GetTitle() const { return title; }
    char GetMiddleInitial() const { return middleInitial; }
    virtual void Print() const;
    virtual void IsA() const;
    virtual void Greeting(const string &) const;
};
```

In the aforementioned class definition, we have augmented our familiar class for Person, adding four virtual functions, namely, the destructor (~Person()), Print(), IsA(), and Greeting(const string &). Notice that we have simply placed the keyword virtual in front of the return type (if any) of each member function. The remainder of the class definition is as we have explored in depth in the previous chapter.

Now, let's examine the non-inline member function definitions for Person:

```
// With in-class initialization, writing the default
// constructor is no longer necessary.
// Also, remember that strings are member objects and will
// be default constructed as empty.


// alternate constructor
Person::Person(const string &fn, const string &ln, char mi,
                const string &t): firstName(fn),
                lastName(ln), middleInitial(mi), title(t)
{
    // dynamically allocate memory for any ptr data members
}


// We are choosing to utilize the default copy constructor.
// If we wanted to prototype/define it, here's the method:
// Person::Person(const Person &p):
```

```
//              firstName(p.firstName), lastName(p.lastName),
//              middleInitial(p.middleInitial), title(p.title)
// {
    // deep copy any pointer data members here
// }

Person::~Person()
{
    // release memory for any dynamically alloc. data mbrs.
    cout << "Person destructor <" << firstName << " "
        << lastName << ">" << endl;
}

void Person::ModifyTitle(const string &newTitle)
{   // assignment between strings ensures a deep assignment
    title = newTitle;
}

void Person::Print() const
{
    cout << title << " " << firstName << " ";
    cout << middleInitial << ". " << lastName << endl;
}

void Person::IsA() const
{
    cout << "Person" << endl;
}

void Person::Greeting(const string &msg) const
{
    cout << msg << endl;
}
```

In the previous segment of code, we have specified all of the non-inline member functions of `Person`. Notice that the four virtual functions – the destructor, `Print()`, `IsA()`, and `Greeting()` – do not include the `virtual` keyword in the methods (that is, member function definitions) themselves.

Next, let's examine the `Student` class definition and its inline functions:

```
class Student: public Person
{
private:
    float gpa = 0.0;    // in-class initialization
    string currentCourse;
    const string studentId;
    static int numStudents;  // static data member
public:
    Student();  // default constructor
    Student(const string &, const string &, char,
            const string &, float, const string &,
            const string &);
    Student(const Student &);  // copy constructor
    ~Student() override;  // virtual destructor
    void EarnPhD();
    // inline function definitions
    float GetGpa() const { return gpa; }
    const string &GetCurrentCourse() const
        { return currentCourse; }
    const string &GetStudentId() const
        { return studentId; }
    void SetCurrentCourse(const string &); // proto. only

    // In the derived class, keyword virtual is optional,
    // and not currently recommended. Use override instead.
    void Print() const final override;
    void IsA() const override;
    // note: we choose not to redefine
    // Person::Greeting(const string &) const
    static int GetNumberStudents(); // static mbr. function
};
```

```
// definition for static data member
int Student::numStudents = 0;   // notice initial value of 0

inline void Student::SetCurrentCourse(const string &c)
{
    currentCourse = c;
}


// Definition for static member function (it's also inline)
inline int Student::GetNumberStudents()
{
    return numStudents;
}
```

In the previous class definition for Student, we again have all of the assorted components we are accustomed to seeing to comprise this class. Additionally, notice that we have overridden and redefined three virtual functions – the destructor, Print(), and IsA() – using the keyword override. These preferred definitions essentially replace or override the default methods specified for these operations in the base class.

Notice, however, that we choose not to redefine void Person::Greeting(const string &), which was introduced as a virtual function in the Person class. Simply inheriting this method is fine if we find the inherited definition acceptable for instances of the Student class. Furthermore, notice the additional qualification on Print() with the final qualifier. This keyword indicates that Print() may not be overridden in derived classes from Student; the method overridden at the Student level will be the final implementation.

Recall that the meaning of override, when paired with a destructor, is unique, in that it does not imply that the derived class destructor replaces the base class destructor. Instead, it means that the derived class (virtual) destructor is the correct beginning point for the *chain of destruction* sequence when initiated by derived class instances (irrespective of how they are stored). The virtual derived class destructor is merely the entry point for the complete destruction sequence.

Also remember, the derived class of Student is not required to override a virtual function that is defined in Person. Should the Student class find the base class method acceptable, it is automatically inherited. Virtual functions merely allow the derived class to redefine an operation with a more appropriate method when so needed.

Next, let's examine the non-inline Student class member functions:

```
Student::Student(): studentId(to_string(numStudents + 100)
                                     + "Id")
{
```

```
    // studentId is const; we need to set at construction.
    // We're using member init list with a unique id based
    // on numStudents + 100), concatenated with string "Id".
    // Remember, string member currentCourse will be default
    // const. with an empty string (it's a member object)
    numStudents++;      // set static data member
}

// Alternate constructor member function definition
Student::Student(const string &fn, const string &ln,
                 char mi, const string &t, float avg,
                 const string &course, const string &id):
                 Person(fn, ln, mi, t), gpa(avg),
                 currentCourse(course), studentId(id)
{
    // dynamically alloc memory for any pointer data members
    numStudents++;
}



// Copy constructor definition
Student::Student(const Student &s) : Person(s),
                 gpa(s.gpa), currentCourse(s.currentCourse),
                 studentId(s.studentId)
{
    // deep copy any pointer data mbrs of derived class here
    numStudents++;
}

// destructor definition
Student::~Student()
{
    // release memory for any dynamically alloc. data mbrs
    cout << "Student destructor <" << GetFirstName() << " "
        << GetLastName() << ">" << endl;
}
```

```
void Student::EarnPhD()
{
    ModifyTitle("Dr.");
}


void Student::Print() const
{   // need to use access functions as these data members
    // are defined in Person as private
    cout << GetTitle() << " " << GetFirstName() << " ";
    cout << GetMiddleInitial() << ". " << GetLastName();
    cout << " with id: " << studentId << " GPA: ";
    cout << setprecision(3) <<   " " << gpa;
    cout << " Course: " << currentCourse << endl;
}


void Student::IsA() const
{
    cout << "Student" << endl;
}
```

In the previously listed section of code, we list the non-inline member function definitions for Student. Again, notice that the keyword override will not appear in any of the virtual member function definitions themselves, only in their respective prototypes.

Lastly, let's examine the main() function:

```
int main()
{
    Person *people[MAX] = { }; // initialize with nullptrs
    people[0] = new Person("Juliet", "Martinez", 'M',
                          "Ms.");
    people[1] = new Student("Hana", "Sato", 'U', "Dr.",
                          3.8, "C++", "178PSU");
    people[2] = new Student("Sara", "Kato", 'B', "Dr.",
                          3.9, "C++", "272PSU");
    people[3] = new Person("Giselle", "LeBrun", 'R',
                          "Miss");
```

```
      people[4] = new Person("Linus", "Van Pelt", 'S',
                             "Mr.");
      // We will soon see a safer and more modern way to loop
      // using a range for loop (starting in Chp. 8).
      // Meanwhile, let's notice mechanics for accessing
      // each element.
      for (int i = 0; i < MAX; i++)
      {
         people[i]->IsA();
         cout << "   ";
         people[i]->Print();
      }
      for (int i = 0; i < MAX; i++)
         delete people[i];    // engage virtual dest. sequence
      return 0;
   }
```

Here, in main(), we declare an array of pointers to Person. Doing so, allows us to collect both Person and Student instances in this set. Of course, the only operations we may apply to instances stored in this generalized fashion are those found in the base class, Person.

Next, we allocate several Person and several Student instances, storing each instance via an element in the generalized set of pointers. When a Student is stored in this fashion, an upcast to the base class type is performed (but the instance is not altered in any fashion). Recall that when we looked at memory layout for derived class instances in *Chapter 6, Implementing Hierarchies with Single Inheritance*, we noticed that a Student instance first includes the memory layout of a Person, followed by the additional memory required for Student data members. This upcast merely points to the starting point of this collective memory.

Now, we proceed through a loop to apply operations as found in the Person class to all instances in this generalized collection. These operations happen to be polymorphic. That is, the virtual functions allow the specific implementation for methods to be called through runtime binding to match the actual object type (irrespective of the fact that the object may be stored in a generalized pointer).

Lastly, we loop through deleting the dynamically allocated instances of Person and Student, again using the generalized Person pointers. Because we know delete() will patch in a call to the destructor, we wisely have made the destructors virtual, enabling dynamic binding to choose the appropriate starting destructor (in the destruction chain) for each object.

When we look at the output for the aforementioned program, we can see that the specific method for each object is appropriately called for each virtual function, including the destruction sequence. Derived class objects have both the derived, then base class destructor invoked and executed. Here is the output for the full program example:

```
Person
   Ms. Juliet M. Martinez
Student
   Dr. Hana U. Sato with id: 178PSU GPA:  3.8 Course: C++
Student
   Dr. Sara B. Kato with id: 272PSU GPA:  3.9 Course: C++
Person
   Miss Giselle R. LeBrun
Person
   Mr. Linus S. Van Pelt
Person destructor <Juliet Martinez>
Student destructor <Hana Sato>
Person destructor <Hana Sato>
Student destructor <Sara Kato>
Person destructor <Sara Kato>
Person destructor <Giselle LeBrun>
Person destructor <Linus Van Pelt>
```

Now that we have competency utilizing the concept of polymorphism and the mechanics of virtual functions, let's take a look at a less usual situation relating to virtual functions, that of function hiding.

Considering function hiding

Function hiding is not an often-used feature of C++. In fact, it is often employed quite by accident! Let's review a key point we know about inherited member functions to get started. When an operation is specified by a base class, it is intended to provide a protocol for usage and redefinition (in the case of virtual functions) for all derived class methods.

Sometimes, a derived class will alter the signature of a method that is intended to redefine an operation specified by a base class (let's think of virtual functions). In this case, the new function, which differs in signature from the operation specified in its ancestor class, will not be considered a virtual redefinition of the inherited operation. In fact, it will *hide* inherited methods for the virtual function that have the same name specified in ancestor classes.

When programs are compiled, the signature of each function is compared against the class definition for correct usage. Typically, when a member function is not found in the class that *seemingly* matches the instance type, the hierarchy is traversed in an upward fashion until such a match is found or until the hierarchy is exhausted. Let us take a closer look at what the compiler contemplates:

- When a function is found with the same name as the function being sought out, the signature is examined to see whether it matches the function call exactly, or if type conversion can be applied. When the function is found, but type conversion cannot be applied, the normal traversal sequence is ended.

- Functions that hide virtual functions normally halt this upward search sequence, thus hiding a virtual function that otherwise may have been invoked. Recall that at compile time, we are just checking syntax (not deciding which version of a virtual function to call). But if we can't find a match, an error is flagged.

- Function hiding is actually considered helpful and was intended by the language. If the class designer provided a specific function with a given signature and interface, that function should be used for instances of that type. Hidden or unsuspected functions above in the hierarchy should not be used in this specific scenario.

Consider the following modification to our previous full program example to, first, illustrate function hiding, and then provide a more flexible solution for managing function hiding:

- Recall that the `Person` class introduces `virtual void Print()` with no parameters. Imagine that `Student`, instead of overriding `Print()` with the same signature, changes the signature to `virtual void Print(const char *)`:

```
class Person  // base class
{
    // data members
public:  // member functions, etc.
    virtual void Print() const;
};


class Student: public Person
{
    // data members
public:  // member functions, etc.
    // Newly introduced virtual fn. --
    // Not a redefinition of Person::Print()
    virtual void Print(const string &) const;
};
```

Notice that the signature of `Print()` has changed from base to derived class. The derived class function does not redefine the `virtual void Print();` of its base class. It is a new function that will in fact hide the existence of `Person::Print()`. This is actually what was intended, since you may not recall that the base class offers such an operation, and tracking upward might cause surprising results in your application if you intended `Print(const string &)` to be called but `Print()` is called instead. By adding this new function, the derived class designer is dictating this interface is the appropriate `Print()` for instances of `Student`.

However, nothing is straightforward in C++. For situations where a `Student` is upcast to a `Person`, the `Person::Print()` with no arguments will be called. The `Student::Print(const string &)` is not a virtual redefinition because it does not have the same signature. Hence, the `Person::Print()` will be called for generalized `Student` instances. And yet `Student::Print(const string &)` will be called for `Student` instances stored in `Student` variables. Unfortunately, this is inconsistent with how an instance will behave if it is stored in its own type versus a generalized type. Though function hiding was meant to work in this fashion, it may inevitably not be what you would like to happen. Programmers, beware!

Let's look at some of the cumbersome code that might ensue:

- Explicit downcasting or use of the scope resolution operator may be required to reveal an otherwise hidden function:

```
constexpr int MAX = 2;
int main()
{
    Person *people[MAX] = { }; // init. with nullptrs
    people[0] = new Person("Jim", "Black", 'M',
                          "Mr.");
    people[1] = new Student("Kim", "Lin", 'Q', "Dr.",
                          3.55, "C++", "334UD");
    people[1]->Print(); // ok, Person::Print() defined
    // people[1]->Print("Go Team!"); // error!
    // explicit downcast to derived type assumes you
    // correctly recall what the object is
    (dynamic_cast<Student *> (people[1]))->
                          Print("I have to study");
    // Student stored in its own type
    Student s1("Jafari", "Kanumba", 'B', "Dr.", 3.9,
            "C++", "845BU");
```

```
    // s1.Print(); // error, base class version hidden
    s1.Print("I got an A!"); // works for type Student
    s1.Person::Print(); // works using scope
                        // resolution to base class type
    return 0;
}
```

In the aforementioned example, we have a generalized set of two `Person` pointers. One entry points to a `Person` and one entry points to a `Student`. Once the `Student` is generalized, the only applicable operations are those found in the `Person` base class. Therefore, a call to `people[1]->Print();` works and a call to `people[1]->Print("Go Team!");` does not work. The latter call to `Print(const char *)` is an error at the generalized base class level, even though the object truly is a `Student`.

If, from a generalized pointer, we wish to call specific functions found at the `Student` level in the hierarchy, we will then need to downcast the instance back to its own type (`Student`). We add a downcast with the call: `(dynamic_cast<Student *> (people[1]))->Print("I have to study");`. Here, we are taking a risk – if `people[1]` was actually a `Person` and not a `Student`, this would generate a runtime error. However, by first checking the result of the dynamic cast to `Student *` prior to invoking `Print()`, we can ensure we have made an appropriate cast.

Next, we instantiate `Student s1;`. Should we try to call `s1.Print()`, we get a compiler error – `Student::Print(const string &)` hides the base class presence of `Person::Print()`. Remember, `s1` is stored in its own type, `Student`, and since `Student::Print(const string &)` is found, the traversal upward to otherwise uncover `Person::Print()` is halted.

Nonetheless, our call to `s1.Print("I got an A!");` is successful because `Print(const string &)` is found at the `Student` class level. Lastly, notice that the call to `s1.Person::Print();` works but requires knowledge of the otherwise hidden function. By using the scope resolution operator (`::`), we can find the base class version of `Print()`. Even though `Print()` is virtual in the base class (implying dynamic binding), using the scope resolution operation reverts this call to a statically bound function call.

Let's propose that we would like to add a new interface to a derived class with a function that would otherwise hide a base class function. Knowing about function hiding, what should we ideally do? We could simply override the virtual function as found in the base class with a new method in the derived class, and then we could overload that function to add the additional interface. Yes, we're now both overriding and overloading. That is, we are overriding the base class function, and overloading the overridden function in the derived class.

Let's take a look at what we would now have:

- Here is the more flexible interface to add the new member function while keeping the existing interface that would otherwise be hidden:

```
class Person   // base class
{
    // data members
public:   // member functions, etc.
    virtual void Print() const;
};


class Student: public Person
{
    // data members
public:   // member functions, etc.
    // Override the base class method so that this
    // interface is not hidden by overloaded fn. below
    void Print() const override;
    // add the additional interface
    // (which is overloaded)
    // Note: this additional Print() is virtual
    // from this point forward in the hierarchy
    virtual void Print(const string &) const;
};


int main()
{
    Student s1("Zack", "Doone", 'A', "Dr.", 3.9,
               "C++", "769UMD");
    s1.Print();   // this version is no longer hidden.
    s1.Print("I got an A!"); // also works
    s1.Person::Print(); // this is no longer necessary
}
```

In the preceding code fragment, the `Student` class both overrides `Person::Print()` with `Student::Print()` and overloads `Student::Print()` with `Student::Print(const string &)` to envelop the additional desired interface. Now, for `Student` objects stored in `Student` variables, both interfaces are available – the base class interface is no longer hidden. Of course, `Student` objects referenced by `Person` pointers only have the `Person::Print()` interface, which is to be expected.

Overall, function hiding does not surface often, but when it does, it is often an unwelcome surprise. Now you understand what may happen and why, which helps in making you a better programmer.

Now that we have looked at all the uses surrounding virtual functions, let's look under the hood to see why virtual functions are able to support dynamic binding of a specific method to an operation. To thoroughly understand runtime binding, we will need to look at the v-table. Let's move forward!

Understanding dynamic binding

Now that we have seen how polymorphism is implemented with virtual functions to allow for dynamic binding of an operation to a specific implementation or method, let's understand why virtual functions allow for runtime binding.

Non-virtual functions are statically bound at compile time. That is, the address of the function in question is determined at compile time, based on the assumed type of the object at hand. For example, if an object is instantiated of type `Student`, a function call would have its prototype verified starting with the `Student` class, and if not found, the hierarchy would be traversed upward to each base class, such as `Person`, to look for the matching prototype. When found, the correct function call would be patched in. This is how static binding works.

However, a virtual function is a type of function in C++ that employs a dynamic binding at runtime. At compile time, any virtual function call is merely replaced with a lookup mechanism to delay binding until runtime. Certainly, each compiler vendor may differ in their implementation of automating virtual functions. However, there is a widely utilized implementation that involves virtual function pointers, a virtual function table, and virtual function table entries for each object type containing virtual functions.

Let's move forward to investigate how dynamic binding is commonly implemented in C++.

Comprehending runtime binding of methods to operations

We know that virtual functions allow for dynamic binding of an operation (specified in a base class) to a specific implementation or method (often specified in a derived class). How does this work?

When a base class specifies one or more new virtual functions (not just redefinitions of an ancestor's virtual functions), a **virtual function pointer (vptr)** is created below the memory comprising a given instance of that type. This happens at runtime when the memory for an instance is created (on the stack, heap, or static/extern area). When the instance in question is constructed, not only will the appropriate constructor be called to initialize the instance, but this vptr will be initialized to point to the **virtual function pointer table (v-table)** entry for that class type.

The v-table entry for a given class type will consist of a set of function pointers. These function pointers are often organized into an array of function pointers. A **function pointer** is a pointer to an actual function. By dereferencing this pointer, you will actually invoke the function to which the pointer points. There is an opportunity to pass arguments to the function, however, in order for this call to be generic through a function pointer, the arguments must be uniform for any version of this function that the pointer in question may point to. The premise of a function pointer gives us the ability to point to different versions of a particular function. That is, we could point to different methods for a given operation. This is the basis for which we can automate dynamic binding in C++ for virtual functions.

Let's consider the particular v-table entry for a specific object type. We know that this table entry will consist of a set of function pointers, such as an array of function pointers. The order in which these function pointers are arranged will be consistent with the order in which the virtual functions are newly introduced by a given class. Functions overriding existing virtual functions that were newly introduced at a higher level in the hierarchy will simply replace table entries with preferred versions of functions to be called, but will not cause an additional entry to be allocated in the array of function pointers.

So, when the program begins running first in global memory (as a hidden external variable), a v-table will be set up. This table will contain entries for each object type that contains virtual functions. The entry for a given object type will contain a set of function pointers (such as an array of function pointers) that organizes and initializes the dynamically-bound functions for that class. The specific order of the function pointers will correspond to the order in which the virtual functions were introduced (possibly by their ancestor class), and the specific function pointers will be initialized to the preferred versions of these functions for the specific class type in question. That is, the function pointers may point to overridden methods as specified at their own class level.

Then, when an object of a given type is instantiated, the vptr within that object (there will be one per subobject level of newly introduced – not redefined – virtual functions) will be set to point to the corresponding v-table entry for that instance.

It will be useful to see this detail with code and a memory diagram. Let's look under the hood to see the code in action!

Interpreting the v-table in detail

In order to detail the memory model and see the underlying C++ mechanics that will be set up at runtime, let's consider our detailed, full program example from this section with base class `Person` and derived class `Student`. As a reminder, we will show the key elements of the program:

- Abbreviated definitions of the `Person` and `Student` classes (we'll omit the data members and most member function definitions to save space):

```
class Person
{
private:    // data members will be as before
protected:  // assume all member funcs. are as before,
public:   // but we will show only virtual funcs. here
    virtual ~Person();        // 4 virt fns introduced
    virtual void Print() const;  // in Person class
    virtual void IsA() const;
    virtual void Greeting(const string &) const;
};

class Student: public Person
{
private:   // data members will be as before
public:    // assume all member funcs. are as before,
    // but we will show only virtual functions here
    ~Student() override;   // 3 virt fns are overridden
    void Print() const override;
    void IsA() const override;
};
```

The `Person` and `Student` class definitions are as expected. Assume that the data members and member functions are as shown in the full program example. For brevity, we have just included the virtual functions introduced or redefined at each level.

- Revisiting key elements of our `main()` function in abbreviated form (reduced to three instances):

```
constexpr int MAX = 3;
int main()
{
    Person *people[MAX] = { }; // init. with nullptrs
    people[0] = new Person("Joy", "Lin", 'M', "Ms.");
    people[1] = new Student("Renee", "Alexander", 'Z',
                     "Dr.", 3.95, "C++", "21-MIT");
    people[2] = new Student("Gabby", "Doone", 'A',
                     "Ms.", 3.95, "C++", "18-GWU");
    // In Chp. 8, we'll upgrade to a range for loop
    for (int i = 0; i < MAX; i++)
    {                      // at compile time, modified to:
        people[i]->IsA();   // *(people[i]->vptr[2])()
        people[i]->Print();
        people[i]->Greeting("Hello");
        delete people[i];
    }
    return 0;
}
```

Notice in our `main()` function that we instantiate one `Person` instance and two `Student` instances. All are stored in a generic array of pointers of the base class type, `Person`. We then iterate through the set calling virtual functions on each instance, namely `IsA()`, `Print()`, `Greeting()`, and the destructor (which is implicitly called when we delete each instance).

Considering the memory model for the previous example, we have the following diagram:

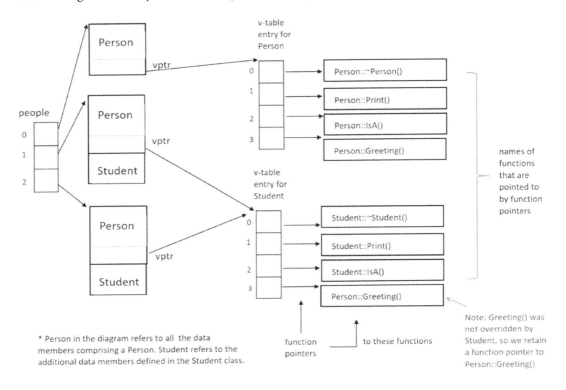

Figure 7.1 – Memory model for the current example

In the aforementioned memory diagram (which follows the preceding program), notice that we have an array of pointers to the genericized instances of Person. The first instance is, in fact, a Person, and the second two instances are of type Student. But, as a Student *Is-A* Person, it is acceptable to upcast a Student to a Person. The top part of the memory layout is in fact, a Person for each of the Student instances. For instances that are in fact of type Student, the additional data members for Student will follow all the memory required for the Person subobject.

Notice that the vptr entries immediately follow the data members for the Person object (or subobject) for each of the three instances. The location for the vptr is the same offset from the top of each object. This is because the virtual functions in question are all introduced at the Person level in the hierarchy. Some may be overridden in the Student class with more appropriate definitions for Student, but the level in which each is introduced is at the Person level, hence the vptr below the Person object (or subobject) will reflect a pointer to the list of operations introduced at the Person level.

As an aside, let's say that Student introduced wholly new virtual functions (and not merely redefinitions of existing virtual functions), such as we saw in the previous function hiding scenario. There would then be a second vptr entry below the Student subobject with those additional (new virtual) operations added.

When each object is instantiated, first the appropriate constructors (proceeding up the hierarchy) will be called for each instance. Additionally, the compiler will patch in a pointer assignment for each instance's vptr to be set to the v-table entry corresponding to the object's type. That is, when a Person is instantiated, its vptr will point to the v-table entry for Person. When a Student is instantiated, its vptr will point to the v-table entry for Student.

Let's assume that the v-table entry for Person or Student contains an array of function pointers to the appropriate virtual functions for that type. The v-table entry for each type actually has more information embedded, such as the size of an instance of that type, and so on. To simplify, we will just look at the portion of the v-table entries that automate the dynamic binding for each class type.

Notice that the v-table entry for Person is an array of four function pointers. Each function pointer will point to the most suitable version of the destructor, Print(), IsA(), and Greeting() for a Person. The order in which these function pointers occur corresponds to the order in which these virtual functions were introduced by this class. That is, vptr[0] will point to the Person destructor, vptr[1] will point to Person::Print(), and so on.

Now, let's look at the v-table entry for Student. The order in which the virtual functions (as function pointers) are laid into the array is the same order as for the Person class. This is because the base class introduced these functions and the ordering in this array of pointers is set by that level. But notice that the actual functions that are pointed to have been overridden for Student instances to mostly be methods that were redefined by the derived class, Student. That is, the Student destructor is specified (as the starting point for destruction), followed by Student::Print(), then Student::IsA(), and then Person::Greeting(). Notice that vptr[3] points to Person::Greeting(). This is because Student did not redefine this function in its class definition; Student found the Person definition, which is inherited, to be acceptable.

Pairing this memory diagram with the code in our main() function, notice that after we instantiate a Person and two Student instances, storing each in the genericized Person array of pointers, we iterate through a loop containing several operations. We uniformly call people[i]->Print();, then people[i]->IsA();, then people[i]->Greeting("Hello");, and then delete people[i]; (which patches in a destructor call).

Because each of these functions is virtual, the decision as to which function should be called is deferred to be looked up at runtime. This is done by accessing each instance's hidden vptr member, indexing into the appropriate v-table entry based on the operation at hand, and then dereferencing the function pointer found at that entry to call the appropriate method. The compiler knows, for example, that vptr[0] will be the destructor, vptr[1] will be the next virtual function introduced in the base class definition, and so on, so that the element position in the v-table that should be activated is easily determined by the name of the polymorphic operation.

Imagine that a call in `main()` to `people[i]->Print();` was replaced with `*(people[i]->vptr[1])();`, which is the syntax for dereferencing a function pointer to call the function at hand. Notice that we are first accessing which function using `people[i]->vptr[1]`, then dereferencing the function pointer using `*`. Notice the parentheses `()` at the end of the statement, which is where any parameters would be passed to the function. Because the code to dereference the function pointer needs to be uniform, the parameters to any such function must also be uniform. That is why any virtual functions overridden in a derived class must use the same signature as specified by the base class. It all makes sense when you look under the hood.

We have now thoroughly examined the OO idea of polymorphism and how it is implemented in C++ using virtual functions. Let's briefly recap what we've covered in this chapter before moving onward to our next chapter.

Summary

In this chapter, we have moved even further along our journey with object-oriented programming by understanding how virtual functions in C++ provide direct language support for the OO idea of polymorphism. We have seen how virtual functions provide dynamic binding of a specific method to an operation in our inheritance hierarchy.

We have seen how, using virtual functions, an operation specified by a base class can be overridden by a derived class, providing a more suitable implementation. We have seen that the correct method for each object can be selected using runtime binding, regardless of whether the object is stored in its own type or in a generalized type.

We have seen that objects are often generalized using base class pointers and how this can allow the uniform processing of related derived class types. We have seen that, regardless of how an instance is stored (as its own type or as that of a base class using a pointer), the correct version of a virtual function will always be applied through dynamic binding. We have seen that in public inheritance hierarchies where upcasting may be routinely done, having a virtual destructor is essential.

We have also seen how dynamic binding works through examining a typical compiler implementation of embedding a vptr into instances, and how these pointers reference v-table entries (containing sets of member function pointers) relevant to each object type.

We have seen that virtual functions allow us to take advantage of dynamic binding of operations to the most appropriate methods, enabling us to use C++ as an OOP language to implement robust designs featuring polymorphism, which promotes easily extensible code.

By extending our OOP knowledge with the utilization of virtual functions, we can now move forward to include additional object-oriented concepts and details relating to inheritance and polymorphism. Continuing to *Chapter 8, Mastering Abstract Classes*, we will next learn how to employ the OO ideal of abstract classes, along with all the assorted OOP considerations surrounding this next object-oriented concept. Let's continue!

Questions

1. Using your *Chapter 6, Implementing Hierarchies with Single Inheritance*, solution, augment your inheritance hierarchy to further specialize `Student` with `GraduateStudent` and `NonDegreeStudent`.

    a. Add necessary data members to your `GraduateStudent` class. Data members to consider might be *dissertation topic* or *graduate advisor*. Include appropriate constructors (default, alternate, and copy), a destructor, access member functions, and a suitable public interface. Be sure to place your data members in the private access region. Do the same for `NonDegreeStudent`.

    b. Add polymorphic operations to `Person`, `Student`, `GraduateStudent`, and `NonDegreeStudent` as necessary. Introduce at the `Person` level virtual functions `IsA()` and `Print()`. Override `IsA()` and `Print()` in your derived classes as necessary. It may be that you override `IsA()` in `Student` and `GraduateStudent`, but choose to override `Print()` only in the `Student()` class. Be sure to include virtual destructors in each of your classes.

    c. Instantiate `Student`, `GraduateStudent`, `NonDegreeStudent`, and `Person` several times and utilize the appropriate `public` interfaces on each. Be sure to dynamically allocate several instances.

    d. Create an array of pointers to `Person` and allocate instances of `Person`, `Student`, `GraduateStudent`, and `NonDegreeStudent` to be members of this array. Once generalized, only call polymorphic operations that are found at the `Person` level (and other public methods of `Person`). Be sure to delete any dynamically allocated instances.

    e. Create an array of pointers to `Student` and allocate only instances of `GraduateStudent` and `NonDegreeStudent` to be members of this array. Now, call operations that are found at the `Student` level to be applied to these generalized instances. Additionally, utilize operations found at the `Person` level – they are inherited and additionally available for generalized `Student` instances. Be sure to delete any dynamically allocated instances pointed to in your array.

8
Mastering Abstract Classes

This chapter will continue expanding our knowledge of object-oriented programming in C++. We will begin by exploring a powerful OO concept, **abstract classes**, and then progress to understanding how this idea is implemented in C++ through *direct language support*.

We will implement abstract classes using pure virtual functions to ultimately support refinements in a hierarchy of related classes. We will understand how abstract classes augment and pair with our understanding of polymorphism. We will also recognize how the OO concept of abstract classes presented in this chapter will support powerful and adaptable designs, allowing us to create easily extensible C++ code.

In this chapter, we will cover the following main topics:

- Understanding the OO concept of an abstract class
- Implementing abstract classes with pure virtual functions
- Creating interfaces using abstract classes and pure virtual functions
- Generalizing derived class objects using abstract classes, and upcasting and downcasting

By the end of this chapter, you will understand the OO concept of an abstract class, and how to implement this idea in C++ through pure virtual functions. You will learn how abstract classes containing only pure virtual functions can define an OOP concept of an interface. You will understand how abstract classes and interfaces contribute to powerful OO designs.

You will see how we can very easily generalize groups of related, specialized objects using sets of abstract types. We will further explore up and downcasting within the hierarchy to understand what is allowed and when such typecasting is reasonable to employ.

By understanding the direct language support of abstract classes in C++ using pure virtual functions, as well as why creating interfaces is useful, you will have more tools available to create an extensible hierarchy of related classes. Let us expand our understanding of C++ as an OOP language by understanding how these concepts are implemented in C++.

Technical requirements

Online code for full program examples can be found in the following GitHub URL: `https://github.com/PacktPublishing/Deciphering-Object-Oriented-Programming-with-CPP/tree/main/Chapter08`. Each full program example can be found in the GitHub under the appropriate chapter heading (subdirectory) in a file that corresponds to the chapter number, followed by a dash, followed by the example number in the chapter at hand. For example, the first full program in this chapter can be found in the subdirectory `Chapter08` in a file named `Chp8-Ex1.cpp` under the aforementioned GitHub directory.

The CiA video for this chapter can be viewed at: `https://bit.ly/3SZvOjy`.

Understanding the OO concept of an abstract class

In this section, we will introduce an essential object-oriented concept, that of an abstract class. This concept will add to your growing knowledge base of key OO ideas including encapsulation, information hiding, generalization, specialization, and polymorphism. You know how to encapsulate a class. You also know how to build inheritance hierarchies using single inheritance, and various reasons to build hierarchies, such as supporting *Is-A* relationships or for the lesser-used reason of supporting implementation inheritance. Furthermore, you know how to employ runtime binding of methods to operations using the concept of polymorphism, implemented by virtual functions. Let's extend our growing OO terminology by exploring **abstract classes**.

An **abstract class** is a base class that is intended to collect commonalities that may exist in derived classes for the purpose of asserting a common interface (that is, a set of operations) on the derived class. An abstract class does not represent a class that is intended for instantiation. Only objects of the derived class types may be instantiated.

Let's start by looking at the C++ language feature that allows us to implement abstract classes, that is, pure virtual functions.

Implementing abstract classes with pure virtual functions

An abstract class is specified by introducing at least one abstract method (that is, a pure virtual function prototype) in the class definition. The OO concept of an **abstract method** is the specification of an operation with only its protocol for usage (that is, with only the *name* and *signature* of the member function), but with no definition for the function. An abstract method will be polymorphic, in that, having no definition, it is expected to be redefined by derived classes.

A **pure virtual** function is used to implement the OO concept of an abstract method in C++. A pure virtual function is specified by a member function whose prototype contains =0 after the arguments to the function. Additionally, it is important to understand the following nuances regarding pure virtual functions:

- Usually, definitions for pure virtual functions are not provided. This equates to the operation (prototype only) being specified at the base class level and all methods (member function definitions) being supplied at the derived class level.

- Derived classes that do not provide methods for all pure virtual functions introduced by their base classes are also considered abstract and are therefore not instantiable.

- The =0 in the prototype is merely an indication to the linker that a definition for this function need not be linked in (or resolved) when creating an executable program.

> **Note**
>
> An abstract class is designated by including one or more pure virtual function prototypes in the class definition. The optional definitions for these methods are not typically provided.

The reason that pure virtual functions will most often not have a definition is because they are meant to provide a protocol of usage for polymorphic operations to be implemented in descendent classes. A pure virtual function designates a class to be abstract; an abstract class cannot be instantiated. Therefore, a definition provided in a pure virtual function will never be selected as the appropriate method for a polymorphic operation because instances of the abstract type will never exist. That being said, a pure virtual function can still provide a definition that could be explicitly called using the scope resolution operator (: :) and base class name. Perhaps, this default behavior might be meaningful as a helper function used by derived class implementations.

Let's begin with a brief overview of the syntax required to specify an abstract class. Remember, a potential keyword of *abstract* is not used to specify an abstract class. Rather, by merely introducing one or more pure virtual functions, we have indicated the class to be an abstract class:

```
class LifeForm    // Abstract class definition
{
private:
    // all LifeForms have a lifeExpectancy
    int lifeExpectancy = 0; // in-class initialization
public:
    LifeForm() = default; // def. ctor, uses in-class init
    LifeForm(int life): lifeExpectancy(life) { }
    // Remember, we get default copy, even w/o proto below
    // LifeForm(const LifeForm &form) = default;
```

```
        // Must include prototype to specify virtual destructor
        virtual ~LifeForm() = default;    // virtual destructor
        // Recall, [[nodiscard]] requires ret. value to be used
        [[nodiscard]] int GetLifeExpectancy() const
            { return lifeExpectancy; }
    virtual void Print() const = 0; // pure virtual fns.
    virtual string IsA() const = 0;
    virtual string Speak() const = 0;
};
```

Notice that in the abstract class definition, we have introduced four virtual functions, three of which are pure virtual functions. The virtual destructor has no memory to release but is indicated as `virtual` so that it will be polymorphic, and so that the correct destruction sequence can be applied to derived class instances stored as pointers to base class types.

The three pure virtual functions, `Print()`, `IsA()`, and `Speak()`, are indicated with `=0` in their prototype. There are no definitions for these operations (though there optionally can be). A pure virtual function can have a default implementation, but not as an inline function. It will be the derived class' responsibility to provide methods for these operations using the interface (that is, signature) specified by this base class definition. Here, the pure virtual functions provide the *interface* for the polymorphic operations that will be defined in derived class definitions.

> **Important note**
> Abstract classes will certainly have derived classes (since we cannot instantiate an abstract class itself). In order to allow the virtual destructor mechanism to work appropriately in the eventual hierarchy, be sure to include a *virtual destructor* in the abstract class definition. This will ensure that all derived class destructors are *virtual*, and can be overridden to provide the correct entry point in an object's destruction sequence.

Now, let's take a deeper look at what it means to have an interface, from an OO perspective.

Creating interfaces

An **interface class** is an OO concept of a class that is a further refinement of an abstract class. Whereas an abstract class can contain generalized attributes and default behaviors (by including data members and default definitions for pure virtual functions or by providing non-virtual member functions), an interface class will only contain abstract methods. An abstract class in C++ containing only abstract methods (that is, pure virtual functions with no optional definitions) can be thought of as an **interface class**.

When considering interface classes as implemented in C++, it is useful to remember the following:

- Abstract classes are not instantiable; they provide (via inheritance) the interfaces (that is, operations) that a derived class must offer.

- Although a pure virtual function may contain an optional implementation (that is, method body) in the abstract class, this implementation should not be provided if the class wishes to be considered an interface class in pure OO terms.

- Although an abstract class may have data members, it should not if the class wishes to be considered an interface class.

- An abstract method, in OO terms, is an operation without a method; it is the interface only and is implemented in C++ as a pure virtual function.

- As a reminder, be sure to include a virtual destructor prototype in the interface class definition; this will ensure that derived class destructors will be virtual. The destructor definition should be empty.

Let's consider various motivations for having interface classes within our OOP arsenal of implementation techniques. Some OOP languages follow very strict OO concepts and only allow for the implementation of very pure OO designs. Other OOP languages, such as C++, offer more flexibility, by allowing more radical OOP ideas to be implemented by the language directly.

For example, in pure object-oriented terms, inheritance should be reserved for *Is-A* relationships. We've seen implementation inheritance, which C++ supports through private and protected base classes. We've seen some acceptable uses of implementation inheritance, that is, to implement a new class in terms of another (with the ability to hide the underlying implementation with the use of protected and public base classes).

Another example of a fringe OOP feature is that of multiple inheritance. We'll see in *Chapter 9, Exploring Multiple Inheritance*, that C++ allows a class to be derived from more than one base class. In some cases, we are truly saying that the derived class has an Is-A relationship with potentially many base classes, but not always.

Some OOP languages do not allow multiple inheritance and those that do not rely more on interface classes to mix in the functionality of (otherwise) multiple base classes. In these situations, the OOP language can allow a derived class to implement the functionality as specified in multiple interface classes without actually using multiple inheritance. Ideally, interfaces are used to *mix-in* functionality from multiple classes. These classes, not surprisingly, are sometimes referred to as **mix-in** classes. In these situations, we are not saying that the Is-A relationship necessarily applies between derived and base classes.

In C++, when we introduce an abstract class with only pure virtual functions, we can think of creating an interface class. When a new class mixes in functionality from multiple interfaces, we can think of this in OO terms as using each interface class as a means to mix-in the desired interfaces for behaviors. Note that the derived classes must override each of the pure virtual functions with their own implementation; we're mixing in only the required API.

C++'s implementation of the OO concept of an interface is merely that of an abstract class containing only pure virtual functions. Here, we're using public inheritance from an abstract class paired with polymorphism to simulate the OO concept of an interface class. Note that other languages (such as Java) implement this idea directly in the language (but then those languages do not support multiple inheritance). In C++, we can do almost anything, yet it remains important to understand how to implement OO ideals (even those not offered with direct language support) in reasonable and meaningful ways.

Let's see an example to illustrate an abstract class used to implement an interface class:

```cpp
class Charitable     // interface class definition
{                    // implemented using an abstract class
public:
    virtual void Give(float) = 0; // interface for 'giving'
    // must include prototype to specify virtual destructor
    virtual ~Charitable() = default; // remember virt. dest
};


class Person: public Charitable    // mix-in an 'interface'
{
    // Assume typical Person class definition w/ data
    // members, constructors, member functions exist.
public:
    virtual void Give(float amt) override
    {   // implement a means for giving here
    }
    ~Person() override;   // virtual destructor prototype
};


// Student Is-A Person which mixes-in Charitable interface
class Student: public Person
{
```

```
    // Assume typical Student class definition w/ data
    // members, constructors, member functions exist.
public:
    virtual void Give(float amt) override
    { // Should a Student have little money to give,
      // perhaps they can donate their time equivalent to
      // the desired monetary amount they'd like to give
    }
    ~Student() override;   // virtual destructor prototype
};
```

In the aforementioned class definitions, we first notice a simple interface class, `Charitable`, implemented using a restricted abstract class. We include no data members and a pure virtual function `virtual void Give(float) = 0;` to define the interface class. We also include a virtual destructor.

Next, `Person` is derived from `Charitable` using public inheritance to implement the `Charitable` interface. We simply override `virtual void Give(float);` to provide a default definition for *giving*. We then derive `Student` from `Person`; note that a *Student Is-A Person that mixes-in (or implements) the Charitable interface*. In our `Student` class, we choose to redefine `virtual void Give(float);` to provide a more suitable `Give()` definition for `Student` instances. Perhaps students have limited finances and opt to donate an amount of their time that is equivalent to a predetermined monetary amount.

Here, we have used an abstract class in C++ to model the OO concept of an interface class.

Let's continue with our discussion relating to abstract classes overall by examining how derived class objects may be collected by abstract class types.

Generalizing derived class objects as abstract types

We've seen in *Chapter 7, Utilizing Dynamic Binding through Polymorphism*, that it is reasonable at times to group related derived class instances in a collection that is stored using base class pointers. Doing so allows uniform processing of related derived class types using polymorphic operations as specified by the base class. We also know that when a polymorphic base class operation is invoked, the correct derived class method will be invoked at runtime by virtue of the virtual functions and internal v-table that implement polymorphism in C++.

You may contemplate, however, whether it is possible to collect a group of related derived class types by a base class type that is an abstract class. Remember, an abstract class is not instantiable, so how might we store a derived class object as an object that cannot be instantiated? The solution is to use *pointers* (or even a reference). Whereas we cannot collect derived class instances in a set of abstract base class instances (those types cannot be instantiated), we can collect derived class instances in a set of pointers of the abstract class type. We may also have a reference of the abstract class type refer to a derived class instance. We've been doing this type of grouping (with base class pointers) since we learned about polymorphism.

Generalized groups of specialized objects employ implicit upcasting. Undoing such an upcast must be done using an explicit downcast, and the programmer will need to be correct as to the derived type that has been previously generalized. An incorrect downcast to the wrong type will cause a runtime error.

When is it necessary to collect derived class objects by base class types, including abstract base class types? The answer is when it makes sense in your application to process related derived class types in a more generic way, that is, when the operations specified in the base class type account for all of the operations you'd like to utilize. Undeniably, you may find just as many situations where keeping derived class instances in their own type (to utilize specialized operations introduced at the derived class level) is reasonable. Now you understand what is possible.

Let's continue by examining a comprehensive example showing abstract classes in action.

Putting all the pieces together

So far in this chapter, we have understood the subtleties of abstract classes, including pure virtual functions, and how to create interface classes using abstract classes and pure virtual functions. It is always important to see our code in action, with all its various components and their various nuances.

Let's take a look at a more complex, full program example to fully illustrate abstract classes, implemented using pure virtual functions in C++. In this example, we will not further designate an abstract class as an interface class, but we will take the opportunity to collect related derived class types using a set of pointers of their abstract base class type. This example will be broken into many segments; the full program can be found in the following GitHub location:

https://github.com/PacktPublishing/Deciphering-Object-Orient-ed-Programming-with-CPP/blob/main/Chapter08/Chp8-Ex1.cpp

```cpp
#include <iostream>
#include <iomanip>
using std::cout;        // preferred to:  using namespace std;
using std::endl;
using std::setprecision;
using std::string;
```

```
using std::to_string;

constexpr int MAX = 5;

class LifeForm    // abstract class definition
{
private:
    int lifeExpectancy = 0;   // in-class initialization
public:
    LifeForm() = default;
    LifeForm(int life): lifeExpectancy(life) { }
    // Remember, we get the default copy ctor included,
    // even without the prototype below:
    // LifeForm(const LifeForm &) = default;
    // Must include prototype to specify virtual destructor
    virtual ~LifeForm() = default;      // virtual destructor
    [[nodiscard]] int GetLifeExpectancy() const
         { return lifeExpectancy; }
    virtual void Print() const = 0;    // pure virtual fns.
    virtual string IsA() const = 0;
    virtual string Speak() const = 0;
};
```

In the aforementioned class definition, we notice that `LifeForm` is an abstract class. It is an abstract class because it contains at least one pure virtual function definition. In fact, it contains three pure virtual function definitions, namely `Print()`, `IsA()`, and `Speak()`.

Now, let's extend `Lifeform` with a concrete derived class, `Cat`:

```
class Cat: public LifeForm
{
private:
    int numberLivesLeft = 9;  // in-class initialization
    string name;
    static constexpr int CAT_LIFE = 15;  // Life exp for cat
public:
    Cat(): LifeForm(CAT_LIFE) { } // note prior in-class init
```

```
    Cat(int lives): LifeForm(CAT_LIFE),
                    numberLivesLeft(lives) { }
    Cat(const string &);
    // Because base class destructor is virtual, ~Cat() is
    // automatically virtual (overridden) whether or not
    // explicitly prototyped. Below prototype not needed:
    // ~Cat() override = default;    // virtual destructor
    const string &GetName() const { return name; }
    int GetNumberLivesLeft() const
        { return numberLivesLeft; }
    void Print() const override; // redef pure virt fns
    string IsA() const override { return "Cat"; }
    string Speak() const override { return "Meow!"; }
};

Cat::Cat(const string &n) : LifeForm(CAT_LIFE), name(n)
{   // numLivesLeft will be set with in-class initialization
}

void Cat::Print() const
{
    cout << "\t" << name << " has " << numberLivesLeft;
    cout << " lives left" << endl;
}
```

In the previous segment of code, we see the class definition for Cat. Notice that Cat has redefined LifeForm's pure virtual functions Print(), IsA(), and Speak() by providing definitions for each of these methods in the Cat class. With the existing methods in place for these functions, any derived class of Cat may optionally choose to redefine these methods with more suitable versions (but they are no longer obligated to do so).

Note that if Cat had failed to redefine even one of the pure virtual functions of LifeForm, then Cat would also be considered an abstract class and therefore not instantiable.

As a reminder, even though virtual functions IsA() and Speak() are written inline to abbreviate the code, virtual functions will almost never be inlined by the compiler, as their correct method must be determined at runtime (except for a few cases involving compiler devirtualization, involving final methods, or when an instance's dynamic type is known).

Notice that in the `Cat` constructors, the member initialization list is used to select the `LifeForm` constructor that takes an integer argument (that is, `:LifeForm(CAT_LIFE)`). A value of 15 (CAT_LIFE) is passed up to the `LifeForm` constructor to initialize `lifeExpectancy`, defined in `LifeForm`, to the value of 15. The member initialization list is additionally used to initialize data members defined in the `Cat` class for the cases when in-class initialization is not used (that is, the value is determined by a parameter to the method).

Now, let's move forward to the class definition for `Person`, along with its inline functions:

```
class Person: public LifeForm
{
private:
    string firstName;
    string lastName;
    char middleInitial = '\0';
    string title;  // Mr., Ms., Mrs., Miss, Dr., etc.
    static constexpr int PERSON_LIFE = 80;  // Life exp of
protected:                                  // a Person
    void ModifyTitle(const string &);
public:
    Person();   // programmer-specified default constructor
    Person(const string &, const string &, char,
          const string &);
    // Default copy constructor prototype is not necessary:
    // Person(const Person &) = default;  // copy const.
    // Because base class destructor is virtual, ~Person()
    // is automatically virtual (overridden) whether or not
    // explicitly prototyped. Below prototype not needed:
    // ~Person() override = default;  // destructor
    const string &GetFirstName() const
        { return firstName; }
    const string &GetLastName() const
        { return lastName; }
    const string &GetTitle() const { return title; }
    char GetMiddleInitial() const { return middleInitial; }
    void Print() const override; // redef pure virt fns
    string IsA() const override;
```

```
      string Speak() const override;
};
```

Notice that `Person` now extends `LifeForm` using public inheritance. In previous chapters, `Person` was a base class at the top of the inheritance hierarchy. `Person` redefines the pure virtual functions from `LifeForm`, namely, `Print()`, `IsA()`, and `Speak()`. As such, `Person` is now a concrete class and can be instantiated.

Now, let's review the member function definitions for `Person`:

```
// select the desired base constructor using mbr. init list
Person::Person(): LifeForm(PERSON_LIFE)
{  // Remember, middleInitial will be set w/ in-class init
   // and the strings will be default constructed to empty
}

Person::Person(const string &fn, const string &ln, char mi,
               const string &t): LifeForm(PERSON_LIFE),
                                  firstName(fn), lastName(ln),
                                  middleInitial(mi), title(t)
{
}

// We're using the default copy constructor. But if we did
// choose to prototype and define it, the method would be:
// Person::Person(const Person &p): LifeForm(p),
//           firstName(p.firstName), lastName(p.lastName),
//           middleInitial(p.middleInitial), title(p.title)
// {
// }

void Person::ModifyTitle(const string &newTitle)
{
    title = newTitle;
}

void Person::Print() const
{
```

```
   cout << "\t" << title << " " << firstName << " ";
   cout << middleInitial << ". " << lastName << endl;
}


string Person::IsA() const
{
   return "Person";
}


string Person::Speak() const
{
   return "Hello!";
}
```

In the `Person` member functions, notice that we have implementations for `Print()`, `IsA()`, and `Speak()`. Additionally, notice that in two of the `Person` constructors, we select `:LifeForm(PERSON_LIFE)` in their member initialization lists to call the `LifeForm(int)` constructor. This call will set the private inherited data member `LifeExpectancy` to 80 (`PERSON_LIFE`) in the `LifeForm` subobject of a given `Person` instance.

Next, let's review the `Student` class definition, along with its inline function definitions:

```
class Student: public Person
{
private:
   float gpa = 0.0;  // in-class initialization
   string currentCourse;
   const string studentId;
   static int numStudents;
public:
   Student();  // programmer-supplied default constructor
   Student(const string &, const string &, char,
           const string &, float, const string &,
           const string &);
   Student(const Student &);  // copy constructor
   ~Student() override;  // virtual destructor
   void EarnPhD();
   float GetGpa() const { return gpa; }
```

```
    const string &GetCurrentCourse() const
       { return currentCourse; }
    const string &GetStudentId() const
       { return studentId; }
    void SetCurrentCourse(const string &);
    // Redefine not all of the virtrtual function; don't
    // override Person::Speak(). Also, mark Print() as
    // the final override
    void Print() const final override;
    string IsA() const override;
    static int GetNumberStudents();
};

int Student::numStudents = 0; // static data mbr def/init

inline void Student::SetCurrentCourse(const string &c)
{
    currentCourse = c;
}

inline int Student::GetNumberStudents()
{
    return numStudents;
}
```

The aforementioned class definition for Student looks much as we've seen in the past. Student extends Person using public inheritance because a Student *Is-A* Person.

Moving forward, we'll recall the non-inline Student class member functions:

```
// default constructor
Student::Student(): studentId(to_string(numStudents + 100)
                                         + "Id")
{   // Set const studentId in mbr init list with unique id
    // (based upon numStudents counter + 100), concatenated
    // with the string "Id". Remember, string member
    // currentCourse will be default constructed with
    // an empty string - it is a member object
```

```
        numStudents++;
}

// Alternate constructor member function definition
Student::Student(const string &fn, const string &ln,
                char mi, const string &t, float avg,
                const string &course, const string &id):
                Person(fn, ln, mi, t), gpa(avg),
                currentCourse(course), studentId(id)
{
    numStudents++;
}

// Copy constructor definition
Student::Student(const Student &s) : Person(s),
                gpa(s.gpa),
                currentCourse(s.currentCourse),
                studentId(s.studentId)
{
    numStudents++;
}

// destructor definition
Student::~Student()
{
    numStudents--;
}

void Student::EarnPhD()
{
    ModifyTitle("Dr.");
}

void Student::Print() const
{
    cout << "\t" << GetTitle() << " " << GetFirstName();
```

```
        cout << " " << GetMiddleInitial() << ". "
            << GetLastName();
        cout << " id: " << studentId << "\n\twith gpa: ";
        cout << setprecision(3) << " " << gpa
            << " enrolled in: " << currentCourse << endl;
}


string Student::IsA() const
{
    return "Student";
}
```

In the previously listed section of code, we see the non-inline member function definitions for
Student. The complete class definition is, at this point, largely familiar to us.

Accordingly, let's examine the main() function:

```
int main()
{
    // Notice that we are creating an array of POINTERS to
    // LifeForms. Since LifeForm cannot be instantiated,
    // we could not create an array of LifeForm(s).
    LifeForm *entity[MAX] = { }; // init. with nullptrs
    entity[0] = new Person("Joy", "Lin", 'M', "Ms.");
    entity[1] = new Student("Renee", "Alexander", 'Z',
                            "Dr.", 3.95, "C++", "21-MIT");
    entity[2] = new Student("Gabby", "Doone", 'A', "Ms.",
                            3.95, "C++", "18-GWU");
    entity[3] = new Cat("Katje");
    entity[4] = new Person("Giselle", "LeBrun", 'R',
                            "Miss");
    // Use range for-loop to process each element of entity
    for (LifeForm *item : entity)  // each item is a
    {                              // LifeForm *
        cout << item->Speak();
        cout << " I am a " << item->IsA() << endl;
        item->Print();
        cout << "\tHas a life expectancy of: ";
```

```
        cout << item->GetLifeExpectancy();
        cout << "\n";
    }

    for (LifeForm *item : entity) // process each element
    {                             // in the entity array
        delete item;
        item = nullptr;    // ensure deleted ptr isn't used
    }
    return 0;
}
```

Here, in `main()`, we declare an array of pointers to `LifeForm`. Recall, `LifeForm` is an abstract class. We could not create an array of `LifeForm` objects, because that would require us to be able to instantiate a `LifeForm`; we can't – `LifeForm` is an abstract class.

However, we can create a set of pointers to an abstract type and this allows us to collect related types, `Person`, `Student`, and `Cat` instances in this set. Of course, the only operations we may apply to instances stored in this generalized fashion are those found in the abstract base class, `LifeForm`.

Next, we allocate a variety of `Person`, `Student`, and `Cat` instances, storing each instance via an element in the generalized set of pointers of type `LifeForm`. When any of these derived class instances is stored in this fashion, an implicit upcast to the abstract base class type is performed (but the instance is not altered in any fashion – we're just pointing to the most base class subobject comprising the entire memory layout).

Now, we proceed through a loop to apply operations as found in the abstract class `LifeForm` to all instances in this generalized collection, such as `Speak()`, `Print()`, and `IsA()`. These operations happen to be polymorphic, allowing each instance's most appropriate implementation to be utilized via dynamic binding. We additionally invoke `GetLifeExpectancy()` on each of these instances, which is a non-virtual function found at the `LifeForm` level. This function merely returns the life expectancy of the `LifeForm` in question.

Lastly, we loop through deleting the dynamically allocated instances of `Person`, `Student`, and `Cat` again using the generalized `LifeForm` pointers. We know that `delete()` will patch in a call to the destructor, and because the destructor is virtual, the appropriate starting level of the destructor and proper destruction sequence will commence. Additionally, by setting `item = nullptr;`, we are ensuring that the deleted pointer will not be used mistakenly as a bonafide address (we are overwriting each relinquished address with a `nullptr`).

The utility of the abstract class `LifeForm` in this example is that its use allows us to generalize common aspects and behaviors of all `LifeForm` objects together in one base class (such as `lifeExpectancy` and `GetLifeExpectancy()`). The common behaviors also extend to a set of pure virtual functions with the desired interfaces that all `LifeForm` objects should have, namely `Print()`, `IsA()`, and `Speak()`.

> **Important reminder**
>
> An abstract class is one that collects common traits of derived classes, yet does not itself represent a tangible entity or object that should be instantiated. In order to specify a class as abstract, it must contain at least one pure virtual function.

Looking at the output for the aforementioned program, we can see that objects of various related derived class types are instantiated and processed uniformly. Here, we've collected these objects by their abstract base class type and have overridden the pure virtual functions in the base class with meaningful definitions in various derived classes.

Here is the output for the full program example:

```
Hello! I am a Person
        Ms. Joy M. Lin
        Has a life expectancy of: 80
Hello! I am a Student
        Dr. Renee Z. Alexander id: 21-MIT
        with gpa:  3.95 enrolled in: C++
        Has a life expectancy of: 80
Hello! I am a Student
        Ms. Gabby A. Doone id: 18-GWU
        with gpa:  3.95 enrolled in: C++
        Has a life expectancy of: 80
Meow! I am a Cat
        Katje has 9 lives left
        Has a life expectancy of: 15
Hello! I am a Person
        Miss Giselle R. LeBrun
        Has a life expectancy of: 80
```

We have now thoroughly examined the OO idea of an abstract class and how it is implemented in C++ using pure virtual functions, as well as how these ideas can extend to creating OO interfaces. Let's briefly recap the language features and OO concepts we've covered in this chapter before moving onward to our next chapter.

Summary

In this chapter, we have continued our progression with object-oriented programming, foremost, by understanding how pure virtual functions in C++ provide direct language support for the OO concept of an abstract class. We have explored how abstract classes without data members that do not contain non-virtual functions can support the OO ideal of an interface class. We've talked about how other OOP languages utilize interface classes, and how C++ may choose to support this paradigm as well by using such restricted abstract classes. We've upcast related derived class types to be stored as pointers of the abstract base class type, as a typical, and overall very useful, programming technique.

We have seen how abstract classes complement polymorphism not only by providing a class to specify common attributes and behaviors that derived classes share, but most notably to provide the interfaces of polymorphic behaviors for the related classes since abstract classes themselves are not instantiable.

By adding abstract classes and potentially the OO concept of interface classes to our programming repertoire in C++, we are able to implement designs that promote easily extensible code.

We are now ready to continue to *Chapter 9*, *Exploring Multiple Inheritance*, to enhance our OOP skills by next learning how and when to appropriately utilize the concept of multiple inheritance, while understanding trade-offs and potential design alternatives. Let's move forward!

Questions

1. Create a hierarchy of shapes using the following guidelines:

    a. Create an abstract base class called `Shape`, which defines an operation to compute the area of a shape. Do not include a method for the `Area()` operation. Hint: use a pure virtual function.

    b. Derive classes `Rectangle`, `Circle`, and `Triangle` from `Shape` using public inheritance. Optionally, derive class `Square` from `Rectangle`. Redefine the operation `Area()` that `Shape` has introduced, in each derived class. Be sure to provide the method to support the operation in each derived class so that you can later instantiate each type of `Shape`.

    c. Add data members and other member functions as necessary to complete the newly introduced class definitions. Remember, only common attributes and operations should be specified in `Shape` – all others belong in their respective derived classes. Don't forget to implement the copy constructor and access functions within each class definition.

    d. Create an array of pointers of the abstract class type, `Shape`. Assign elements in this array point to instances of type `Rectangle`, `Square`, `Circle`, and `Triangle`. Since you are now treating derived class objects as generalized `Shape` objects, loop through the array of pointers and invoke the `Area()` function for each. Be sure to `delete()` any dynamically allocated memory you have allocated.

    e. Is your abstract `Shape` class also an interface class in conceptual OO terms? Why, or why not?

Exploring Multiple Inheritance

This chapter will continue broadening our knowledge of object-oriented programming in C++. We will begin by examining a controversial OO concept, **multiple inheritance** (**MI**), understanding why it is controversial, how it can reasonably be used to support OO designs, as well as when alternative designs may be more appropriate.

Multiple inheritance can be implemented in C++ with *direct language support*. In doing so, we will be faced with several OO design issues. We will be asked to critically evaluate an inheritance hierarchy, asking ourselves whether we are using the best design possible to represent a potential set of object relationships. Multiple inheritance can be a powerful OOP tool; using it wisely is paramount. We will learn when to use MI to sensibly extend our hierarchies.

In this chapter, we will cover the following main topics:

- Understanding multiple inheritance mechanics
- Examining reasonable uses for multiple inheritance
- Creating diamond-shaped hierarchies and exploring issues arising from their usage
- Using virtual base classes to resolve diamond-shaped hierarchy duplication
- Applying discriminators to evaluate the worthiness of a diamond-shaped hierarchy and MI in a design, as well as considering design alternatives

By the end of this chapter, you will understand the OO concept of multiple inheritance, and how to implement this idea in C++. You will understand not only the simple mechanics of MI but the reasons for its usage (mix-in, Is-A or controversially, Has-A).

You will see why MI is controversial in OOP. Having more than one base class can lead to oddly shaped hierarchies, such as diamond-shaped; these types of hierarchies come with potential implementation issues. We will see how C++ incorporates a language feature (virtual base classes) to solve these conundrums, but the solution is not always ideal.

Once we understand the complexities caused by multiple inheritance, we will use OO design metrics, such as discriminators, to evaluate whether a design using MI is the best solution to represent a set of object relationships. We'll look at alternative designs, and you will then be better equipped to understand not only what multiple inheritance is, but when it's best utilized. Let us expand our understanding of C++ as a *"you can do anything"* OOP language by moving forward with MI.

Technical requirements

Online code for full program examples can be found in the following GitHub URL: `https://github.com/PacktPublishing/Deciphering-Object-Oriented-Programming-with-CPP/tree/main/Chapter09`. Each full program example can be found in the GitHub under the appropriate chapter heading (subdirectory) in a file that corresponds to the chapter number, followed by a dash, followed by the example number in the chapter at hand. For example, the first full program in this chapter can be found in the subdirectory `Chapter09` in a file named `Chp9-Ex1.cpp` under the aforementioned GitHub directory.

The CiA video for this chapter can be viewed at: `https://bit.ly/3Cbqt7y`.

Understanding multiple inheritance mechanics

In C++, a class can have more than one immediate base class. This is known as **multiple inheritance**, and is a very controversial topic in both OO designs and OOP. Let's begin with the simple mechanics; we will then move forward to the design issues and programming logistics surrounding MI during the progression of this chapter.

With multiple inheritance, the derived class specifies who each of its immediate ancestors or base classes are, using the base class list in its class definition.

In a similar fashion to single inheritance, the constructors and destructors are invoked all the way up the hierarchy as objects of the derived class type are instantiated and destroyed. Reviewing and expanding upon the subtleties of construction and destruction for MI, we are reminded of the following logistics:

- The calling sequence for a constructor starts with the derived class, but immediately passes control to a base constructor, and so on up the hierarchy. Once the calling sequence passes control to the top of the hierarchy, the execution sequence begins. All the highest-level base class constructors at the same level are first executed, and so on down the hierarchy until we arrive at the derived class constructor, whose body is executed last in the construction chain.
- The derived class destructor is invoked and executed first, followed by all the immediate base class destructors and so on, as we progress up the inheritance hierarchy.

The member initialization list in the derived class constructor may be used to specify which constructor for each immediate base class should be invoked. In the absence of this specification, the default constructor will be used for that base class' constructor.

Let's take a look at a typical multiple inheritance example to implement a quintessential application of MI from an OO design, as well as to understand basic MI syntax in C++. This example will be broken into many segments; the full program can be found in the following GitHub location:

https://github.com/PacktPublishing/Deciphering-Object-Oriented-Programming-with-CPP/blob/main/Chapter09/Chp9-Ex1.cpp

```cpp
#include <iostream>
using std::cout;      // preferred to: using namespace std;
using std::endl;
using std::string;
using std::to_string;

class Person
{
private:
    string firstName;
    string lastName;
    char middleInitial = '\0';  // in-class initialization
    string title;  // Mr., Ms., Mrs., Miss, Dr., etc.
protected:
    void ModifyTitle(const string &);
public:
    Person() = default;   // default constructor
    Person(const string &, const string &, char,
           const string &);
    Person(const Person &) = delete;  // prohibit copies
    virtual ~Person();   // destructor prototype
    const string &GetFirstName() const
        { return firstName; }
    const string &GetLastName() const
        { return lastName; }
    const string &GetTitle() const { return title; }
    char GetMiddleInitial() const { return middleInitial; }
};
```

In the previous code segment, we have an expected class definition for Person, containing the class elements that we are accustomed to defining.

Next, let's see the accompanying member functions for this class:

```
// With in-class initialization, writing the default
// constructor is no longer necessary.
// Also, remember strings are member objects and will
// be default constructed as empty.

Person::Person(const string &fn, const string &ln, char mi,
               const string &t): firstName(fn),
               lastName(ln), middleInitial(mi), title(t)
{
}


// Simple destructor so we can trace the destruction chain
Person::~Person()
{
    cout << "Person destructor <" << firstName << " " <<
            lastName << ">" << endl;
}

void Person::ModifyTitle(const string &newTitle)
{
    title = newTitle;
}
```

In the previous segment of code, the member function definitions for `Person` are as expected. Nonetheless, it is useful to see the `Person` class defined, as this class will serve as a building block, and portions of it will be directly accessed in upcoming code segments.

Now, let's define a new class, `BillableEntity`:

```
class BillableEntity
{
private:
    float invoiceAmt = 0.0;    // in-class initialization
public:
    BillableEntity() = default;
    BillableEntity(float amt) invoiceAmt(amt) { }
```

```
    // prohibit copies with prototype below
    BillableEntity(const BillableEntity &) = delete;
    virtual ~BillableEntity();
    void Pay(float amt) { invoiceAmt -= amt; }
    float GetBalance() const { return invoiceAmt; }
    void Balance() const;
};


// Simple destructor so we can trace destruction chain
BillableEntity::~BillableEntity()
{
    cout << "BillableEntity destructor" << endl;
}

void BillableEntity::Balance() const
{
    if (invoiceAmt)
       cout << "Owed amount: $ " << invoiceAmt << endl;
    else
       cout << "Credit: $ " << 0.0 - invoiceAmt << endl;
}
```

In the previous `BillableEntity` class, we define a class containing simple functionality to encapsulate a billing structure. That is, we have an invoice amount and methods such as `Pay()` and `GetBalance()`. Notice that the copy constructor indicates = `delete` in its prototype; this will prohibit copies, which seems appropriate given the nature of this class.

Next, let's combine the two aforementioned base classes, `Person` and `BillableEntity`, to serve as base classes for our `Student` class:

```
class Student: public Person, public BillableEntity
{
private:
    float gpa = 0.0;   // in-class initialization
    string currentCourse;
    const string studentId;
    static int numStudents;
```

```
public:
    Student();   // default constructor
    Student(const string &, const string &, char,
            const string &, float, const string &,
            const string &, float);
    Student(const Student &) = delete;   // prohibit copies
    ~Student() override;
    void Print() const;
    void EarnPhD();
    float GetGpa() const { return gpa; }
    const string &GetCurrentCourse() const
        { return currentCourse; }
    const string &GetStudentId() const
        { return studentId; }
    void SetCurrentCourse(const string &);
    static int GetNumberStudents();
};

// definition for static data member
int Student::numStudents = 0;   // notice initial value of 0

inline void Student::SetCurrentCourse(const string &c)
{
    currentCourse = c;
}

inline int Student::GetNumberStudents()
{
    return numStudents;
}
```

In the preceding class definition for Student, two public base classes, Person and BillableEntity, are specified in the base class list for Student. These two base classes are merely comma-separated in the Student base class list. We have also included the inline function definitions with the class definition, as these are usually bundled together in a header file.

Let's further see what accommodations must be made in the remainder of the `Student` class by examining its member functions:

```cpp
// Due to non-specification in the member init list, this
// constructor calls the default base class constructors
Student::Student() : studentId(to_string(numStudents + 100)
                                            + "Id")
{
    // Note: since studentId is const, we need to set it at
    // construction using member init list. Remember, string
    // members are default constructed w an empty string.
    numStudents++;
}


// The member initialization list specifies which versions
// of each base class constructor should be utilized.
Student::Student(const string &fn, const string &ln,
        char mi, const string &t, float avg,
        const string &course, const string &id, float amt):
        Person(fn, ln, mi, t), BillableEntity(amt),
        gpa(avg), currentCourse(course), studentId(id)
{
    numStudents++;
}


// Simple destructor so we can trace destruction sequence
Student::~Student()
{
    numStudents--;
    cout << "Student destructor <" << GetFirstName() << " "
        << GetLastName() << ">" << endl;
}


void Student::Print() const
{
    cout << GetTitle() << " " << GetFirstName() << " ";
    cout << GetMiddleInitial() << ". " << GetLastName();
```

```
        cout << " with id: " << studentId << " has a gpa of: ";
        cout << " " << gpa << " and course: " << currentCourse;
        cout << " with balance: $" << GetBalance() << endl;
    }


    void Student::EarnPhD()
    {
        ModifyTitle("Dr.");
    }
```

Let's consider the previous code segment. In the default constructor for Student, due to the lack of base class constructor specification in the member initialization list, the default constructors will be called for both the Person and BillableEntity base classes.

However, notice that in the alternate Student constructor, we merely comma-separate our two base class constructor choices in the member initialization list – that is, Person(const string &, const string &, char, const string &) and BillableEntity(float) – and then pass various parameters from the Student constructor to the base class constructors using this list.

Finally, let's take a look at our main() function:

```
    int main()
    {
        float tuition1 = 1000.00, tuition2 = 2000.00;
        Student s1("Gabby", "Doone", 'A', "Ms.", 3.9, "C++",
                   "178GWU", tuition1);
        Student s2("Zack", "Moon", 'R', "Dr.", 3.9, "C++",
                   "272MIT", tuition2);
        // public mbrs. of Person, BillableEntity, Student are
        // accessible from any scope, including main()
        s1.Print();
        s2.Print();
        cout << s1.GetFirstName() << " paid $500.00" << endl;
        s1.Pay(500.00);
        cout << s2.GetFirstName() << " paid $750.00" << endl;
        s2.Pay(750.00);
        cout << s1.GetFirstName() << ": ";
        s1.Balance();
        cout << s2.GetFirstName() << ": ";
```

```
        s2.Balance();
        return 0;
}
```

In our `main()` function in the previous code, we instantiate several `Student` instances. Notice that `Student` instances can utilize any methods in the public interface of `Student`, `Person`, or `BillableEntity`.

Let's look at the output for the aforementioned program:

```
Ms. Gabby A. Doone with id: 178GWU has a gpa of:  3.9 and
course: C++ with balance: $1000
Dr. Zack R. Moon with id: 272MIT has a gpa of:  3.9 and course:
C++ with balance: $2000
Gabby paid $500.00
Zack paid $750.00
Gabby: Owed amount: $ 500
Zack: Owed amount: $ 1250
Student destructor <Zack Moon>
BillableEntity destructor
Person destructor <Zack Moon>
Student destructor <Gabby Doone>
BillableEntity destructor
Person destructor <Gabby Doone>
```

Notice the destruction sequence in the aforementioned output. We can see each `Student` instance invokes the `Student` destructor, as well as the destructors for each base class (`BillableEntity` and `Person`).

We have now seen the language mechanics for MI with a typically implemented OO design. Now, let's move forward by looking at the typical reasons for employing multiple inheritance in OO designs, some of which are more widely accepted than others.

Examining reasonable uses for multiple inheritance

Multiple inheritance is a controversial concept that arises when creating OO designs. Many OO designs avoid MI; other designs embrace it with strict usage. Some OOP languages, such as Java, do not explicitly provide direct language support for multiple inheritance. Instead, they offer interfaces, such as we've modeled in C++ by creating interface classes using abstract classes (restricted to containing only pure virtual functions) in *Chapter 8, Mastering Abstract Classes*.

Of course, in C++, inheriting from two interface classes is still a use of multiple inheritance. Though C++ does not include interface classes within the language, this concept can be simulated by employing a more restrictive use of MI. For example, we can programmatically streamline abstract classes to include only pure virtual functions (no data members, and no member functions with definitions) to mimic the OO design idea of an interface class.

Typical MI conundrums form the basis of why MI is contentious in OOP. Classic MI quandaries will be detailed in this chapter and can be avoided by restricting MI to the usage of interface classes only, or through a redesign. This is why some OOP languages only support interface classes versus allowing unrestricted MI. In C++, you can carefully consider each OO design and choose when to utilize MI, when to utilize a restrictive form of MI (interface classes), or when to employ a redesign eliminating MI.

C++ is a *"you can do anything"* programming language. As such, C++ allows multiple inheritance without restrictions or reservations. As an OO programmer, we will look more closely at typical reasons to embrace MI. As we move further into this chapter, we will evaluate issues that arise by using MI and see how C++ solves these issues with additional language features. These MI issues will allow us to then apply metrics to understand more reasonably when we should use MI and when a redesign may be more appropriate.

Let's begin our pursuit of reasonable uses of MI by considering Is-A and mix-in relationships, and then move to examining the controversial use of MI to implement Has-A relationships.

Supporting Is-A and mix-in relationships

As we have learned with single inheritance, an Is-A relationship is most often used to describe the relationship between two inherited classes. For example, a `Student` *Is-A* `Person`. The same desired ideal continues with MI; Is-A relationships are the primary motivations to specify inheritance. In pure OO designs and programming, inheritance should be used only to support Is-A relationships.

Nonetheless, as we have learned when we looked at interface classes (a concept modeled in C++ using abstract classes with the restriction of containing only pure virtual functions), mix-in relationships often apply when we inherit from an interface. Recall that a mix-in relationship is when we use inheritance to mix-in the functionality of another class, simply because that functionality is useful or meaningful for the derived class to have. The base class need not be an abstract or interface class, but employing an ideal OO design, it would be as such.

The mix-in base class represents a class in which an Is-A relationship does not apply. Mix-ins exist more so with multiple inheritance, at least as the reason supporting the necessity of one of the (many) base classes. Since C++ has direct language support for multiple inheritance, MI can be used to support implementing mix-ins (whereas languages such as Java may only use interface classes). In practice, MI is often used to inherit from one class to support an Is-A relationship and to also inherit from another class to support a mix-in relationship. In our last example, we saw that a `Student` *Is-A* `Person`, and a `Student` chooses to *mix-in* `BillableEntity` capabilities.

Reasonable uses of MI in C++ include supporting both Is-A and mix-in relationships; however, our discussion would not be complete without next considering an unusual use of MI – implementing Has-A relationships.

Supporting Has-A relationships

Less commonly, and much more controversially, MI can be used to implement a Has-A relationship, that is, to model containment, or a whole versus part relationship. We will see in *Chapter 10, Implementing Association, Aggregation, and Composition*, a more widely accepted implementation for Has-A relationships; however, MI provides a very simple implementation. Here, the parts serve as the base classes. The whole inherits from the parts, automatically including the parts in its memory layout (and also automatically inheriting the parts' members and functionality).

For example, a `Student` *Is-A* `Person` and a `Student` *Has-A(n)* `Id`; the usage of the second base class (`Id`) is for containment. `Id` will serve as a base class and `Student` will be derived from `Id` to factor in all that an `Id` offers. The `Id` public interface is immediately usable to `Student`. In fact, any class that inherits from `Id` will inherit a uniform interface when utilizing its `Id` parts. This simplicity is a driving reason why inheritance is sometimes used to model containment.

However, using inheritance to implement Has-A relationships can cause unnecessary usage of MI, which can then complicate an inheritance hierarchy. Unnecessary usage of MI is the primary reason why using inheritance to model Has-A relationships is very controversial and is quite frankly frowned upon in pure OO designs. Nonetheless, we mention it because you will see some C++ applications using MI for Has-A implementation.

Let's move forward to explore other controversial designs employing MI, namely that of a diamond-shaped hierarchy.

Creating a diamond-shaped hierarchy

When using multiple inheritance, sometimes it is tempting to utilize sibling (or cousin) classes as base classes for a new derived class. When this happens, the hierarchy is no longer a tree in shape, but rather, a graph containing a *diamond*.

Whenever an object of the derived class type is instantiated in such a situation, two copies of the common base class will be present in the instance of the derived class. Duplication of this sort obviously wastes space. Additional time is also wasted by calling duplicate constructors and destructors for this repeated subobject and by maintaining two parallel copies of a subobject (most likely unnecessarily). Ambiguities also result when trying to access members from this common base class.

Let's see an example detailing this issue, starting with abbreviated (and simplified) class definitions of LifeForm, Horse, and Person. Though only portions of the full program example are shown, the program in its entirety can be found in our GitHub as follows:

https://github.com/PacktPublishing/Deciphering-Object-Orient-ed-Programming-with-CPP/blob/main/Chapter09/Chp9-Ex2.cpp

```cpp
class Lifeform
{   // abbreviated class definition - see full code online
private:
    int lifeExpectancy = 0;   // in-class initialization
public:
    LifeForm(int life): lifeExpectancy(life) { }
    [[nodiscard]] int GetLifeExpectancy() const
        { return lifeExpectancy; }
    // additional constructors, destructor, etc.
    virtual void Print() const = 0; // pure virtual funcs.
    virtual string IsA() const = 0;
    virtual string Speak() const = 0;
};

class Horse: public LifeForm
{   // abbreviated class definition
private:
    string name;
    static constexpr int HORSE_LIFE = 35; // life exp Horse
public:
    Horse(): LifeForm(HORSE_LIFE) { }
    // additional constructors, destructor, etc …
    void Print() const override { cout << name << endl; }
    string IsA() const override { return "Horse"; }
    string Speak() const override { return "Neigh!"; }
};

class Person: public LifeForm
{   // abbreviated class definition
private:
```

```
        string firstName;
        string lastName;
        static constexpr int PERSON_LIFE = 80; // life expect.
                                               // of Person
        // additional data members (imagine them here)
    public:
        Person(): LifeForm(PERSON_LIFE) { }
        // additional constructors, destructor, etc.
        const string &GetFirstName() const
            { return firstName; }
        // additional access methods, etc.
        void Print() const override
            { cout << firstName << " " << lastName << endl; }
        string IsA() const override { return "Person"; }
        string Speak() const override { return "Hello!"; }
    };
```

The previous fragment of code shows skeleton class definitions for `LifeForm`, `Person`, and `Horse`. Each class shows a default constructor, which merely serves as an example to show how `lifeExpectancy` is set for each class. In the default constructors for `Person` and `Horse`, the member initialization list is used to pass a value of 35 (HORSE_LIFE) or 80 (PERSON_LIFE) to the `LifeForm` constructor to set this value.

Though the previous class definitions are abbreviated (that is, purposely incomplete) to save space, let's assume that each class has appropriate additional constructors defined, an appropriate destructor, and other necessary member functions.

We notice that `LifeForm` is an abstract class, in that it offers pure virtual functions: `Print()`, `IsA()`, and `Speak()`. Both `Horse` and `Person` are concrete classes and will be instantiable because they override these pure virtual functions with virtual functions. These virtual functions are shown inline, only to make the code compact for viewing (virtual functions will almost never be inlined by the compiler as their methods are nearly always determined at runtime).

Next, let's look at a new derived class that will introduce the graph, or diamond shape, in our hierarchy:

```
class Centaur: public Person, public Horse
{   // abbreviated class definition
public:
    // constructors, destructor, etc …
    void Print() const override
        { cout << GetFirstName() << endl; }
```

```
    string IsA() const override { return "Centaur"; }
    string Speak() const override
        { return "Neigh! and Hello!"; }
};
```

In the previous fragment, we define a new class, Centaur, using multiple inheritance. At first glance, we truly do mean to assert the Is-A relationship between Centaur and Person, and also between Centaur and Horse. However, we'll soon challenge our assertion to test whether it is more of a combination than a true Is-A relationship.

We will assume that all of the necessary constructors, the destructor, and member functions exist to make Centaur a well-defined class.

Now, let's move forward to look at a potential main() function we might utilize:

```
int main()
{
    Centaur beast("Wild", "Man");
    cout << beast.Speak() << " I'm a " << beast.IsA();
    cout << endl;

    // Ambiguous method call - which LifeForm sub-object?
    // cout << beast.GetLifeExpectancy();
    cout << "It is unclear how many years I will live: ";
    cout << beast.Person::GetLifeExpectancy() << " or ";
    cout << beast.Horse::GetLifeExpectancy() << endl;
    return 0;
}
```

Here, in main(), we instantiate a Centaur and we name the instance beast. We easily call two polymorphic operations on beast, namely Speak() and IsA(). Then we try to call the public inherited GetLifeExpectancy(), which is defined in LifeForm. Its implementation is included in Lifeform so that Person, Horse, or Centaur do not need to provide a definition (nor should they – it's not a virtual function meant to be redefined).

Unfortunately, calls to GetLifeExpectancy() via Centaur instances are ambiguous. This is because there are two LifeForm subobjects in the beast instance. Remember, Centaur is derived from Horse, which is derived from LifeForm, providing the memory layout for all the aforementioned base class data members (Horse and LifeForm). And Centaur is also derived from Person, which is derived from Lifeform, which contributes the memory layout for Person and LifeForm within Centaur. The LifeForm piece is duplicated.

There are two copies of the inherited data member int lifeExpectancy;. There are two subobjects of LifeForm within the Centaur instance. Therefore, when we try to call GetLifeExpectancy() through the Centaur instance, the method call is ambiguous. Which lifeExpectancy are we trying to initialize? Which LifeForm subobject will serve as the this pointer when GetLifeExpectancy() is called? It is simply not clear, so the compiler will not choose for us.

To disambiguate the GetLifeExpectancy() function call, we must use the scope resolution operator. We precede the :: operator with the intermediate base class from which we want the LifeForm subobject. Notice that we call, for example, beast.Horse::GetLifeExpectancy() to choose the lifeExpectancy from the Horse subobject's path, which will include LifeForm. This is awkward, as neither Horse nor Person includes the ambiguous member; lifeExpectancy is found in LifeForm.

Let's consider the output for the aforementioned program:

```
Neigh! and Hello! I'm a Centaur.
It is unclear how many years I will live: 80 or 35.
```

We can see that designing a hierarchy that includes a diamond shape has drawbacks. These conundrums include programming ambiguities that need to be resolved in an awkward fashion, duplication in memory of repeated subobjects, plus time to construct and destruct these duplicate subobjects.

Luckily, C++ has a language feature to alleviate these hardships with diamond-shaped hierarchies. After all, C++ is a language that will allow us to do anything. Knowing when and whether we should utilize these features is another concern. Let's first take a look at the C++ language solution to deal with diamond-shaped hierarchies and their inherent problems by looking at virtual base classes.

Utilizing virtual base classes to eliminate duplication

We have just seen the MI implementation issues that quickly arise when a diamond shape is included in an OO design – duplication in memory for a repeated subobject, ambiguity accessing that subobject (even through inherited member functions), and the duplication of construction and destruction. For these reasons, pure OO designs will not include graphs in a hierarchy (that is, no diamond shapes). Yet, we know C++ is a powerhouse of a language and anything is possible. As such, C++ will provide us with a solution to these issues.

Virtual base classes are a C++ language feature that can alleviate duplication of a common base class when using multiple inheritance. The keyword virtual is placed in the base class list between the access label and the base class name of the sibling or cousin class that may *later* be used as a base class for the same derived class. Note that knowing two sibling classes may later be combined as common base classes for a new derived class can be difficult. It is important to note that sibling classes that do not specify a virtual base class will demand their own copy of the (otherwise) shared base class.

Virtual base classes should be used sparingly in implementation because they place restrictions and overhead on instances that have such a class as an ancestor class. Restrictions to be aware of include the following:

- An instance having a virtual base class can use more memory than its non-virtual counterpart (the instance contains a pointer to the potentially shared base class component).

- Casting from an object of a base class type to a derived class type is prohibited when a virtual base class is in the ancestor hierarchy.

- The member initialization list of the most derived class must be used to specify which constructor of the shared object type should be used for initialization. If this specification is ignored, the default constructor will be used to initialize this subobject.

Let us now look at a full program example that employs virtual base classes. As usual, the full program can be found in our GitHub as follows:

https://github.com/PacktPublishing/Deciphering-Object-Orient-ed-Programming-with-CPP/blob/main/Chapter09/Chp9-Ex3.cpp

```cpp
#include <iostream>
using std::cout;    // preferred to: using namespace std;
using std::endl;
using std::string;
using std::to_string;

class LifeForm
{
private:
    int lifeExpectancy = 0;  // in-class initialization
public:
    LifeForm() = default;
    LifeForm(int life): lifeExpectancy(life) { }
    // We're accepting default copy constructor, but if we
    // wanted to write it, it would look like:
    // LifeForm(const LifeForm &form):
    //          lifeExpectancy(form.lifeExpectancy) { }
    // prototype necessary to specify virtual dest. below
    virtual ~LifeForm() = default;
```

```
    [[nodiscard]] int GetLifeExpectancy() const
        { return lifeExpectancy; }
    virtual void Print() const = 0;
    virtual string IsA() const = 0;
    virtual string Speak() const = 0;
};
```

In the previous segment of code, we see the full class definition of `LifeForm`. Notice that the member functions with bodies are inlined in the class definition. Of course, the compiler will not actually make inline substitutions for constructors or the virtual destructor; knowing this, it is convenient to write the methods as inline to make the class compact for reviewing.

Next, let's see the class definition for `Horse`:

```
class Horse: public virtual LifeForm
{
private:
    string name;
    static constexpr int HORSE_LIFE = 35; // Horse life exp
public:
    Horse() : LifeForm(HORSE_LIFE) { }
    Horse(const string &n);
    // Remember, it isn't necessary to proto def. copy ctor
    // Horse(const Horse &) = default;
    // Because base class destructor is virtual, ~Horse()
    // is automatically virtual (overridden) even w/o proto
    // ~Horse() override = default;
    const string &GetName() const { return name; }
    void Print() const override
        { cout << name << endl; }
    string IsA() const override { return "Horse"; }
    string Speak() const override { return "Neigh!"; }
};


Horse::Horse(const string &n) : LifeForm(HORSE_LIFE),
                                name(n)
{
}
```

```
// We are using the default copy constructor, but if we
// wanted to write it, this is what it would look like:
// Horse::Horse(const Horse &h): LifeForm (h), name(h.name)
// {
// }
```

In the previous segment of code, we have the full class definition for Horse. Keep in mind that though certain methods are written as inline for compactness, the compiler will never actually inline a constructor or destructor. Nor can a virtual function be inlined, as its whole point is to have the appropriate method determined at runtime (except rare scenarios involving devirtualization).

Here, LifeForm is a virtual base class of Horse. This means that if Horse ever has a sibling (or cousin) that also inherits from LifeForm using a virtual base class, and those siblings serve as base classes for a derived class, then those siblings will *share* their copy of LifeForm. The virtual base class will reduce storage, reduce extra constructor and destructor calls, and eliminate ambiguity.

Notice the Horse constructors that specify a constructor specification of LifeForm (HORSE_LIFE) in their member initialization lists. This base class initialization will be ignored if LifeForm actually is a shared virtual base class, though these constructor specifications are certainly valid for instances of Horse or for instances of descendants of Horse in which the diamond shape hierarchy does not apply. In hierarchies where Horse is combined with a sibling class to truly serve as a virtual base class, the LifeForm (HORSE_LIFE) specification will be ignored and, instead, either the default LifeForm constructor will be called, or another will be selected at a lower (and unusual) level in the hierarchy.

Next, let's see more of this program by looking at additional class definitions, beginning with Person:

```
class Person: public virtual LifeForm
{
private:
    string firstName;
    string lastName;
    char middleInitial = '\0';  // in-class initialization
    string title;  // Mr., Ms., Mrs., Miss, Dr., etc.
    static constexpr int PERSON_LIFE = 80; // Life expect.
protected:
    void ModifyTitle(const string &);
public:
    Person();   // default constructor
    Person(const string &, const string &, char,
           const string &);
```

```
    // Default copy constructor prototype is not necessary
    // Person(const Person &) = default;  // copy ctor.
    // Because base class destructor is virtual, ~Person()
    // is automatically virtual (overridden) even w/o proto
    // ~Person() override = default;  // destructor
    const string &GetFirstName() const
        { return firstName; }
    const string &GetLastName() const
        { return lastName; }
    const string &GetTitle() const { return title; }
    char GetMiddleInitial() const { return middleInitial; }
    void Print() const override;
    string IsA() const override;
    string Speak() const override;
};
```

In the prior segment of code, we see that Person has a public virtual base class of LifeForm. Should Person and a sibling of Person ever be combined using multiple inheritance to be base classes for a new derived class, those siblings that have indicated a virtual base class of LifeForm will agree to share a single subobject of LifeForm.

Moving onward, let's review the member functions of Person:

```
Person::Person(): LifeForm(PERSON_LIFE)
{  // Note that the base class init list specification of
   // LifeForm(PERSON_LIFE) is ignored if LifeForm is a
   // shared, virtual base class.
}  // This is the same in all Person constructors.


Person::Person(const string &fn, const string &ln, char mi,
               const string &t): LifeForm(PERSON_LIFE),
               firstName(fn), lastName(ln),
               middleInitial(mi), title(t)
{
}


// We're using the default copy constructor, but if we
// wrote/prototyped it, here's what the method would be:
```

```cpp
// Person::Person(const Person &p): LifeForm(p),
//               firstName(p.firstName), lastName(p.lastName),
//               middleInitial(p.middleInitial), title(p.title)
// {
// }

void Person::ModifyTitle(const string &newTitle)
{
    title = newTitle;
}

void Person::Print() const
{
    cout << title << " " << firstName << " ";
    cout << middleInitial << ". " << lastName << endl;
}

string Person::IsA() const
{
    return "Person";
}

string Person::Speak() const
{
    return "Hello!";
}
```

In the aforementioned methods of Person, we see few details that surprise us; the methods are largely as expected. However, as a reminder, note that the LifeForm(PERSON_LIFE) specifications in the member initialization lists of the Person constructor will be ignored if Person is combined in a diamond-shaped hierarchy where the LifeForm subobject becomes shared, rather than duplicated.

Next, let's take a look at where multiple inheritance comes into play, with the definition of the Centaur class:

```cpp
class Centaur: public Person, public Horse
{
private:
```

```
        // no additional data members required, but the below
        // static constexpr eliminates a magic number of 1000
        static constexpr int CENTAUR_LIFE = 1000; //life expect
public:
        Centaur(): LifeForm(CENTAUR_LIFE) { }
        Centaur(const string &, const string &, char = ' ',
                const string & = "Mythological Creature");
        // We don't want default copy constructor due to the
        // needed virtual base class in the mbr init list below
        Centaur(const Centaur &c):
                Person(c), Horse(c), LifeForm(CENTAUR_LIFE) { }
        // Because base class' destructors are virt, ~Centaur()
        // is automatically virtual (overridden) w/o prototype
        // ~Centaur() override = default;
        void Print() const override;
        string IsA() const override;
        string Speak() const override;
};


// Constructors for Centaur need to specify how the shared
// base class LifeForm will be initialized
Centaur::Centaur(const string &fn, const string &ln,
                char mi, const string &title):
                Person(fn, ln, mi, title), Horse(fn),
                LifeForm(CENTAUR_LIFE)
{
    // All initialization has been taken care of in
}   // member initialization list

void Centaur::Print() const
{
    cout << "My name is " << GetFirstName();
    cout << ".  I am a " << GetTitle() << endl;
}
```

```
string Centaur::IsA() const
{
    return "Centaur";
}

string Centaur::Speak() const
{
    return "Neigh! Hello! I'm a master of two languages.";
}
```

In the aforementioned Centaur class definition, we can see that Centaur has public base classes of Horse and Person. We are implying that a Centaur *Is-A* Horse and Centaur *Is-A* Person.

Notice, however, that the keyword virtual is not used in the base class list with the Centaur class definition. Yet, Centaur is the level in the hierarchy where the diamond shape is introduced. This means that we must plan ahead in our design stage to know to utilize the virtual keyword in the base class list for our Horse and Person class definitions. This is an example of why a proper design session is critical versus just jumping into implementation.

Also, quite unusually, notice the base class list of Person(fn, ln, mi, title), Horse(fn), LifeForm(CENTAUR_LIFE) in the Centaur alternate constructor. Here, we not only specify the preferred constructor of our immediate base classes of Person and Horse, but also the preferred constructor for *their* common base class of LifeForm. This is highly unusual. Without LifeForm as a virtual base class for Horse and Person, Centaur would not be able to specify how to construct the shared LifeForm piece (that is, by choosing a constructor for other than its immediate base classes). You also will notice the base class constructor specification of :LifeForm(CENTAUR_LIFE) in the member initialization list of the default as well as copy constructors for the same purposes. The virtual base class usage makes the Person and Horse classes less reusable for other applications, for reasons outlined at the beginning of this subsection.

Let's take a look at what our main() function entails:

```
int main()
{
    Centaur beast("Wild", "Man");
    cout << beast.Speak() << endl;
    cout << " I'm a " << beast.IsA() << ". ";
    beast.Print();
    cout << "I will live: ";
    cout << beast.GetLifeExpectancy();// no longer ambiguous
    cout << " years" << endl;
```

```
    return 0;
}
```

Similar to the `main()` function in our non-virtual base class example, we can see that `Centaur` is likewise instantiated and that virtual functions such as `Speak()`, `IsA()`, and `Print()` are easily called. Now, however, when we call `GetLifeExpectancy()` through our `beast` instance, the call is no longer ambiguous. There is only one subobject of `LifeForm`, whose `lifeExpectancy` (an integer) has been initialized to `1000` (`CENTAUR_LIFE`).

Here is the output for the full program example:

```
Neigh! Hello! I'm a master of two languages.
I am a Centaur. My name is Wild. I am a Mythological Creature.
I will live: 1000 years.
```

Virtual base classes have solved a difficult MI conundrum. But we have also seen that the code required to do so is less flexible for future expansion and reuse. As such, virtual base classes should be carefully and sparingly used only when the design truly supports a diamond-shaped hierarchy. With that in mind, let's consider an OO concept of a discriminator, and consider when alternate designs may be more appropriate.

Considering discriminators and alternate designs

A **discriminator** is an object-oriented concept that helps outline the reasons why a given class is derived from its base class. **Discriminators** tend to characterize the types of groupings of specializations that exist for a given base class.

For example, in the aforementioned program examples with diamond-shaped hierarchies, we have the following discriminators (shown in parentheses), outlining our purpose for specializing a new class from a given base class:

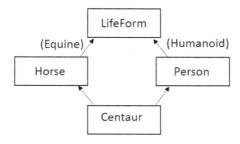

* Discriminators are in ()'s

Figure 9.1 – Multiple inheritance diamond-shaped design shown with discriminators

Whenever temptation leads to the creation of a diamond-shaped hierarchy, examining the discriminators can help us decide whether the design is reasonable, or whether perhaps an alternate design would prove better. Here are some good design metrics to consider:

- If the discriminators for the sibling classes that are being brought back together are the same, then the diamond-shaped hierarchy is better off redesigned.

- When the sibling classes do not have unique discriminators, the attributes and behaviors they will introduce will consist of duplications stemming from having a *like-discriminator*. Consider making the discriminator a class to house those commonalities.

- If the discriminators for the sibling classes are unique, then the diamond-shaped hierarchy may be warranted. In this case, virtual base classes will prove helpful and should be added to the appropriate locations in the hierarchy.

In the previous example, the discriminator detailing why `Horse` specializes `LifeForm` is `Equine`. That is, we are specializing `LifeForm` with equine characteristics and behaviors (hooves, galloping, neighing, etcetera). Had we derived classes such as `Donkey` or `Zebra` from `LifeForm`, the discriminator for these classes would also be `Equine`. Considering the same aforementioned example, the `Person` class would have a `Humanoid` discriminator when specializing `LifeForm`. Had we derived classes such as `Martian` or `Romulan` from `LifeForm`, these classes would also have `Humanoid` as a discriminator.

Bringing `Horse` and `Person` together as base classes for `Centaur` is combining two base classes with different discriminators, `Equine` and `Humanoid`. As such, wholly different types of characteristics and behaviors are factored in by each base class. Though an alternate design may be possible, this design is acceptable (except to OO design purists), and virtual base classes may be used in C++ to eliminate duplication of the otherwise-replicated `LifeForm` piece. Bringing two classes together that share a common base class and that each specializes the base class using distinct discriminators is an example of how MI and virtual base classes are reasonable in C++.

However, bringing together two classes such as `Horse` and `Donkey` (both derived from `LifeForm`) together in a derived class such as `Mule` also creates a diamond-shaped hierarchy. Examining the discriminators for `Horse` and `Donkey` reveals that both have the discriminator of `Equine`. In this case, bringing together these two classes using a diamond-shaped design is not the optimal design choice. Another design choice is possible and preferred. In this case, a preferred solution would be to make the discriminator, `Equine`, its own class, and then derive `Horse`, `Donkey`, and `Mule` from `Equine`. This would avoid MI and a diamond-shaped hierarchy. Let's take a look at the two design options:

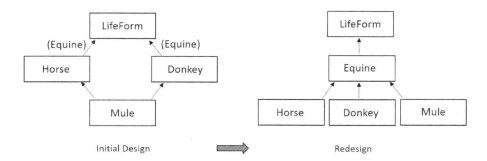

Figure 9.2 – Diamond-shaped multiple inheritance redesigned without MI

> **Reminder**
>
> In a diamond-shaped hierarchy, if the discriminators for the *combined* classes are the same, a better design is possible (by making the discriminator its own class). However, if the discriminators are different, consider keeping the diamond-shaped MI hierarchy and then use virtual base classes to avoid duplication of the common base class subobject.

We have now thoroughly examined the OO concept of a discriminator and have seen how discriminators can be used to help evaluate the reasonableness of a design. In many cases, designs using diamond-shaped hierarchies can be redesigned to not only eliminate the diamond shape but to eliminate multiple inheritance altogether. Let's briefly recap the MI issues and OO concepts we've covered in this chapter before moving onward to our next chapter.

Summary

In this chapter, we have marched onward with our quest for understanding object-oriented programming by exploring a controversial OOP topic, that of multiple inheritance. First, in this chapter, we have understood the simple mechanics of multiple inheritance. Equally important, we have reviewed reasons for building inheritance hierarchies and possible reasons for using MI (that is, specifying Is-A, mix-in, and Has-A relationships). We have been reminded that using inheritance to specify Is-A relationships supports pure OO designs. We have also seen the utility of using MI to implement mix-in relationships. We have also taken a look at the controversial use of MI to quickly implement a Has-A relationship; we'll see in *Chapter 10, Implementing Association, Aggregation, and Composition*, a preferred implementation for Has-A.

We have seen how having multiple inheritance in our OO design toolkit can lead to diamond-shaped hierarchies. We have seen the inevitable issues arising from diamond-shaped hierarchies, such as duplication in memory, duplication in construction/destruction, and ambiguity in accessing a replicated subobject. We have also seen that C++ provides a language-supported mechanism to correct these issues, using virtual base classes. We know that virtual base classes solve a tedious problem, yet they themselves are not perfect solutions.

In an effort to critique diamond-shaped hierarchies, we have looked at an OO concept of a discriminator to help us weigh the validity of an OO design using MI in a diamond shape. This has also led us to understand that alternate designs can apply to a set of objects; sometimes a redesign is a more elegant approach in which the solution will yield easier, long-term use.

C++ is a *"you can do anything"* OOP language, and multiple inheritance is a controversial OO concept. Knowing when certain MI designs may be warranted and understanding language features to help with those MI issues will make you a better programmer. Knowing when a redesign is in order is also critical.

We are now ready to continue to *Chapter 10, Implementing Association, Aggregation, and Composition*, to further enhance our OOP skills by next learning how to represent association, aggregation, and composition with programming techniques. These upcoming concepts will *not* have direct language support, but the concepts are instrumental in our OO arsenal of skills. Let's move onward!

Questions

1. Type in (or use the online code) for the diamond-shaped hierarchy example in this chapter that uses virtual base classes. Run it as is. Hint: you may want to add explicit destructors with `cout` statements to trace the destruction sequence:

    a. How many `LifeForm` subobjects exist for the instance of `Centaur`?

    b. How many times is the `LifeForm` constructor (and destructor) invoked? Hint: you may want to place trace statements using `cout` in each of your constructors and destructor.

    c. Which `LifeForm` constructor would be invoked if the constructor selections for `LifeForm` in the member initialization list of the `Centaur` constructors were omitted?

2. Now, remove the keyword `virtual` from the base class list of `Person` and `Horse` (that is, `LifeForm` will no longer be a virtual base class of `Person` and `Horse`. `LifeForm` will just be a typical base class of `Person` and `Horse`.) Also, remove the `LifeForm` constructor selection from the member initialization list of the `Centaur` constructors. Now, instantiate `Centaur`:

    a. How many `LifeForm` subobjects exist for the instance of `Centaur`?

    b. Now, how many times is the `LifeForm` constructor (and destructor) invoked? Hint: you may want to add trace statements to your constructors and destructor.

10

Implementing Association, Aggregation, and Composition

This chapter will continue advancing our knowledge of object-oriented programming in C++. We will augment our understanding of object relationships by exploring the object-oriented concepts of association, aggregation, and composition. None of these OO concepts have direct language support in C++; we will instead learn multiple programming techniques to implement these ideas. We will also understand which implementation techniques are preferred for various concepts, as well as the advantages and pitfalls of various practices.

Association, aggregation, and composition occur copiously in OO designs. It is crucial to understand how to implement these important object relationships.

In this chapter, we will cover the following main topics:

- Understanding the OO concepts of aggregation and composition, and various implementations
- Understanding the OO concept of association and its implementation, including the importance of backlink maintenance and the utility of reference counting

By the end of this chapter, you will understand the OO concepts of association, aggregation, and composition, and how to implement these relationships in C++. You will also understand many housekeeping approaches necessary to keep these relationships up to date, such as reference counting and backlink maintenance. Though the concepts are relatively straightforward, you will see why there is a substantial amount of bookkeeping required to maintain accuracy for these types of object relationships.

Let's broaden our understanding of C++ as an OOP language by exploring these core object relationships.

Technical requirements

Online code for full program examples can be found in the following GitHub URL: `https://github.com/PacktPublishing/Deciphering-Object-Oriented-Programming-with-CPP/tree/main/Chapter10`. Each full program example can be found in the GitHub under the appropriate chapter heading (subdirectory) in a file that corresponds to the chapter number, followed by a dash, followed by the example number in the chapter at hand. For example, the first full program in this chapter can be found in the subdirectory `Chapter10` in a file named `Chp10-Ex1.cpp` under the aforementioned GitHub directory.

The CiA video for this chapter can be viewed at: `https://bit.ly/3clgvGe`.

Understanding aggregation and composition

The object-oriented concept of aggregation arises in many OO designs. It comes up as frequently as inheritance does to specify object relationships. **Aggregation** is used to specify Has-A, whole-part, and in some cases, containment relationships. A class may contain aggregates of other objects. Aggregation can be broken into two categories – *composition* as well as a less strict and *generalized* form of aggregation.

Both **generalized aggregation** and **composition** imply a Has-A or whole-part relationship. However, the two differ in the existence requirements between the two related objects. With generalized aggregation, the objects can exist independently from one another, yet with composition, the objects cannot exist without one another.

Let's take a look at each variety of aggregation, starting with composition.

Defining and implementing composition

Composition is the most specialized form of aggregation and is often what most OO designers and programmers think of when they consider aggregation. Composition implies containment, and is most often synonymous with a whole-part relationship – that is, the whole entity is composed of one or more parts. The whole *contains* the parts. The Has-A relationship will also apply to composition.

The outer object, or *whole*, can be made up of *parts*. With composition, parts do not exist without the whole. Implementation is usually an embedded object – that is, a data member of the contained object type. On rare occasions, the outer object will contain a pointer or reference to the contained object type; however, when this occurs, the outer object will be responsible for the creation and destruction of the inner object. The contained object has no purpose without its outer layer. Likewise, the outer layer is not *ideally* complete without its inner, contained pieces.

Let's take a look at a composition as typically implemented. The example will illustrate containment – a Student *Has-A(n)* Id. More so, we will imply that an Id is a necessary part of a Student and will not exist without a Student. Id objects on their own serve no purpose. Id objects simply do not need to exist if they are not a part of a primary object that gives them purpose. Likewise, you might argue that a Student is incomplete without an Id, though this is a bit subjective! We will implement the *part* using an embedded object within the *whole*.

The composition example will be broken into many segments. Though only portions of the example are shown, the full program can be found in the following GitHub location:

https://github.com/PacktPublishing/Deciphering-Object-Orient-ed-Programming-with-CPP/blob/main/Chapter10/Chp10-Ex1.cpp

```
#include <iostream>
#include <iomanip>
using std::cout;
using std::endl;
using std::setprecision;
using std::string;
using std::to_string;

class Id final  // the contained 'part'
{          // this class is not intended to be extended
private:
    string idNumber;
public:
    Id() = default;
    Id(const string &id): idNumber(id) { }
    // We get default copy constructor, destructor
    // without including without including prototype
    // Id(const Id &id) = default;
    // ~Id() = default;
    const string &GetId() const { return idNumber; }
};
```

In the previous code fragment, we have defined an Id class. An Id will be a class that can be contained by other classes needing a fully functional Id capability. The Id will become a *part* of any *whole* object that may choose to contain it.

Let's move onward to build a set of classes that will ultimately contain this Id. We will start with a class we are familiar with, Person:

```
class Person
{
private:
    string firstName;
    string lastName;
    char middleInitial = '\0';   // in-class initialization
    string title;  // Mr., Ms., Mrs., Miss, Dr., etc.
protected:
    void ModifyTitle(const string &);
public:
    Person() = default;   // default constructor
    Person(const string &, const string &, char,
           const string &);
    // We get default copy constructor w/o prototype
    // Person(const Person &) = default;  // copy ctor.
    // But, we need prototype destructor to add 'virtual'
    virtual ~Person() = default;  // virtual destructor
    const string &GetFirstName() const
        { return firstName; }
    const string &GetLastName() const { return lastName; }
    const string &GetTitle() const { return title; }
    char GetMiddleInitial() const { return middleInitial; }
    // virtual functions
    virtual void Print() const;
    virtual void IsA() const;
    virtual void Greeting(const string &) const;
};
//  Assume the member functions for Person exist here
//  (they are the same as in previous chapters)
```

In the previous segment of code, we have defined the Person class, as we are accustomed to describing it. To abbreviate this example, let us assume that the accompanying member functions exist as prototyped in the aforementioned class definition. You can reference these member functions in the earlier provided GitHub link for online code.

Now, let's define our Student class. Though it will contain elements that we are accustomed to seeing, Student will also contain an Id as an embedded object:

```cpp
class Student: public Person  // 'whole' object
{
private:
    float gpa = 0.0;    // in-class initialization
    string currentCourse;
    static int numStudents;
    Id studentId;   // is composed of a 'part'
public:
    Student();  // default constructor
    Student(const string &, const string &, char,
            const string &, float, const string &,
            const string &);
    Student(const Student &);  // copy constructor
    ~Student() override;  // destructor
    // various member functions (many are inline)
    void EarnPhD() { ModifyTitle("Dr."); }
    float GetGpa() const { return gpa; }
    const string &GetCurrentCourse() const
        { return currentCourse; }
    void SetCurrentCourse(const string &); // proto. only
    void Print() const override;
    void IsA() const override
        { cout << "Student" << endl; }
    static int GetNumberStudents() { return numStudents; }
    // Access function for embedded Id object
    const string &GetStudentId() const;    // prototype only
};
int Student::numStudents = 0;  // static data member

inline void Student::SetCurrentCourse(const string &c)
{
    currentCourse = c;
}
```

In the preceding Student class, we routinely notice that Student is derived from Person. As we already know, this means that a Student instance will include the memory layout of a Person, as a Person subobject.

However, notice the data member, Id studentId;, in the Student class definition. Here, studentId is of type Id. It is not a pointer, nor is it a reference to an Id. Data member studentId is an embedded (that is, an aggregate or member) object. This means that when a Student class is instantiated, not only will the memory from inherited classes be included but also the memory for any embedded objects. We will need to provide a means to initialize the embedded object, studentId. Note, we have seen member objects before, such as data members of type string; that is, data members that are of another class type.

Let's move forward with the Student member functions to understand how we may initialize, manipulate, and access the embedded object:

```
Student::Student(): studentId(to_string(numStudents + 100)
                                        + "Id")
{
    numStudents++;    // increment static counter
}


Student::Student(const string &fn, const string &ln,
                 char mi, const string &t, float avg,
                 const string &course, const string &id):
                 Person(fn, ln, mi, t), gpa(avg),
                 currentCourse(course), studentId(id)
{
    numStudents++;
}


Student::Student(const Student &s): Person(s),
                 gpa(s.gpa), currentCourse(s.currentCourse),
                 studentId(s.studentId)
{
    numStudents++;
}


Student::~Student()    // destructor definition
{
```

```
        numStudents--;       // decrement static counter
        // embedded object studentId will also be destructed
    }


    void Student::Print() const
    {
        cout << GetTitle() << " " << GetFirstName() << " ";
        cout << GetMiddleInitial() << ". " << GetLastName();
        cout << " with id: " << studentId.GetId() << " GPA: ";
        cout << setprecision(3) <<  " " << gpa;
        cout << " Course: " << currentCourse << endl;
    }


    const string &GetStudentId() const
    {
        return studentId.GetId();
    }
```

In the previously listed member functions of Student, let's begin with our constructors. Notice in the default constructor, we utilize the member initialization list (:) to specify studentId(to_ string(numStudents + 100) + "Id"). Because studentId is a member object, we have the opportunity to select (via the member initialization list) which constructor should be used for its initialization. Here, we merely select the one with the Id(const string &) signature. In the absence of a specific value to use to initialize Id, we manufacture a string value to serve as the needed ID.

Similarly, in the alternate constructor for Student, we use the member initialization list to specify studentId(id), which will also select the Id(const string &) constructor, passing the parameter id to this constructor.

The copy constructor for Student additionally specifies how to initialize the studentId member object with the studentId(s.studentId) specification in the member initialization list. Here, we simply call the copy constructor for Id.

In our destructor for Student, we do not need to deallocate studentId. As this data member is an embedded (aggregate) object, its memory will go away when the memory for the outer object goes away. Of course, because studentId is an object itself, its own destructor will first be called before its memory is released. Under the hood, the compiler will (covertly) patch in a call to the Id destructor for studentId as the last line of code in the Student destructor. Actually, this will be the penultimate (next to last) implicit line in the destructor – the last line that will be covertly patched in will be a call to the Person destructor (to continue the destruction sequence).

Lastly, in the previous segment of code, let's notice the call to `studentId.GetId()`, which occurs in both `Student::Print()` and `Student::GetStudentId()`. Here, the embedded object `studentId` calls its own public function `Id::GetId()` to retrieve its private data member in the scope of the `Student` class. Because `studentId` is private in `Student`, this embedded object may only be accessed within the scope of `Student` (that is, member functions of `Student`). However, the addition of `Student::GetStudentId()` provides a public wrapper for `Student` instances in other scopes to retrieve this information.

Finally, let's take a look at our `main()` function:

```
int main()
{
    Student s1("Cyrus", "Bond", 'I', "Mr.", 3.65, "C++",
               "6996CU");
    Student s2("Anne", "Brennan", 'M', "Ms.", 3.95, "C++",
               "909EU");
    cout << s1.GetFirstName() << " " << s1.GetLastName();
    cout << " has id #: " << s1.GetStudentId() << endl;
    cout << s2.GetFirstName() << " " << s2.GetLastName();
    cout << " has id #: " << s2.GetStudentId() << endl;
    return 0;
}
```

In the aforementioned `main()` function, we instantiate two `Student` instances: `s1` and `s2`. When the memory is created (in this case, on the stack) for each `Student`, memory for any inherited classes will also be included as subobjects. Additionally, memory for any embedded objects, such as `Id`, will also be laid out as a subobject within `Student`. The memory for the contained object, or *part*, will be allocated along with the allocation for the outer object, or *whole*.

Next, let's notice the access to the contained piece, the embedded `Id` object. We start with a call to `s1.GetStudentId()`; `s1` accesses a `Student` member function, `GetStudentId()`. That student member function will utilize the member object of `studentId` to call `Id::GetId()` on this inner object of type `Id`. The member function `Student::GetStudentId()` can implement this desired public access by simply returning the value that `Id::GetId()` returns on the embedded object.

Let's look at the output for the aforementioned program:

```
Cyrus Bond has id #: 6996CU
Anne Brennan has id #: 909EU
```

This example details composition with its typical implementation, an embedded object. Let's now take a look at a much less used, alternate implementation – that of inheritance.

Considering an alternate implementation for composition

It is useful to understand that composition can alternatively be implemented using inheritance, however, this is extremely controversial. Remember, inheritance is most often used to implement *Is-A* and not *Has-A* relationships. We briefly described using inheritance to implement Has-A relationships in *Chapter 9, Exploring Multiple Inheritance*.

To recap, you would simply inherit from the *part*, rather than embed the part as a data member. When doing so, you no longer need to provide *wrapper* functions to the *part*, such as we saw in the previous program, with the `Student::GetStudentId()` method calling `studentId.GetId()` to provide access to its embedded part. The wrapper function was necessary with the embedded object example, as the part (`Id`) was private in the whole (`Student`). Programmers could not have accessed the private `studentId` data member of `Student` outside the scope of `Student`. Of course, member functions of `Student` (such as `GetStudentId()`) can access their own class' private data members and in doing so, can implement the `Student::GetStudentId()` wrapper function to provide such (safe) access.

Had inheritance been used, the public interface of `Id::GetId()` would have been simply inherited as a public interface in `Student`, providing simple access without the need to first go through the embedded object explicitly.

Nonetheless, though inheriting a *part* is simple in some ways, it vastly compounds multiple inheritance. We know multiple inheritance can provide many potential complications. Also, using inheritance, the *whole* can only contain one of each *part* – not multiples of a *part*.

Additionally, implementing a whole-part relationship with inheritance may be confusing when you compare the implementation to the OO design. Remember, inheritance usually means *Is-A* and not *Has-A*. For these reasons, the most typical and appreciated implementation of an aggregate is through an embedded object.

Next, let's move onward by looking at a more general form of aggregation.

Defining and implementing a generalized aggregation

We have looked at the most commonly used form of aggregation in OO designs, that of composition. Most notably, with composition, we have seen that the part does not have a reason to exist without the whole. Nonetheless, a more generalized (but less common) form of aggregation exists and is sometimes specified in OO designs. We will now consider this less common form of aggregation.

In a **generalized aggregation**, a *part* may exist without the *whole*. A part will be created separately and then attached to the whole at a later point in time. When the *whole* goes away, a *part* may then remain to be salvaged for use with another outer or *whole* object.

In a generalized aggregation, the Has-A relationship certainly applies, as does the whole-part designation. The difference is that the *whole* object will not create nor destroy a *part* subobject. Consider the straightforward example that a Car *Has-A(n)* Engine. A Car object also *Has-A* set of four Tire objects. The Engine or Tire objects can be manufactured separately and then passed to the constructor of the Car to provide these parts to the whole. Yet should an Engine be destroyed, a new Engine can easily be swapped out (using a member function), without requiring the entire Car to be destroyed and then reconstructed.

A generalized aggregation is equivalent to a Has-A relationship, yet we think of this with more flexibility and permanence of the individual parts as we did with composition. We consider this relationship as an aggregation simply because we wish to equate the objects with a Has-A meaning. The Has-A relationship in the Car, Engine, and Tire example is strong; the Engine and Tires are necessary parts, required to make the whole Car.

Here, implementation typically is with the *whole* containing a pointer (or set of pointers) to the *part(s)*. It is important to note that the parts will be passed into a constructor (or another member function) of the outer object to establish the relationship. The critical marker is that the whole will not create (nor destroy) the parts, and the parts will never destroy the whole.

Incidentally, the longevity of the individual pieces (and the basic implementation) of a generalized aggregation will be similar to our next topic – association. Let's move forward to our next section to understand the similarities, as well as the OO conceptual differences (sometimes subtle) between generalized aggregation and association.

Understanding association

An **association** models a relationship that exists between otherwise unrelated class types. An association can provide ways for objects to interact to fulfill these relationships. Associations are not used for Has-A relationships, however, in some cases, there are shades of gray as to whether we're describing a bonafide Has-A relationship, or whether we are merely using the phrase Has-A because it sounds appropriate linguistically.

Multiplicity for associations exists: one-to-one, one-to-many, many-to-one, or many-to-many. For example, a Student may be associated with a single University, and that University may be associated with many Student instances; this is a one-to-many association.

Associated objects have an independent existence. That is, two or more objects may be instantiated and exist independently for a portion of the application. At some point, one object may wish to assert a dependency or relationship with the other object. Later in the application, the associated objects may part ways and continue on their own unrelated paths.

For example, consider the relationship between a `Course` and an `Instructor`. A `Course` is associated with an `Instructor`. A `Course` requires an `Instructor`; an `Instructor` is integral to the `Course`. An `Instructor` may be associated with many `Course`(s). Yet each part exists independently – one will not create nor destroy the other. Instructors may also exist independently without courses; perhaps an instructor is taking time to write a book, is taking a sabbatical, or is a professor conducting research.

In this example, the association is very similar to a generalized aggregation. In both cases, the related objects also exist independently. In this case, whether one says that `Course` Has-A(n) `Instructor` or that a `Course` has a dependency on an `Instructor`, can be a shade of gray. You may ask yourself – is it just spoken language that makes me choose the wording of Has-A? Do I instead mean that there is a necessary link between the two? Perhaps the relationship is an association, and its descriptive adornment (to further describe the nature of the association) is *teaches*. You may have arguments supporting either choice. For this reason, generalized aggregations can be considered specialized types of associations; we will see that their implementations are the same using independently existing objects. Nonetheless, we will distinguish a typical association as being a relationship between objects that decisively do not support a true Has-A relationship.

For example, consider the relationship between `University` and `Instructor`. Rather than thinking of this as a Has-A relationship, we may instead consider the relationship between the two as that of association; we can think of the adornment describing this relationship as *employs*. Likewise, `University` ascertains a relationship with many `Student` objects. The association here may be described by the adornment *educates*. The distinction can be made that `University` is made up of `Department` objects, `Building` objects, and components of this nature to support any of its Has-A relationships through containment, yet its relationships with `Instructor` objects, `Student` objects, and so on are made using associations.

Now that we have distinguished typical associations from generalized aggregations, let's take a look at how we can implement associations and some of the complexities involved.

Implementing association

Typically, an association between two or more objects is implemented using pointers or sets of pointers. The *one* side is implemented using a pointer to the associated object, whereas the *many* side of the relationship is implemented as a set of pointers to the associated objects. A set of pointers may be an array of pointers, a linked list of pointers, or truly any collection of pointers. Each type of collection will have its own set of advantages and drawbacks. For example, arrays of pointers are easy to use, have direct access to specific members, yet have a fixed number of items. Linked lists of pointers can accommodate any quantity of items, yet accessing a specific element requires traversing past others to find the desired item.

Occasionally, a reference may be used to implement the *one* side of an association. Recall that a reference must be initialized and cannot at a later date be reset to reference another object. Using a reference to model an association implies that one instance will be associated with a precise other instance for the duration of the primary object's existence. This is extremely restrictive, so references are used very infrequently to implement associations.

Regardless of the implementation, when the primary object goes away, it will not interfere with (that is, delete) the associated object.

Let's see a typical example illustrating the preferred implementation of a one-to-many association, utilizing a pointer on the *one* side, and a set of pointers on the *many* side. In this example, a University will be associated with many Student instances. And, for simplicity, a Student will be associated with a single University.

To save space, some portions of this program that are the same as in our last example will not be shown; however, the program in its entirety can be found in our GitHub as follows:

https://github.com/PacktPublishing/Deciphering-Object-Orient-ed-Programming-with-CPP/blob/main/Chapter10/Chp10-Ex2.cpp

```cpp
#include <iostream>
#include <iomanip>
using std::cout;
using std::endl;
using std::setprecision;
using std::string;
using std::to_string;


// classes Id and Person are omitted here to save space.
// They will be as shown in previous example: Chp10-Ex1.cpp
class Student; // forward declaration

class University
{
private:
    string name;
    static constexpr int MAX = 25; // max students allowed
    // Notice: each studentBody element is set to a nullptr
    // using in-class initialization
    Student *studentBody[MAX] = { }; // Association to
                                     // many students
```

```
        int currentNumStudents = 0;   // in-class initialization
public:
    University();
    University(const string &);
    University(const University &) = delete; // no copies
    ~University();
    void EnrollStudent(Student *);
    const string &GetName() const { return name; }
    void PrintStudents() const;
};
```

In the preceding segment, let's first notice the forward declaration of class Student;. This declaration will allow our code to reference the Student type prior to the Student class definition. In the University class definition, we see that there is an array of pointers to Student. We also see that the EnrollStudent() method takes a Student * as an argument. The forward declaration enables such usage of Student prior to its definition.

We also notice that the University has a simple interface with constructors, a destructor, and a few member functions.

Next, let's take a look at the University member function definitions:

```
// Remember, currentNumStudents will be set w in-class init
// and name, as a string member object, will be init to
// empty. And studentBody (array of ptrs) will also set w
// in-class initialization.
University::University()
{
    // in-lieu of in-class init, we could alternatively set
    // studentBody[i] to nullptr iteratively in a loop:
    // (the student body will start out empty)
    // for (int i = 0; i < MAX; i++)
    //     studentBody[i] = nullptr;
}


University::University(const string &n) : name(n)
{
    // see default constructor for alt init of studentBody
}
```

```cpp
University::~University()
{
    // The University will not delete the students
    for (int i = 0; i < MAX; i++)    // only null out
        studentBody[i] = nullptr;     // their link
}


void University::EnrollStudent(Student *s)
{
    // set an open slot in the studentBody to point to the
    // Student passed in as an input parameter
    studentBody[currentNumStudents++] = s;
}


void University::PrintStudents() const
{
    cout << name << " has the following students:" << endl;
    // Simple loop to process set of students, however we
    // will soon see safer, more modern ways to iterate
    // over partial arrays w/o writing explicit 'for' loops
    for (int i = 0; i < currentNumStudents; i++)
    {
        cout << "\t" << studentBody[i]->GetFirstName();
        cout << " " << studentBody[i]->GetLastName();
        cout << endl;
    }
}
```

Taking a closer look at the aforementioned University methods, we can see that in both constructors for University, we could alternatively null out the pointers to the elements comprising our studentBody using nullptr (versus our choice of using in-class initialization, which similarly initializes each element). Likewise, in the destructor, we similarly null out our links to the associated Student instances. Shortly in this section, we will see that there will be some additional backlink maintenance required, but for now, the point is that we will not delete the associated Student objects.

Since `University` objects and `Student` objects will exist independently, neither will create nor destroy instances of the other type.

We also come across an interesting member function, `EnrollStudent(Student *)`. In this method, a pointer to a specific `Student` will be passed in as an input parameter. We merely index into our array of pointers to `Student` objects, namely `studentBody`, and set an unused array element point to the newly enrolled `Student`. We keep track of how many current `Student` objects exist using a `currentNumStudents` counter, which is incremented with a post increment after the pointer assignment to the array is made.

We also notice that the `University` class has a `Print()` method that prints the university's name, followed by its current student body. It does so by simply accessing each associated `Student` object in `studentBody` and asking each `Student` instance to invoke the `Student::GetFirstName()` and `Student::GetLastName()` methods.

Next, let's now take a look at our `Student` class definition, along with its inline functions. Recall that we're assuming that the `Person` class is the same as seen earlier in this chapter:

```
class Student: public Person
{
private:
    // data members
    float gpa = 0.0;  // in-class initialization
    string currentCourse;
    static int numStudents;
    Id studentId;  // part, Student Has-A studentId
    University *univ = nullptr;  // Assoc. to Univ object
public:
    // member function prototypes
    Student();  // default constructor
    Student(const string &, const string &, char,
            const string &, float, const string &,
            const string &, University *);
    Student(const Student &);  // copy constructor
    ~Student() override;  // destructor
    void EarnPhD() { ModifyTitle("Dr."); }
    float GetGpa() const { return gpa; }
    const string &GetCurrentCourse() const
        { return currentCourse; }
    void SetCurrentCourse(const string &); // proto. only
```

```
        void Print() const override;
        void IsA() const override
            { cout << "Student" << endl; }
        static int GetNumberStudents() { return numStudents; }
        // Access functions for aggregate/associated objects
        const string &GetStudentId() const
            { return studentId.GetId(); }
        const string &GetUniversity() const
            { return univ->GetName(); }
};
int Student::numStudents = 0;  // def. of static data mbr.

inline void Student::SetCurrentCourse(const string &c)
{
    currentCourse = c;
}
```

Here, in the previous code segment, we see the Student class definition. Notice that we have an association with a University with the pointer data member University *univ = nullptr; and that this member is initialized to nullptr using in-class initialization.

In the class definition for Student, we can also see there is a wrapper function to encapsulate access to the student's university's name with Student::GetUniversity(). Here, we allow the associated object, univ, to call its public method University::GetName() and return that value as the result of Student::GetUniversity().

Now, let's take a look at the non-inline member functions of Student:

```
Student::Student(): studentId(to_string(numStudents + 100)
                                        + "Id")
{
    // no current University association (set to nullptr
    // with in-class initialization)
    numStudents++;
}

Student::Student(const string &fn, const string &ln,
        char mi, const string &t, float avg,
        const string &course, const string &id,
```

```
            University *univ): Person(fn, ln, mi, t),
            gpa(avg), currentCourse(course), studentId(id)
{
    // establish link to University, then back link
    // note: forward link could also be set in the
    // member initialization list
    this->univ = univ;   // required use of <this>
    univ->EnrollStudent(this);   // another required 'this'
    numStudents++;
}


Student::Student(const Student &s): Person(s),
            gpa(s.gpa), currentCourse(s.currentCourse),
            studentId(s.studentId)
{
    // Notice, these three lines of code are the same as
    // in the alternate constructor - we could instead make
    // a private helper method with this otherwise
    // duplicative code as a means to simplify code
    // maintenance.
    this->univ = s.univ;
    univ->EnrollStudent(this);
    numStudents++;
}


Student::~Student()   // destructor
{
    numStudents--;
    univ = nullptr;   // a Student does not delete its Univ
    // embedded object studentId will also be destructed
}


void Student::Print() const
{
    cout << GetTitle() << " " << GetFirstName() << " ";
    cout << GetMiddleInitial() << ". " << GetLastName();
```

```
        cout << " with id: " << studentId.GetId() << " GPA: ";
        cout << setprecision(3) <<   " " << gpa;
        cout << " Course: " << currentCourse << endl;
    }
```

In the preceding code segment, notice that the default `Student` constructor and the destructor both only null out their link to the `University` object (using `nullptr`). The default constructor has no way to set this link to an existing object, and should certainly not create a `University` instance to do so. Likewise, the `Student` destructor should not delete the `University` merely because the `Student` object's life expectancy is complete.

The most interesting part of the preceding code happens in both the alternate constructor and copy constructor of `Student`. Let's examine the alternate constructor. Here, we establish the link to the associated `University` as well as the backlink from the `University` back to the `Student`.

In the `this->univ = univ;` line of code, we are assigning the data member, `univ` (as pointed to by the `this` pointer) by setting it to point to where the input parameter, `univ`, points. Look closely at the previous class definition – the identifier for the `University *` is named `univ`. Additionally, the input parameter for the `University *` in the alternate constructor is named `univ`. We cannot simply assign `univ = univ;` in the body of this constructor (or in the member initialization list). The `univ` identifier that is in the most local scope is the input parameter, `univ`. Assigning `univ = univ;` would set this parameter to itself. Instead, we disambiguate the `univ` on the left-hand side of this assignment using the `this` pointer. The statement `this->univ = univ;` sets the data member `univ` to the input parameter `univ`. Could we merely have renamed the input parameter something different, such as `u`? Sure, but it is important to understand how to disambiguate an input parameter and data member with the same identifier when the need arises to do so.

Now, let's examine the next line of code, `univ->EnrollStudent(this);`. Now that `univ` and `this->univ` point to the same object, it does not matter which is used to set the backlink. Here, `univ` calls `EnrollStudent()`, which is a public member function in the `University` class. No problem, `univ` is of type `University`. `University::EnrollStudent(Student *)` expects to be passed a pointer to a `Student` to complete the linkage on the `University` side. Luckily, the `this` pointer in our `Student` alternate constructor (the scope of the calling function) is a `Student *`. The `this` pointer (in the alternate constructor) is literally the `Student *` that we need to create the backlink. Here is another example where the explicit use of the `this` pointer is required to complete the task at hand.

Let's move forward to our `main()` function:

```
int main()
{
    University u1("The George Washington University");
    Student s1("Gabby", "Doone", 'A', "Miss", 3.85, "C++",
```

```
                   "4225GWU", &u1);
    Student s2("Giselle", "LeBrun", 'A', "Ms.", 3.45,
                   "C++", "1227GWU", &u1);
    Student s3("Eve", "Kendall", 'B', "Ms.", 3.71, "C++",
                   "5542GWU", &u1);
    cout << s1.GetFirstName() << " " << s1.GetLastName();
    cout << " attends " << s1.GetUniversity() << endl;
    cout << s2.GetFirstName() << " " << s2.GetLastName();
    cout << " attends " << s2.GetUniversity() << endl;
    cout << s3.GetFirstName() << " " << s3.GetLastName();
    cout << " attends " << s3.GetUniversity() << endl;
    u1.PrintStudents();
    return 0;
}
```

Finally, in the previous code fragment in our `main()` function, we can create several independently existing objects, create an association between them, and then view that relationship in action.

First, we instantiate a `University`, namely `u1`. Next, we instantiate three `Student` objects, `s1`, `s2`, and `s3`, and associate each to `University u1`. Note that this association can be set when we instantiate a `Student`, or later on, for example, if the `Student` class supported a `SelectUniveristy(University *)` interface to do so.

We then print out each `Student`, along with the name of the `University` each `Student` attends. Then, we print out the student body for our `University`, `u1`. We notice that the link built between the associated objects is complete in both directions.

Let's look at the output for the aforementioned program:

```
Gabby Doone attends The George Washington University
Giselle LeBrun attends The George Washington University
Eve Kendall attends The George Washington University
The George Washington University has the following students:
        Gabby Doone
        Giselle LeBrun
        Eve Kendall
```

We've seen how easily associations can be set up and utilized between related objects. However, a lot of housekeeping will arise from implementing associations. Let's move forward to understanding the necessary and related issues of reference counting and backlink maintenance, which will help with these housekeeping endeavors.

Utilizing backlink maintenance and reference counting

In the previous subsection, we have seen how to implement associations using pointers. We've seen how to link an object with a pointer to an object in an associated instance. And we've seen how to complete the circular, two-sided relationship by establishing a backlink.

However, as is typical for associated objects, the relationships are fluid and change over time. For example, the given `Student` body will change quite often for a given `University`, or the various `Course` set an `Instructor` will teach will change each semester. It will be typical, then, to remove a particular object's association to another object, and perhaps associate, instead, to a different instance of that class. But, that also means that the associated object must know to remove its link to the first mentioned object. This becomes complicated.

For example, consider the `Student` and `Course` relationship. A `Student` is enrolled in many `Course` instances. A `Course` contains an association to many `Student` instances. This is a many-to-many association. Let's imagine that the `Student` wishes to drop a `Course`. It is not enough for a specific `Student` instance to remove a pointer to a specific `Course` instance. Additionally, the `Student` must let the particular `Course` instance know that the `Student` in question should be removed from that `Course`'s roster. This is known as backlink maintenance.

Consider what would happen in the above scenario if a `Student` were to simply null out its link to the `Course` it was dropping, and do nothing further. The `Student` instance in question would be fine. However, the formerly associated `Course` instance would still contain a pointer to the `Student` in question. Perhaps this would equate to the `Student` receiving a failing grade in the `Course` as the `Instructor` still thinks the `Student` in question is enrolled, yet hasn't been turning in their homework. In the end, the `Student` has been affected after all, with the failing grade.

Remember, with associated objects, one object will not delete the other when it is done with the other object. For example, when a `Student` drops a `Course`, they will not delete that `Course` – only remove their pointer to the `Course` in question (and definitely also handle the required backlink maintenance).

One idea to help us with overall link maintenance is to consider **reference counting**. The purpose of reference counting is to keep track of how many pointers may be pointing to a given instance. For example, if other objects point to a given instance, that instance should not be deleted. Otherwise, the pointers in the other object will point to deallocated memory, which will lead to numerous runtime errors.

Let's consider an association with multiplicity, such as the relationship between a `Student` and a `Course`. A `Student` should keep track of how many `Course` pointers are pointing to the `Student`, that is, how many `Courses` the `Student` is taking. A `Student` should not be deleted while various `Courses` point to that `Student`. Otherwise, `Courses` will point to deleted memory. One way to handle this situation is to check within the `Student` destructor whether the object (`this`) contains any non-null pointers to `Course` instances. If the object does, it then needs to call a method through each of the active `Course` instances to request links to the `Student` be removed from each such `Course`.

After each link is removed, the reference counter corresponding to the set of Course instances can be decremented.

Likewise, link maintenance should occur in the Course class in favor of Student instances. Course instances should not be deleted until all Student instances enrolled in that Course have been notified. Keeping a counter of how many Student instances point to a particular instance of a Course through reference counting is helpful. In this example, it is as simple as maintaining a variable to reflect the current number of Student instances enrolled in the Course.

We can meticulously conduct link maintenance ourselves, or we may choose to use smart pointers to manage the lifetime of an associated object. **Smart pointers** can be found in the C++ Standard Library. They encapsulate a pointer (that is, wrap a pointer within a class) to add smart features, including reference counting and memory management. Because smart pointers utilize templates, which we will not cover until *Chapter 13, Working with Templates*, we will just mention their potential utility here.

We have now seen the importance of backlink maintenance and the utility of reference counting to fully support associations and their successful implementation. Let's now briefly recap the OO concepts we've covered in this chapter – association, aggregation, and composition – before moving onward to our next chapter.

Summary

In this chapter, we have pressed forward with our pursuit of object-oriented programming by exploring various object relationships – association, aggregation, and composition. We have understood the various OO design concepts representing these relationships and have seen that C++ does not offer direct language support through keywords or specific language features to implement these concepts.

Nonetheless, we have learned several techniques for implementing these core OO relationships, such as embedded objects for composition and generalized aggregation, or using pointers to implement association. We have looked at the typical longevity of object existence with these relationships; for example, with aggregation, by creating and destroying its inner part (through an embedded object, or more rarely, by allocating and deallocating a pointer member). Or through the independent existence of associated objects that neither create nor destroy one another. We have also looked under the hood at the housekeeping required to implement association, particularly associations with multiplicity, by examining backlink maintenance and reference counting.

We have added key features to our OOP skills through understanding how to implement association, aggregation, and composition. We have seen examples of how these relationships may even be more prolific in OO designs than inheritance. By mastering these skills, we have completed our core skillset of implementing essential OO concepts in C++.

We are now ready to continue to *Chapter 11, Handling Exceptions*, which will begin our quest to expand our C++ programming repertoire. Let's continue forward!

Questions

1. Add an additional `Student` constructor to the `University`/`Student` example in this chapter to accept the `University` constructor argument by reference, rather than by pointer. For example, in addition to the constructor with the signature `Student::Student(const string &fn, const string &ln, char mi, const string &t, float avg, const string &course, const string &id, University *univ);`, overload this function with a similar one, but with `University &univ` as the last parameter. How does this change the implicit call to this constructor?

    Hint: within your overloaded constructor, you will now need to take the address-of (`&`) the `University` reference parameter to set the association (which is stored as a pointer). You may need to switch to object notation (`.`) to set the backlink (if you use parameter `univ`, versus data member `this->univ`).

2. Write a C++ program to implement a many-to-many association between objects of type `Course` and of type `Student`. You may choose to build on your previous programs that encapsulate `Student`. The many-to-many relationship should work as follows:

    a. A given `Student` may take zero to many `Courses`, and a given `Course` will associate to many `Student` instances. Encapsulate the `Course` class to minimally contain a course name, a set of pointers to associated `Student` instances, and a reference count to keep track of the number of `Student` instances that are in the `Course` (this will equate to how many `Student` instances point to a given instance of a `Course`). Add the appropriate interface to reasonably encapsulate this class.

    b. Add to your `Student` class a set of pointers to the `Course` instances in which that `Student` is enrolled. Additionally, keep track of how many `Course` instances a given `Student` is currently enrolled. Add appropriate member functions to support this new functionality.

    c. Model your many-sided associations using either a linked list of pointers (that is, the data part is a pointer to the associated object) or as an array of pointers to the associated objects. Note that an array will enforce a limit on the number of associated objects you can have, however, this may be reasonable because a given `Course` can only accommodate a maximum number of `Students` and a `Student` may only enroll up to a maximum number of `Courses` per semester. If you choose the array of pointers approach, make sure your implementation includes error checking to accommodate exceeding the maximum number of associated objects in each array.

    d. Be sure to check for simple errors, such as trying to add `Students` to a `Course` that is full, or adding too many `Courses` to a `Student`'s schedule (assume there is an upper bound to five courses per semester).

    e. Make sure your destructors do not delete the associated instances.

f.   Introduce at least three Student objects, each of which takes two or more Courses. Additionally, make sure each Course has multiple Students enrolled. Print each Student, including each Course in which they are enrolled. Likewise, print each Course, showing each Student enrolled in the Course.

3.   (Optional) Enhance your program in *Exercise 2* to gain experience with backlink maintenance and reference counting as follows:

a.   Implement a DropCourse() interface for Student. That is, create a Student:: DropCourse(Course *) method in Student. Here, find the Course the Student wishes to drop in their course list, but before removing the Course, call a method on that Course to remove the aforementioned Student (that is, this) from the Course. Hint: you can make a Course::RemoveStudent(Student *)) method to help with backlink removal.

b.   Now, fully implement proper destructors. When a Course is destructed, have the Course destructor first tell each remaining associated Student to remove their link to that Course. Likewise, when a Student is destructed, loop through the Student's course list to ask those Courses to remove the aforementioned Student (that is, this) from their student list. You may find reference counting in each class (that is, by checking numStudents or numCourses) helpful to see whether these tasks must be engaged.

Part 3: Expanding Your C++ Programming Repertoire

The goal of this part is to expand your C++ programming skills, beyond the OOP skills, to encompass other critical features of C++.

The initial chapter in this section explores exception handling in C++ through understanding the mechanisms of `try`, `throw`, and `catch`, and through examining many examples to explore exception mechanics by delving into various exception handling scenarios. Additionally, this chapter expands exception class hierarchies with new exception classes.

The next chapter digs into the topics of the proper usage of friend functions and friend classes, as well as operator overloading (which may sometimes require friends), to make operations polymorphic between built-in and user defined types.

The subsequent chapter explores using C++ templates to help make code generic and usable for a variety of data types using template functions and template classes. Additionally, this chapter explains how operator overloading will assist in making template code extensible for virtually any data type.

In the next chapter, the Standard Template Library in C++ is introduced, and core STL containers such as `list`, `iterator`, `deque`, `stack`, `queue`, `priority_queue`, and `map` (including one using a functor) are examined. Additionally, STL algorithms and functors are introduced.

The final chapter in this section surveys testing OO programs and components by exploring canonical class form, creating drivers for component testing, testing classes related through inheritance, association, aggregation, and testing exception-handling mechanisms.

This part comprises the following chapters:

- *Chapter 11, Handling Exceptions*
- *Chapter 12, Friends and Operator Overloading*
- *Chapter 13, Working with Templates*
- *Chapter 14, Understanding STL Basics*
- *Chapter 15, Testing Classes and Components*

11
Handling Exceptions

This chapter will begin our quest to expand your C++ programming repertoire beyond OOP concepts, with the goal of enabling you to write more robust and extensible code. We will begin this endeavor by exploring exception handling in C++. Adding language-prescribed methods in our code to handle errors will allow us to achieve less buggy and more reliable programs. By using the formal exception handling mechanisms *built into the language*, we can achieve a uniform handling of errors, which leads to more easily maintainable code.

In this chapter, we will cover the following main topics:

- Understanding exception handling basics – `try`, `throw`, and `catch`

- Exploring exception handling mechanics – trying code that may raise exceptions, raising (throwing), catching, and handling exceptions using several variations

- Utilizing exception hierarchies with standard exception objects or by creating customized exception classes

By the end of this chapter, you will understand how to utilize exception handling in C++. You will see how to identify an error to raise an exception, transfer control of the program to a designated area by throwing an exception, and then handle the error by catching the exception and hopefully repairing the problem at hand.

You will also learn how to utilize standard exceptions from the C++ Standard Library, as well as create customized exception objects. A hierarchy of exception classes can be designed to add robust error detection and handling capabilities.

Let's increase our understanding of C++ by expanding our programming repertoire by exploring the built-in language mechanisms of exception handling.

Technical requirements

Online code for full program examples can be found in the following GitHub URL: `https://github.com/PacktPublishing/Deciphering-Object-Oriented-Programming-with-CPP/tree/main/Chapter11`. Each full program example can be found in the GitHub under the appropriate chapter heading (subdirectory) in a file that corresponds to the chapter number, followed by a dash, followed by the example number in the chapter at hand. For example, the first full program in this chapter can be found in the subdirectory `Chapter11` in a file named `Chp11-Ex1.cpp` under the aforementioned GitHub directory.

The CiA video for this chapter can be viewed at: `https://bit.ly/3QZi638`.

Understanding exception handling

Error conditions may occur within an application that would prevent a program from continuing correctly. Such error conditions may include data values that exceed application limits, necessary input files or databases have become unavailable, heap memory has become exhausted, or any other imaginable issue. C++ exceptions provide a uniform, language-supported manner for handling program anomalies.

Prior to the introduction of language supported exception handling mechanisms, each programmer would handle errors in their own manner, and sometimes not at all. Program errors and exceptions that are not handled imply that somewhere further in the application, an unexpected result will occur and the application will most often terminate abnormally. These potential outcomes are certainly undesirable!

C++ **exception handling** provides a language supported mechanism to detect and correct program anomalies so that an application can remain running, rather than ending abruptly.

Let's take a look at the mechanics, starting with the language supported keywords `try`, `throw`, and `catch`, which comprise exception handling in C++.

Utilizing exception handling with try, throw, and catch

Exception handling detects a program anomaly, as defined by the programmer or by a class library, and passes control to another portion of the application where the specific problem may be handled. Only as a last resort will it be necessary to exit an application.

Let's begin by taking a look at the keywords that support exception handling. The keywords are as follows:

- `try`: Allows programmers to *try* a portion of code that might cause an exception.
- `throw`: Once an error is found, `throw` raises the exception. This will cause a jump to the catch block below the associated try block; `throw` will allow an argument to be returned to the associated catch block. The argument thrown may be of any standard or user defined type.

- catch: Designates a block of code designed to seek exceptions that have been thrown, to attempt to correct the situation. Each catch block in the same scope will handle an exception of a different type.

When utilizing exception handling, it is useful to review the idea of backtracking. When a sequence of functions is called, we build up, on the stack, state information applicable to each successive function call (parameters, local variables, and return value space), as well as the return address for each function. When an exception is thrown, we may need to unwind the stack to the point of origination where this sequence of function calls (or try blocks) began, resetting the stack pointer as well. This process is known as **backtracking** and allows a program to return to an earlier sequence in the code. Backtracking applies not only to function calls but to nested blocks including nested try blocks.

Here is a simple example to illustrate basic exception handling syntax and usage. Though portions of the code are not shown to save space, the complete example can be found in our GitHub as follows:

https://github.com/PacktPublishing/Deciphering-Object-Orient-ed-Programming-with-CPP/blob/main/Chapter11/Chp11-Ex1.cpp

```
// Assume Student class is as seen before, but with one
// additional virtual mbr function. Assume usual headers.

void Student::Validate() // defined as virtual in class def
{                        // so derived classes may override
    // check constructed Student; see if standards are met
    // if not, throw an exception
    throw string("Does not meet prerequisites");
}


int main()
{
    Student s1("Sara", "Lin", 'B', "Dr.", 3.9,
               "C++", "23PSU");
    try       // Let's 'try' this block of code --
    {         // Validate() may raise an exception
        s1.Validate(); // does s1 meet admission standards?
    }
    catch (const string &err)
    {
        cout << err << endl;
        // try to fix problem here…
```

```
        exit(1); // only if you can't fix error,
    }               // exit as gracefully as possible
    cout << "Moving onward with remainder of code.";
    cout << endl;
    return 0;
}
```

In the previous code fragment, we can see the keywords `try`, `throw`, and `catch` in action. First, let's notice the `Student::Validate()` member function. Imagine, in this virtual method, we verify that a `Student` meets admission standards. If so, the function ends normally. If not, an exception is thrown. In this example, a simple `string` is thrown encapsulating the message `"Does not meet prerequisites"`.

In our `main()` function, we first instantiate a `Student`, namely `s1`. Then, we nest our call to `s1.Validate()` within a try block. We are literally saying that we'd like to *try* this block of code. Should `Student::Validate()` work as expected, error-free, our program completes the try block, skips the catch block(s) below the try block, and merely continues with the code below any catch blocks.

However, should `Student::Validate()` throw an exception, we will skip any remaining code in our try block and seek an exception matching the type of `const string &` in a subsequently defined catch block. Here, in the matching catch block, our goal is to correct the error if at all possible. If we are successful, our program will continue with the code below the catcher. If not, our job is to end the program gracefully.

Let's look at the output for the aforementioned program:

Student does not meet prerequisites

Next, let us summarize the overall flow of exception handling with the following logistics:

- When a program completes a try block without encountering any thrown exceptions, the code sequence continues with the statement following the catch block. Multiple catch blocks (with different argument types) may follow a try block.

- When an exception is thrown, the program must backtrack and return to the try block containing the originating function call. The program may have to backtrack past multiple functions. When backtracking occurs, the objects encountered on the stack will be popped off, and hence destructed.

- Once a program (with an exception raised) backtracks to the function where the try block was executed, the program will continue with the catch block (following the try block) whose signature matches the type of the exception that was thrown.

- Type conversion (with the exception of upcasting objects related through public inheritance) will not be done to match potential catch blocks. However, a catch block with ellipses (...) may be used as the most general type of catch block and can catch any type of exception.

- If a matching catch block does not exist, the program will call `terminate()` from the C++ Standard Library. Note that `terminate()` will call `abort()`; however, the programmer may instead register another function for `terminate()` to call via the `set_terminate()` function.

Now, let's see how to register a function with `set_terminate()`. Though we only show key portions of the code here, the complete program can be found in our GitHub:

```
https://github.com/PacktPublishing/Deciphering-Object-Orient-
ed-Programming-with-CPP/blob/main/Chapter11/Chp11-Ex2.cpp
```

```cpp
void AppSpecificTerminate()
{   // first, do what's necessary to end program gracefully
    cout << "Uncaught exception. Program terminating";
    cout << endl;
    exit(1);
}

int main()
{
    set_terminate(AppSpecificTerminate);   // register fn.
    return 0;
}
```

In the previous code fragment, we define our own `AppSpecificTerminate()` function. This is the function we wish to have the `terminate()` function call rather than its default behavior of calling `abort()`. Perhaps we use `AppSpecificTerminate()` to end our application a bit more gracefully, saving key data structures or database values. Of course, we would also then `exit()` (or `abort()`) ourselves.

In `main()`, we merely call `set_terminate(AppSpecificTerminate)` to register our terminate function with `set_terminate()`. Now, when `abort()` would otherwise be called, our function will be called instead.

It is interesting to note that `set_terminate()` returns a function pointer to the previously installed `terminate_handler` (which upon its first call will be a pointer to `abort()`). Should we choose to save this value, we can use it to reinstate previously registered terminate handlers. Notice that we have not opted to save this function pointer in this example.

Here is what the output would look like for an uncaught exception using the aforementioned code:

```
Uncaught exception. Program terminating
```

Keep in mind that functions such as `terminate()`, `abort()`, and `set_terminate()` are from the Standard Library. Though we may precede their names with the library name using the scope resolution operator, such as `std::terminate()`, this is not necessary.

> **Note**
> Exception handling is not meant to take the place of simple programmer error checking; exception handling has greater overhead. Exception handling should be reserved to handle more severe programmatic errors in a uniform manner and in a common location.

Now that we have seen the basic mechanics for exception handling, let's take a look at slightly more complex exception handling examples.

Exploring exception handling mechanics with typical variations

Exception handling can be more sophisticated and flexible than the basic mechanics previously illustrated. Let's take a look at various combinations and variations of exception handling basics, as each may be applicable to different programming situations.

Passing exceptions to outer handlers

Caught exceptions may be passed up to outer handlers for processing. Alternatively, exceptions may be partially handled and then thrown to outer scopes for further handling.

Let's build on our previous example to demonstrate this principle. The full program can be seen in the following GitHub directory:

https://github.com/PacktPublishing/Deciphering-Object-Oriented-Programming-with-CPP/blob/main/Chapter11/Chp11-Ex3.cpp

```cpp
// Assume Student class is as seen before, but with
// two additional member fns. Assume usual header files.

void Student::Validate()  // defined as virtual in class def
{                         // so derived classes may override
    // check constructed student; see if standards are met
    // if not, throw an exception
    throw string("Does not meet prerequisites");
}
```

```cpp
bool Student::TakePrerequisites()
{
    // Assume this function can correct the issue at hand
    // if not, it returns false
    return false;
}

int main()
{
    Student s1("Alex", "Ren", 'Z', "Dr.", 3.9,
               "C++", "89CU");
    try    // illustrates a nested try block
    {
        // Assume another important task occurred in this
        // scope, which may have also raised an exception
        try
        {
            s1.Validate();  // may raise an exception
        }
        catch (const string &err)
        {
            cout << err << endl;
            // try to correct (or partially handle) error.
            // If you cannot, pass exception to outer scope
            if (!s1.TakePrerequisites())
                throw;      // re-throw the exception
        }
    }
    catch (const string &err) // outer scope catcher
    {                         // (e.g. handler)
        cout << err << endl;
        // try to fix problem here…
        exit(1); // only if you can't fix, exit gracefully
    }
    cout << "Moving onward with remainder of code. ";
    cout << endl;
```

```
        return 0;
}
```

In the aforementioned code, let's assume that we have our usual header files included and the usual class definition for `Student` defined. We will now augment the `Student` class by adding the `Student::Validate()` method (virtual, so that it may be overridden) and the `Student::TakePrerequisites()` method (not virtual, descendants should use it as-is).

Notice that our `Student::Validate()` method throws an exception, which is merely a string literal containing a message indicating the issue at hand. We can imagine the complete implementation of the `Student::TakePrerequisites()` method verifies that the `Student` has met the appropriate prerequisites, and returns a boolean value of `true` or `false` accordingly.

In our `main()` function, we now notice a set of nested try blocks. The purpose here is to illustrate an inner try block that may call a method, such as `s1.Validate()`, which may raise an exception. Notice that the same level handler as the inner try block catches this exception. Ideally, an exception is handled at the level equal to the try block from which it originates, so let's assume that the catcher in this scope tries to do so. For example, our innermost catch block presumably tries to correct the error and tests whether the correction has been made using a call to `s1.TakePrerequisites()`.

But perhaps this catcher is only able to process the exception partially. Perhaps there is the knowledge that an outer level handler knows how to do the remaining corrections. In such cases, it is acceptable to re-throw this exception to an outer (nested) level. Our simple `throw;` statement in the innermost catch block does just this. Notice that there is a catcher at the outer level. Should the thrown exception match, type-wise, this outer level will now have the opportunity to further handle the exception and hopefully correct the problem so that the application can continue. Only if this outer catch block is unable to correct the error should the application be exited. In our example, each catcher prints out the string representing the error message; therefore, this message occurs twice in the output.

Let's look at the output for the aforementioned program:

```
Student does not meet prerequisites
Student does not meet prerequisites
```

Now that we have seen how to use nested try and catch blocks, let us move forward to see how a variety of thrown types and a variety of catch blocks can be used together.

Adding an assortment of handlers

Sometimes, a variety of exceptions may be raised from an inner scope, creating the necessity to craft handlers for a variety of data types. Exception handlers (that is, catch blocks) can receive an exception of any data type. We can minimize the number of catchers we introduce by utilizing catch blocks with base class types; we know that derived class objects (related through public inheritance) can always be upcast to their base class type. We can also use the ellipses (...) in a catch block to allow us to catch anything not previously specified.

Let's build on our initial example to illustrate assorted handlers in action. Though abbreviated, our full program example can be found in our GitHub as follows:

https://github.com/PacktPublishing/Deciphering-Object-Orient-ed-Programming-with-CPP/blob/main/Chapter11/Chp11-Ex4.cpp

```cpp
// Assume Student class is as seen before, but with one
// additional virtual member function, Graduate(). Assume
// a simple Course class exists. All headers are as usual.

void Student::Graduate()
{   // Assume the below if statements are fully implemented
    if (gpa < 2.0) // if gpa doesn't meet requirements
        throw gpa;
    // if Student is short credits, throw number missing
        throw numCreditsMissing;  // assume this is an int
    // or if Student is missing a Course, construct, then
    // throw the missing Course as a referenceable object
    // Assume appropriate Course constructor exists
        throw Course("Intro. To Programming", 1234);
    // or if another issue, throw a diagnostic message
        throw string("Does not meet requirements");
}


int main()
{
    Student s1("Ling", "Mau", 'I', "Ms.", 3.1,
               "C++", "55UD");
    try
    {
        s1.Graduate();
    }
    catch (float err)
    {
        cout << "Too low gpa: " << err << endl;
        exit(1); // only if you can't fix, exit gracefully
    }
```

```cpp
    catch (int err)
    {
        cout << "Missing " << err << " credits" << endl;
        exit(2);
    }
    catch (const Course &err)
    {
        cout << "Need to take: " << err.GetTitle() << endl;
        cout << "Course #: " << err.GetCourseNum() << endl;
        // Ideally, correct the error, and continue program
        exit(3); // Otherwise, exit, gracefully if possible
    }
    catch (const string &err)
    {
        cout << err << endl;
        exit(4);
    }
    catch (...)
    {
        cout << "Exiting" << endl;
        exit(5);
    }
    cout << "Moving onward with remainder of code.";
    cout << endl;
    return 0;
}
```

In the aforementioned segment of code, we first examine the Student::Graduate() member function. Here, we can imagine that this method runs through many graduation requirements, and as such, can potentially raise a variety of different types of exceptions. For example, should the Student instance have too low of a gpa, a float is thrown as the exception, indicating the student's poor gpa. Should the Student have too few credits, an integer is thrown, indicating how many credits the Student still needs to earn their degree.

Perhaps the most interesting potential error that Student::Graduate() might raise would be if a required Course is missing from a student's graduation requirements. In this scenario, Student::Graduate() would instantiate a new Course object, filling it with the Course name and number via construction. This anonymous object would then be thrown from Student::Graduate(), much as an anonymous `string` object may be alternatively thrown in this method. The handler may then catch the `Course` (or `string`) object by reference.

In the `main()` function, we merely wrap the call to `Student::Graduate()` within a try block, as this statement may raise an exception. A sequence of catchers follows the try block – one `catch` statement per type of object that may be thrown. The last catch block in this sequence uses ellipses (...), indicating that this catcher will handle any other type of exception thrown by `Student::Graduate()` that has not been caught by the other catchers.

The catch block that is actually engaged is the one in which a `Course` is caught using `const Course &err`. With the `const` qualifier, we may not modify the `Course` in the handler, so we may only apply `const` member functions to this object.

Note that though each earlier catcher shown merely prints out an error and then exits, ideally, a catcher would try to correct the error so that the application would not need to terminate, allowing code below the catch blocks to continue onward.

Let's look at the output for the aforementioned program:

```
Need to take: Intro. to Programming
Course #: 1234
```

Now that we have seen a variety of thrown types and a variety of catch blocks, let us move forward to understand what we should group together within a single try block.

Grouping together dependent items within a try block

It is important to remember that when a line of code in a try block encounters an exception, the remainder of the try block is ignored. Instead, the program continues with a matching catcher (or calls `terminate()` if no suitable catcher exists). Then, if the error is repaired, the code beyond the catcher commences. Note that we never return to complete the remainder of the initial try block. The implication of this behavior is that you should only group together elements within a try block that go together. That is, if one item causes an exception, it is no longer important to complete the other item in that grouping.

Keep in mind that the goal of a catcher is to correct an error if at all possible. This means that the program may continue forward after the applicable catch block. You may ask: Is it now acceptable that an item was skipped in the associated try block? Should the answer be no, then rewrite your code. For example, you may want to add a loop around the `try-catch` grouping such that if an error is corrected by a catcher, the whole enterprise is retried starting with the initial try block.

Alternatively, make smaller, successive `try-catch` groupings. That is, *try* only one important task in its own try block (followed by applicable catchers). Then *try* the next task in its own try block with its associated catchers and so on.

Next, let's take a look at a way to include in a function's prototype the type of exceptions it may throw.

Examining exception specifications in function prototypes

We can optionally specify the types of exceptions a C++ function may throw by extending the signature of that function to include the object types of what may be thrown. However, because a function may throw more than one type of exception (or none at all), checking which type is actually thrown must be done at runtime. For this reason, these augmented specifiers in the function prototype are also known as **dynamic exception specifications**. Though deprecated, dynamic exceptions will lay the groundwork for the noexcept specifier, which we'll see shortly. Uses of dynamic exceptions also occur in existing code bases and libraries, so let's briefly examine its usage.

Let's see an example using exception types in the extended signature of a function:

```cpp
void Student::Graduate() throw(float, int,
                               Course &, string)
{
   // this method might throw any type included in
   // its extended signature
}

void Student::Enroll() throw()
{
   // this method might throw any type of exception
}
```

In the aforementioned code fragment, we see two member functions of Student. Student::Graduate() includes the throw keyword after its parameter list and then, as part of this method's extended signature, includes the types of objects that may be thrown from this function. Notice that the Student::Enroll() method merely has an empty list following throw() in its extended signature. This means that Student::Enroll() might throw any type of exception.

In both cases, by adding the throw() keyword with optional data types to the signature, we are providing a means to announce to the user of this function what types of objects might be thrown. We are then asking programmers to include any calls to this method within a try block followed by appropriate catchers.

We will see that though the idea of an extended signature seems very helpful, it has unfavorable issues in practice. For this reason, dynamic exception specifications have been *deprecated*. Because you may still see these specifications used in existing code, including Standard Library prototypes (such as with exceptions), this deprecated feature is still supported by compilers, and you will need to understand their usage.

Though dynamic exceptions (extended function signatures as previously described) have been deprecated, a specifier with a similar purpose has been added to the language, the noexcept keyword.

This specifier can be added after the extended signature as follows:

```
void Student::Graduate() noexcept    // will not throw()
{              // same as  noexcept(true) in extended signature
}              // same as deprecated throw() in ext. signature

void Student::Enroll() noexcept(false)   // may throw()
{                                         // an exception
}
```

Nonetheless, let's investigate why unfavorable issues exist relating to dynamic exceptions by looking at what happens when our application throws exceptions that are not part of a function's extended signature.

Dealing with unexpected types of dynamic exceptions

Should an exception be thrown of a type other than that specified in the extended function prototype, unexpected(), from the C++ Standard Library, will be called. You can register your own function with unexpected(), much as we registered our own function with set_terminate() earlier in this chapter.

You can allow your AppSpecificUnexpected() function to rethrow an exception of the type that the originating function should have thrown; however, if that does not occur, terminate() will then be called. Furthermore, if no possible matching catcher exists to handle what is correctly thrown from the originating function (or rethrown by AppSpecificUnexpected()), then terminate() will be called.

Let's see how to use set_unexpected() with our own function:

```
void AppSpecificUnexpected()
{
    cout << "An unexpected type was thrown" << endl;
    // optionally re-throw the correct type, or
    // terminate() will be called.
}

int main()
{
    set_unexpected(AppSpecificUnexpected)
}
```

Registering our own function with `set_unexpected()` is very simple, as illustrated in the aforementioned code fragment.

Historically, one motivating reason for employing exception specification in a function's extended signature was to provide a documentative effect. That is, you could see which exceptions a function might possibly throw simply by examining its signature. You could then plan to enclose that function call within a try block and provide appropriate catchers to handle any potential situation.

Nonetheless, regarding dynamic exceptions, it is useful to note that compilers do not check that the types of exceptions *actually* thrown in a function body match the types specified in the function's extended signature. It is up to the programmer to ensure that they are in sync. Therefore, this deprecated feature can be error-prone and, overall, less useful than its original intention.

Though well intended, dynamic exceptions are currently unused, except in large quantities of library code such as the Standard C++ Library. Since you will inevitably utilize these libraries, it is important to understand these anachronisms.

> **Important note**
> Dynamic exception specifications (that is, the ability to specify exception types in a method's extended signature) have been *deprecated* in C++. This is because compilers are not able to validate their use, which must then be delayed until runtime. Their use, though still supported (many libraries have such specifications), is now deprecated.

Now that we have seen an assortment of exception handling detection, raising, catching, and (hopefully) correction schemes, let's take a look at how we might create a hierarchy of exception classes to add sophistication to our error handling abilities.

Utilizing exception hierarchies

Creating a class to encapsulate the details relating to a program error seems like a useful endeavor. In fact, the C++ Standard Library has created one such generic class, `exception`, to provide the basis for building an entire hierarchy of useful exception classes.

Let's take a look at the `exception` class with its Standard Library descendants, and then how we may extend `exception` with our own classes.

Using standard exception objects

The **exception** class is defined in the C++ Standard Library and is available merely by including the `<exception>` header. The `exception` class includes virtual functions with the following signatures: `virtual const char *what() const noexcept` and `virtual const char *what() const throw()`. These signatures indicate that derived classes should redefine `what()` to return a `const char *` with a description of the error at hand. The `const` keyword after what() indicates that these are const member functions; they will not change any members of the derived class. The noexcept usage in the first prototype indicates that what() is non-throwing. The throw() in the extended signature of the second prototype indicates that this function may throw any type. The usage of throw() in the second signature is a deprecated anachronism and should not be used in new code.

The std::exception class is the base class of a variety of predefined C++ exception classes, including bad_alloc, bad_cast, bad_exception, bad_function_call, bad_typeid, bad_weak_ptr, logic_error, runtime_error, and nested class ios_base::failure. Many of these derived classes have descendants themselves, adding additional standard exceptions to the predefined hierarchy of exceptions.

Should a function throw any of the aforementioned exceptions, these exceptions may be caught by either catching the base class type, exception, or by catching an individual derived class type. Depending on what course of action your handler will take, you can decide whether you wish to catch one such exception as its generalized base class type or as its specific type.

Just as the Standard Library has set up a hierarchy of classes based on the exception class, so may you. Let's next take a look at how we might do just this!

Creating customized exception classes

As a programmer, you may decide that it is advantageous to establish your own specialized exception types. Each type can pack useful information into an object detailing just what went wrong with the application. Additionally, you may be able to pack clues into the object (which will be thrown) as to how to correct the error at hand. Simply derive your class from the Standard Library exception class.

Let's take a look at how easily this may be done by examining the critical portions of our next example, which can be found as a full program in our GitHub:

https://github.com/PacktPublishing/Deciphering-Object-Oriented-Programming-with-CPP/blob/main/Chapter11/Chp11-Ex5.cpp

```
#include <iostream>
#include <exception>
// See online code for many using std:: inclusions

class StudentException: public exception
{
private:
    int errCode = 0;   // in-class init, will be over-
    // written with bonified value after successful
    // alternate constructor completion
```

```cpp
    string details;
public:
    StudentException(const string &det, int num):
                     errCode(num), details(det) { }
    // Base class destructor (exception class) is virtual.
    // Override at this level if there's work to do.
    // We can omit the default destructor prototype.
    // ~StudentException() override = default;
    const char *what() const noexcept override
    {   // overridden function from exception class
        return "Student Exception";
    }
    int GetCode() const { return errCode; }
    const string &GetDetails() const { return details; }
};

// Assume Student class is as we've seen before, but with
// one additional virtual member function, Graduate()
void Student::Graduate() // fn. may throw StudentException
{
    // if something goes wrong, construct a
    // StudentException, packing it with relevant data,
    // and then throw it as a referenceable object
    throw StudentException("Missing Credits", 4);
}

int main()
{
    Student s1("Alexandra", "Doone", 'G', "Miss", 3.95,
               "C++", "231GWU");
    try
    {
        s1.Graduate();
    }
```

```
    catch (const StudentException &e)   // catch exc. by ref
    {
        cout << e.what() << endl;
        cout << e.GetCode() << " " << e.GetDetails();
        cout << endl;
        // Grab useful info from e and try to fix problem
        // so that the program can continue.
        exit(1);   // only exit if we can't fix the problem!
    }
    return 0;
}
```

Let's take a few minutes to examine the previous segment of code. Foremost, notice that we define our own exception class, StudentException. It is a derived class from the C++ Standard Library exception class.

The StudentException class contains data members to hold an error code as well as alphanumeric details describing the error condition using data members errCode and details, respectively. We have two simple access functions, StudentException::GetCode() and StudentException::GetDetails(), to easily retrieve these values. As these methods do not modify the object, they are const member functions.

We notice that the StudentException constructor initializes the two data members – one through the member initialization list and one in the body of the constructor. We also override the virtual const char *what() const noexcept method (as introduced by the exception class) in our StudentException class to return the string of characters "Student Exception".

Next, let's examine our Student::Graduate() method. This method may throw a StudentException. If an exception must be thrown, we instantiate one, constructing it with diagnostic data, and then throw the StudentException from this function. Note that the object thrown has no local identifier in this method – there's no need, as any such local variable name would soon be popped off the stack after the throw occurred.

In our main() function, we wrap our call to s1.Graduate() within a try block, and it is followed by a catch block that accepts a reference (&) to a StudentException, which we treat as const. Here, we first call our overridden what() method and then print out the diagnostic details from within the exception, e. Ideally, we would use this information to try to correct the error at hand and only exit the application if truly necessary.

Let's look at the output for the aforementioned program:

```
Student Exception
4 Missing Credits
```

Though the most usual way to create a customized exception class is to derive a class from the Standard exception class, you may also wish to utilize a different technique, that of an embedded exception class.

Creating a nested exception class

As an alternative implementation, exception handling may be embedded into a class by adding a nested class definition in the public access region for a particular outer class. The inner class will represent the customized exception class.

Objects of nested, user defined types may be created and thrown to catchers anticipating such types. These nested classes are built into the public access region of the outer class, making them easily available for derived class usage and specialization. In general, exception classes built into an outer class must be public so that the instances of nested types thrown can be caught and handled outside the scope of the outer class (that is, in the scope where the primary, outer instance exists).

Let's take a look at this alternate implementation of an exception class by examining key segments of code, which can be found as a full program in our GitHub:

https://github.com/PacktPublishing/Deciphering-Object-Orient-ed-Programming-with-CPP/blob/main/Chapter11/Chp11-Ex6.cpp

```
// Assume Student class is as before, but with the addition
// of a nested exception class. All headers are as usual.

class Student: public Person
{
private:   // assume usual data members
public:    // usual constructors, destructor, and methods
    virtual void Graduate();
    class StudentException    // nested exception class
    {
    private:
        int number = 0;   // will be over-written after
        // successful alternate constructor
        // note: there is no default constructor
    public:
        StudentException(int num): number(num) { }
        // Remember, it is unnecessary to proto. default ~
        // ~StudentException() = default;
        int GetNum() const { return number; }
```

```
        };
    };

    void Student::Graduate()
    {    // assume we determine an error and wish to throw
         // the nested exception type
         throw StudentException(5);
    }

    int main()
    {
        Student s1("Ling", "Mau", 'I', "Ms.", 3.1,
                   "C++", "55UD");
        try
        {
            s1.Graduate();
        }
        // following is one of many catch blocks in online code
        catch (const Student::StudentException &err)
        {
            cout << "Error: " << err.GetNum() << endl;
            // If you correct error, continue the program
            exit(5);   // Otherwise, exit application
        }
        cout << "Moving onward with remainder of code.";
        cout << endl;
        return 0;
    }
```

In the previous code fragments, we expanded our Student class to include a private, nested class called StudentException. Though the class shown is overly simplified, the nested class ideally should define a means to catalog the error in question as well as collect any useful diagnostic information.

In our main() function, we instantiate a Student, namely s1. In a try block, we then call s1.Graduate();. Our Student::Graduate() method presumably checks that the Student has met graduation requirements, and if not, throws an exception of the nested class type, Student::StudentException (which will be instantiated as needed).

Notice that our corresponding catch block utilizes scope resolution to specify the inner class type for `err`, the referenced object (that is, `const Student::StudentException &err`). Though we ideally would like to correct the program error within the handler, if we cannot, we simply print a message and `exit()`.

Let's look at the output for the aforementioned program:

```
Error: 5
```

Understanding how to create our own exception class (both as a nested class or derived from `std::exception`) is useful. We may additionally wish to create a hierarchy of application-specific exceptions. Let's move ahead to see how to do so.

Creating hierarchies of user defined exception types

An application may wish to define a series of classes that support exception handling to raise specific errors, and hopefully, also provide a means to collect diagnostics for an error so that the error may be addressed in an appropriate segment of the code.

You may wish to create a subhierarchy, derived from the C++ Standard Library `exception`, of your own exception classes. Be sure to use public inheritance. When utilizing these classes, you will instantiate an object of your desired exception type (filling it with valuable diagnostic information), and then throw that object.

Also, if you create a hierarchy of exception types, your catchers can catch specific derived class types or more general base class types. The option is yours, depending on how you will plan to handle the exception. Keep in mind, however, that if you have a catcher for both the base and derived class types, place the derived class types first – otherwise your thrown object will first match to the base class type catcher without realizing that a more appropriate derived class match is available.

We have now seen both the hierarchy of C++ Standard Library exception classes, as well as how to create and utilize your own exception classes. Let's now briefly recap the exception features we've learned in this chapter, before moving forward to our next chapter.

Summary

In this chapter, we have begun expanding our C++ programming repertoire beyond OOP language features to include features that will enable us to write more robust programs. User code can inevitably be error-prone by nature; using language supported exception handling can help us achieve less buggy and more reliable code.

We have seen how to utilize the core exception handling features with `try`, `throw`, and `catch`. We've seen a variety of uses of these keywords – throwing exceptions to outer handlers, using an assortment of handlers featuring various types, and selectively grouping program elements together within a single try block, for example. We have seen how to register our own functions with `set_terminate()` and `set_unexpected()`. We have seen how to utilize the existing C++ Standard Library `exception` hierarchy. We have additionally explored defining our own exception classes to extend this hierarchy.

We have added key features to our C++ skills by exploring exception handling mechanisms. We are now ready to move forward to *Chapter 12, Friends and Operator Overloading*, so that we can continue expanding our C++ programming repertoire with useful language features that will make us better programmers. Let's move forward!

Questions

1. Add exception handling to your previous `Student` / `University` exercise from *Chapter 10, Implementing Association, Aggregation, and Composition*, as follows:

    a. Should a `Student` try to enroll in more than the MAX defined number of allowable courses per `Student`, throw a `TooFullSchedule` exception. This class may be derived from the Standard Library `exception` class.

    b. Should a `Student` try to enroll in a `Course` that is already full, have the `Course::AddStudent(Student *)` method throw a `CourseFull` exception. This class may be derived from the Standard Library `exception` class.

    c. There are many other areas in the `Student` / `University` application that could utilize exception handling. Decide which areas should employ simple error checking and which are worthy of exception handling.

12
Friends and Operator Overloading

This chapter will continue our pursuit of expanding your C++ programming repertoire beyond OOP concepts, with the goal of writing more extensible code. We will next explore **friend functions**, **friend classes**, and **operator overloading** in C++. We will understand how operator overloading can extend operators beyond their usage with standard types to behave uniformly with user defined types, and why this is a powerful OOP tool. We will learn how friend functions and classes can be safely used to achieve this goal.

In this chapter, we will cover the following main topics:

- Understanding friend functions and friend classes, appropriate reasons to utilize, and measures to add safety to their usage

- Learning about operator overloading essentials – how and why to overload operators, ensuring operators are polymorphic between standard and user defined types

- Implementing operator functions and knowing when friends may be necessary

By the end of this chapter, you will unlock the proper usage of friends and understand their utility in harnessing C++'s ability to overload operators. Though the usage of friend functions and classes can be exploited, you will instead insist on their contained usage only within two tightly coupled classes. You will understand how the proper usage of friends can enhance operator overloading, allowing operators to be extended to support user defined types so they may work associatively with their operands.

Let's increase our understanding of C++ by expanding your programming repertoire through exploring friend functions, friend classes, and operator overloading.

Technical requirements

Online code for full program examples can be found in the following GitHub URL: `https://github.com/PacktPublishing/Deciphering-Object-Oriented-Programming-with-CPP/tree/main/Chapter12`. Each full program example can be found in the GitHub repository under the appropriate chapter heading (subdirectory) in a file that corresponds to the chapter number, followed by a dash, followed by the example number in the chapter at hand. For example, the first full program in this chapter can be found in the subdirectory `Chapter12` in a file named `Chp12-Ex1.cpp` under the aforementioned GitHub directory.

The CiA video for this chapter can be viewed at: `https://bit.ly/3K0f4tb`.

Understanding friend classes and friend functions

Encapsulation is a valuable OOP feature that C++ offers through the proper usage of classes and access regions. Encapsulation offers uniformity in the manner in which data and behaviors are manipulated. In general, it is unwise to forfeit the encapsulated protection that a class offers.

There are, however, selected programming situations in which breaking encapsulation slightly is considered more acceptable than the alternative of providing an *overly public* interface to a class. That is, when a class needs to provide methods for two classes to cooperate, yet, in general, those methods are inappropriate to be publicly accessible.

Let's consider a scenario that may lead us to consider slightly forfeiting (that is, breaking) the sacred OOP concept of encapsulation:

- Two tightly coupled classes may exist that are not otherwise related to one another. One class may have one or more associations with the other class and need to manipulate the other class's members. Yet, a public interface to allow access to such members would make these internals *overly public* and subject to manipulation well beyond the needs of the pair of tightly coupled classes.

- In this situation, it is a better choice to allow one class in the tightly coupled pair to have access to the other class's members versus providing a public interface in the other class that allows for more manipulation of these members than is generally safe. We will see, momentarily, how to minimize this prospective loss of encapsulation.

- Selected operator overloading situations, which we will soon see, may require an instance to have access to its members while in a function that is outside of its class scope. Again, a fully accessible public interface may be considered dangerous.

Friend functions and **friend classes** allow this selective breaking of encapsulation to occur. Breaking encapsulation is serious and should not be done to simply override access regions. Instead, friends can be used – with added safety measures – when the choices are slightly breaking encapsulation between two tightly coupled classes or providing an overly public interface that would yield greater and potentially unwanted access to another class's members from various scopes in the application.

Let us take a look at how each may be used, and then we will add the relevant safety measures we should insist on employing. Let's start with friend functions and friend classes.

Using friend functions and friend classes

Friend functions are functions that are individually granted *extended scope* to include the class with which they are associated. Let's examine the implications and logistics:

- In the scope of friend functions, an instance of the associated type can access its own members as if it were in the scope of its own class.
- A friend function needs to be prototyped as a friend in the class definition of the class relinquishing access (that is, extending its scope).
- The keyword `friend` is used in front of the prototype that provides access.
- Functions overloading friend functions are not considered friends.

Friend classes are classes in which every member function of the class is a friend function of the associated class. Let's examine the logistics:

- A friend class should have a forward declaration in the class definition of the class that is providing it with access to its members (that is, scope).
- The keyword `friend` should precede the forward declaration of the class gaining access (that is, whose scope has been extended).

> **Important note**
> Friend classes and friend functions should be utilized sparingly, only when breaking encapsulation selectively and slightly is it a better choice than offering an *overly public* interface (that is, a public interface that would universally offer undesired access to selected members within any scope of the application).

Let's begin by examining the syntax for friend classes and friend function declarations. The following classes do not represent complete class definitions; however, the complete program can be found in our online in our GitHub repository as follows:

https://github.com/PacktPublishing/Deciphering-Object-Oriented-Programming-with-CPP/blob/main/Chapter12/Chp12-Ex1.cpp

```
class Student;  // forward declaration of Student class

class Id  // Partial class - full class can be found online
{
```

```
private:
    string idNumber;
    Student *student = nullptr;   // in-class initialization
public:   // Assume constructors, destructor, etc. exist
    void SetStudent(Student *);
    // all member fns. of Student are friend fns to/of Id
    friend class Student;
};


// Note: Person class is as often defined; see online code

class Student : public Person
{
private:
    float gpa = 0.0;     // in-class initialization
    string currentCourse;
    static int numStudents;
    Id *studentId = nullptr;
public:    // Assume constructors, destructor, etc. exist
    // only the following mbr fn. of Id is a friend fn.
    friend void Id::SetStudent(Student *); // to/of Student
};
```

In the preceding code fragment, we first notice a friend class definition within the Id class. The statement friend class Student; indicates that all member functions in Student are friend functions to Id. This all-inclusive statement is in lieu of naming every function of the Student class as a friend function of Id.

Also, in the Student class, notice the declaration of friend void Id::SetStudent(Student *);. This friend function declaration indicates that only this specific member function of Id is a friend function of Student.

The implication of the friend function prototype friend void Id::SetStudent(Student *); is that if a Student finds itself in the scope of the Id::SetStudent() method, that Student may manipulate its own members as though it is in its own scope, namely that of Student. You may ask, which Student may find itself in the scope of Id::SetStudent(Student *)? That's easy, it is the one passed to the method as an input parameter. The result is that the input parameter of type Student * in the Id::SetStudent() method may access its own private and protected members as if the Student instance were in its own class scope – it is in the scope of a friend function.

Similarly, the implication of the friend class forward declaration: `friend class Student;` found in the `Id` class is that if any `Id` instance finds itself in a `Student` method, that `Id` instance can access its own private or protected methods as if it were in its own class. The `Id` instance may be in any member function of its friend class, `Student`; it is as though those methods have been augmented to also have the scope of the `Id` class.

Notice that the class giving up access – that is, the class widening scope – is the one to announce friendship. That is, the `friend class Student;` statement in `Id` says: If any `Id` happens to be in any member function of `Student`, allow that `Id` to have full access to its members as if it is in its own scope. Likewise, the friend function statement in `Student` indicates that if a `Student` instance is found (via the input parameter) in this particular method of `Id`, it may have access to its elements fully, as though it were in a member function of its own class. Think in terms of friendship as a means of augmenting scope.

Now that we have seen the basic mechanics of friend functions and friend classes, let's employ a simple contract to make it a bit more appealing to selectively break encapsulation.

Making access safer when using friends

We have seen that two tightly coupled classes, such as those related through an association, may need to extend their scope somewhat to selectively include one another through the use of **friend functions** or **friend classes**. The alternative is offering a public interface to select elements of each class. However, consider that you may not want the public interface to those elements to be uniformly accessible to be used in any scope of the application. You are truly facing a tough choice: utilize friends or provide an *overly public* interface.

Though it may make you initially cringe to utilize friends, it may be safer than the alternative of providing an undesired public interface to class elements.

To lessen the panic that you feel towards the selective breaking of encapsulation that friends allow, consider adding the following contract to your usage of friends:

- When utilizing friends, to lessen the loss of encapsulation, one class can provide private access methods to the other class' data members. Consider making these methods inline for efficiency, as they are simple access methods (typically single line methods not likely to add software bloat through their expansion).

- The instance in question should agree to only utilize the private access methods created to appropriately access its desired members while in the scope of the friend function (even though it could actually unrestrictedly access any data or methods of its own type in the scope of the friend function). This informal understanding is, of course, a gentleman's agreement, and not language imposed.

Here is a simple example to illustrate two tightly coupled classes appropriately using a **friend class**. Though the main() function and several methods are not shown to save space, the complete example can be found in our GitHub repository as follows:

https://github.com/PacktPublishing/Deciphering-Object-Orient-ed-Programming-with-CPP/blob/main/Chapter12/Chp12-Ex2.cpp

```cpp
using Item = int;
class LinkList;   // forward declaration

class LinkListElement
{
private:
    void *data = nullptr;    // in-class initialization
    LinkListElement *next = nullptr;
    // private access methods to be used in scope of friend
    void *GetData() const { return data; }
    LinkListElement *GetNext() const { return next; }
    void SetNext(LinkListElement *e) { next = e; }
public:
    // All member functions of LinkList are friend
    // functions of LinkListElement
    friend class LinkList;
    LinkListElement() = default;
    LinkListElement(Item *i): data(i), next(nullptr) { }
    ~LinkListElement() { delete static_cast<Item *>(data);
                         next = nullptr; }
};


// LinkList should only be extended as a protected/private
// base class; it does not contain a virtual destructor. It
// can be used as-is, or as implementation for another ADT.
class LinkList
{
private:
    LinkListElement *head = nullptr, *tail = nullptr,
                    *current = nullptr;  // in-class init.
```

```
public:
    LinkList() = default;
    LinkList(LinkListElement *e)
        { head = tail = current = e; }
    void InsertAtFront(Item *);
    LinkListElement *RemoveAtFront();
    void DeleteAtFront()   { delete RemoveAtFront(); }
    bool IsEmpty() const { return head == nullptr; }
    void Print() const;      // see online definition
    ~LinkList() { while (!IsEmpty()) DeleteAtFront(); }
};
```

Let's examine the preceding class definitions for `LinkListElement` and `LinkList`. Notice that in the `LinkListElement` class, we have three private member functions: `void *GetData();`, `LinkListElement *GetNext();`, and `void SetNext(LinkListElement *);`. These three member functions should not be part of the public class interface. It is only appropriate for these methods to be used within the scope of `LinkList`, a class that is tightly coupled with `LinkListElement`.

Next, notice the `friend class LinkList;` forward declaration in the `LinkListElement` class. This declaration means that all member functions of `LinkList` are friend functions of `LinkListElement`. As a result, any `LinkListElement` instances that find themselves in `LinkList` methods may access their own aforementioned private `GetData()`, `GetNext()`, and `SetNext()` methods simply because they will be in the scope of a friend class.

Next, let's take a look at the `LinkList` class in the preceding code. The class definition itself does not have any unique declarations with respect to friendship. After all, it is the `LinkListElement` class that has widened its scope to include methods of the `LinkedList` class, not the other way around.

Now, let's take a look at two selected member functions of the `LinkList` class. The full complement of these methods may be found online, at the previously mentioned URL:

```
void LinkList::InsertAtFront(Item *theItem)
{
    LinkListElement *newHead = new LinkListElement(theItem);
    // Note: temp can access private SetNext() as if it were
    // in its own scope - it is in the scope of a friend fn.
    newHead->SetNext(head);// same as: newHead->next = head;
    head = newHead;
}
```

```
LinkListElement *LinkList::RemoveAtFront()
{
    LinkListElement *remove = head;
    head = head->GetNext();   // head = head->next;
    current = head;      // reset current for usage elsewhere
    return remove;
}
```

As we examine the aforementioned code, we can see that in a sampling of `LinkList` methods, a `LinkListElement` can call private methods on itself because it is in the scope of a friend function (which is essentially its own scope, widened). For example, in `LinkList::InsertAtFront()`, `LinkListElement *temp` sets its next member to head using `temp->SetNext(head)`. Certainly, we could have also directly accessed the private data member here using `temp->next = head;`. However, we maintained a modicum of encapsulation by `LinkListElement` providing private access functions, such as `SetNext()`, and asking `LinkList` methods (friend functions) to have `temp` utilize private method `SetNext()`, rather than just directly manipulating the data member itself.

Because `GetData()`, `GetNext()`, and `SetNext()` in `LinkListElement` are inline functions, we do not forfeit performance by providing a sense of encapsulated access to members `data` and `next`.

We can similarly see that other member functions of `LinkList`, such as `RemoveAtFront()` (and `Print()` which appears in the online code) have `LinkListElement` instances utilizing its private access methods, rather than allowing the `LinkListElement` instances to grab their private `data` and `next` members directly.

`LinkListElement` and `LinkList` are iconic examples of two tightly coupled classes in which it may be better to extend one class to include the other's scope for access, rather than providing an *overly public* interface. After all, we wouldn't want users in `main()` to get their hands on a `LinkListElement` and apply `SetNext()`, for example, which could change an entire `LinkedList` without the `LinkList` class' knowledge.

Now that we have seen the mechanics as well as suggested usage for friend functions and classes, let's explore another language feature that may need to utilize friends – that of operator overloading.

Deciphering operator overloading essentials

C++ has a variety of operators in the language. C++ allows most operators to be redefined to include usage with user defined types; this is known as **operator overloading**. In this way, user defined types may utilize the same notation as standard types to perform these well-understood operations. We can view an overloaded operator as polymorphic in that its same form can be used with a variety of types – standard and user defined.

Not all operators may be overloaded in C++. The following operators cannot be overloaded: the member access operator (.), the ternary conditional operator (? :), the scope resolution operator (::), the pointer-to-member operator (.*), the sizeof() operator, and the typeid() operator. All the rest may be overloaded, provided at least one operand is a user defined type.

When overloading an operator, it is important to promote the same meaning that the operator has for standard types. For example, the extraction operator (<<) is defined when used in conjunction with cout to print to standard output. This operator can be applied to various standard types, such as integers, floating-point numbers, character strings, and so on. Should the extraction operator (<<) be overloaded for a user defined type, such as Student, it should also mean to print to standard output. In this fashion, operator << is polymorphic when used in the context of an output buffer, such as cout; that is, it has the same meaning but different implementation for all types.

It is important to note that when overloading an operator in C++, we may not change the predefined precedence of the operators as they occur in the language. This makes sense – we are not rewriting the compiler to parse and interpret expressions differently. We are merely extending the meaning of an operator from its usage with standard types to include usage with user defined types. Operator precedence will remain unchanged.

An **operator function** is utilized to redefine or *overload* an operator in C++. The name of the function is simply the name operator, followed by the symbol representing the operator which you wish to overload.

Let's take a look at the simple syntax of an operator function prototype:

```
Student &operator+(float gpa, const Student &s);
```

Here, we intend to provide a means to add a floating-point number and a Student instance using the C++ addition operator (+). The meaning of this addition might be to average the new floating-point number with the student's existing grade point average. Here, the name of the operator function is operator+().

In the aforementioned prototype, the operator function is not a member function of any class. The left expected operand will be a float and the right operand will be a Student. The return type of the function (Student &) allows us to cascade the use of + with multiple operands or be paired with multiple operators, such as s1 = 3.45 + s2;. The overall concept is that we can define how to use + with multiple types, provided at least one operand is a user defined type.

There's actually a lot more involved than the simple syntax shown in the previous prototype. Before we fully examine a detailed example, let's first take a look at more logistics relating to implementing operator functions.

Implementing operator functions and knowing when friends might be necessary

An **operator function**, the mechanism to overload an operator, may be implemented as a member function or as a regular, external function. Let's summarize the mechanics of implementing operator functions with the following key points:

- Operator functions that are implemented as member functions will receive an implicit argument (the `this` pointer), plus, at most, one explicit argument. If the left operand in the overloaded operation is a user defined type in which modifications to the class can easily be made, implementing the operator function as a member function is reasonable and preferred.

- Operator functions that are implemented as external functions will receive one or two explicit arguments. If the left operand in the overloaded operation is a standard type or a class type that is not modifiable, then an external (non-member) function must be used to overload this operator. This external function may need to be a `friend` of any object type that is used as the right-hand function argument.

- Operator functions should most often be implemented reciprocally. That is, when overloading a binary operator, ensure that it has been defined to work no matter what order the data types (should they differ) appear in the operator.

Let's take a look at a full program example to illustrate the mechanics of operator overloading, including member and non-member functions, as well as scenarios requiring the usage of friends. Though some well-known portions of the program have been excluded to save space, the full program example can be found online in our GitHub repository:

https://github.com/PacktPublishing/Deciphering-Object-Orient-ed-Programming-with-CPP/blob/main/Chapter12/Chp12-Ex3.cpp

```
// Assume usual header files and std namespace inclusions
class Person
{
private:
    string firstName, lastname;
    char middleInitial = '\0';
    char *title = nullptr; // use ptr member to demonstrate
                           // deep assignment
protected:
    void ModifyTitle(const string &); // converts to char *
public:
    Person() = default;    // default constructor
```

```
    Person(const string &, const string &, char,
        const char *);
    Person(const Person &);  // copy constructor
    virtual ~Person();  // virtual destructor
    const string &GetFirstName() const
        { return firstName; }
    const string &GetLastName() const { return lastName; }
    const char *GetTitle() const { return title; }
    char GetMiddleInitial() const { return middleInitial; }
    virtual void Print() const;
    virtual void IsA() const;
    // overloaded operator functions
    Person &operator=(const Person &); // overloaded assign
    bool operator==(const Person &);   // overloaded
                                       // comparison
    Person &operator+(const string &); // overloaded plus
    // non-mbr friend fn. for op+ (to make associative)
    friend Person &operator+(const string &, Person &);
};
```

Let's begin our code examination by first looking at the preceding class definition for `Person`. In addition to the class elements that we are accustomed to seeing, we have four operator functions prototyped: `operator=()`, `operator==()`, and `operator+()` (which is implemented twice so that the operands to + can be reversed).

Functions for `operator=()`, `operator==()`, and one version of `operator+()` will be implemented as member functions of this class, whereas the other `operator+()`, with `const char *` and `Person` parameters, will be implemented as a non-member function and will additionally necessitate the use of a friend function.

Overloading the assignment operator

Let's move forward to examine the applicable operator function definitions for this class, starting by overloading the assignment operator:

```
// Assume the required constructors, destructor and basic
// member functions prototyped in the class def. exist.


// overloaded assignment operator
Person &Person::operator=(const Person &p)
```

```
    {
        if (this != &p)   // make sure we're not assigning an
        {                  // object to itself
            // delete any previously dynamically allocated data
            // from the destination object
            delete title;
            // Also, remember to reallocate memory for any
            // data members that are pointers
            // Then, copy from source to destination object
            // for each data member
            firstName = p.firstName;
            lastName = p.lastName;
            middleInitial = p.middleInitial;
            // Note: a pointer is used for title to demo the
            // necessary steps to implement a deep assignment -
            // otherwise, we would implement title with string
            title = new char[strlen(p.title) + 1]; // mem alloc
            strcpy(title, p.title);
        }
        return *this;  // allow for cascaded assignments
    }
```

Let us now review the overloaded assignment operator in the preceding code. It is designated by the member function `Person &Person::operator=(const Person &p);`. Here, we will be assigning memory from a source object, which will be input parameter p, to a destination object, which will be pointed to by `this`.

Our first order of business will be to ensure that we are not assigning an object to itself. Should this be the case, there is no work to be done! We make this check by testing `if (this != &p)` to see whether both addresses point to the same object. If we're not assigning an object to itself, we continue.

Next, within the conditional statement (`if`), we first deallocate the existing memory for the dynamically allocated data members pointed to by `this`. After all, the object on the left-hand of the assignment pre-exists and undoubtedly has allocations for these data members.

Now, we notice that the core piece of code within the conditional statement looks very similar to that of the copy constructor. That is, we carefully allocate space for pointer data members to match the sizes needed from their corresponding data members of input parameter p. We then copy the applicable data members from input parameter p to the data members pointed to by `this`. For the `char` data member, `middleInitial`, a memory allocation is not necessary; we merely use an assignment.

This is also true for the `string` data members, `firstName` and `lastName`. In this segment of code, we ensure that we have performed a deep assignment for any pointer data members. A shallow (pointer) assignment, where the source and destination object would otherwise share memory for the data portions of data members that are pointers, would be a disaster waiting to happen.

Lastly, at the end of our implementation of `operator=()`, we return `*this`. Notice that the return type from this function is a reference to a `Person`. Since `this` is a pointer, we merely dereference it so that we may return a referenceable object. This is done so that assignments between `Person` instances can be cascaded; that is, `p1 = p2 = p3;` where `p1`, `p2`, and `p3` are each an instance of `Person`.

> **Note**
>
> When overloading `operator=`, always check for self-assignment. That is, make sure you are not assigning an object to itself. Not only is there no work to be done in the case of self-assignment, but proceeding with an unnecessary self-assignment can actually create unexpected errors! For example, if we have dynamically allocated data members, we will be releasing destination object memory and re-allocating those data members based on the details of the source object's memory (which, when being the same object, will have been released). The resulting behavior can be unpredictable and error-prone.

Should the programmer wish to disallow assignment between two objects, the keyword `delete` can be used in the prototype of the overloaded assignment operator as follows:

```
// disallow assignment
Person &operator=(const Person &) = delete;
```

It is useful to remember that an overloaded assignment operator shares many similarities with the copy constructor; the same care and cautions apply to both language features. Keep in mind, however, that the assignment operator will be invoked when conducting an assignment between two pre-existing objects, whereas the copy constructor is implicitly invoked for initialization following the creation of a new instance. With the copy constructor, the new instance uses the existing instance as its basis for initialization; similarly, the left-hand object of the assignment operator uses the right-hand object as its basis for the assignment.

> **Important note**
>
> An overloaded assignment operator is not inherited by derived classes; therefore, it must be defined by each class in the hierarchy. Neglecting to overload `operator=` for a class will force the compiler to provide you with a default, shallow assignment operator for that class; this is dangerous for any classes containing data members that are pointers.

Overloading the comparison operator

Next, let's take a look at our implementation of the overloaded comparison operator:

```
// overloaded comparison operator
bool Person::operator==(const Person &p)
{
    // if the objects are the same object, or if the
    // contents are equal, return true. Otherwise, false.
    if (this == &p)
        return true;
    else if ( (!firstName.compare(p.firstName)) &&
              (!lastName.compare(p.lastName)) &&
              (!strcmp(title, p.title)) &&
              (middleInitial == p.middleInitial) )
        return true;
    else
        return false;
}
```

Continuing with a segment from our previous program, we overload the comparison operator. It is designated by the member function `int Person::operator==(const Person &p);`. Here, we will be comparing a `Person` object on the right-hand side of the operator, which will be referenced by input parameter p, to a `Person` object on the left-hand side of the operator, which will be pointed to by `this`.

Similarly, our first order of business will be to test whether the object on the **right-hand side (rhs)** is the same as the object on the **left-hand side (lhs)**. We make this check by testing `if (this != &p)` to see whether both addresses point to the same object. If both addresses point to the same object, we return the boolean (`bool`) value of `true`.

Next, we check whether the two `Person` objects contain identical values. They may be separate objects in memory, yet if they contain identical values, we can likewise choose to return a `bool` value of `true`. If there is no match, we then return a `bool` value of `false`.

Overloading the addition operator as a member function

Now, let's take a look at how to overload `operator+` for `Person` and a `string`:

```
// overloaded operator + (member function)
Person &Person::operator+(const string &t)
{
```

```
        ModifyTitle(t);
        return *this;
}
```

Moving forward with the preceding program, we overload the addition operator (+) to be used with a Person and a string. The operator function is designated by the member function prototype Person& Person::operator+(const string &t);. The parameter, t, will represent the right operand of operator+, which is a character string (which will bind to a reference to a string). The left-hand operand will be pointed to by this. An example use would be p1 + "Miss", where we wish to add a title to the Person p1 using operator+.

In the body of this member function, we merely use the input parameter t as an argument to ModifyTitle(), that is, ModifyTitle(t);. We then return *this so that we may cascade the use of this operator (notice the return type is a Person &).

Overloading the addition operator as a non-member function (using friends)

Now, let's reverse the order of operands with operator+ to allow for a string and a Person:

```
// overloaded + operator (not a mbr function)
Person &operator+(const string &t, Person &p)
{
    p.ModifyTitle(t);
    return p;
}
```

Continuing forward with the preceding program, we would ideally like operator+ to work not only with a Person and a string but also with the operands reversed, that is, with a string and a Person. There is no reason this operator should work one way and not the other.

To implement operator+ fully, we next overload operator+() to be used with const string & and Person. The operator function is designated by the non-member function Person& operator+(const string &t, Person &p);, which has two explicit input parameters. The first parameter, t, will represent the left operand of operator+, which is a character string (binding this parameter to a reference to a string as the first formal parameter in the operator function). The second parameter, p, will be a reference to the right operand used in operator+. An example use might be "Miss" + p1, where we wish to add a title to the Person p1 using operator+. Note that "Miss" will be constructed as a string using the std::string(const char *) constructor—the string literal is simply the initial value for the string object.

In the body of this non-member function, we merely take input parameter p and apply the protected method ModifyTitle() using the string of characters specified by parameter t, that is, p.ModifyTitle(t). However, because Person::ModifyTitle() is protected, Person

&p may not invoke this method outside of member functions of `Person`. We are in an external function; we are not in the scope of `Person`. Therefore, unless this member function is a `friend` of `Person`, p may not invoke `ModifyTitle()`. Luckily, `Person &operator+(const string &, Person &);` has been prototyped as a friend function in the `Person` class, providing the necessary scope to p to allow it to invoke its protected method. It is as if p is in the scope of `Person`; it is in the scope of a friend function of `Person`!

Let us now move forward to our `main()` function, tying together our many aforementioned code segments, so we may see how to invoke our operator functions utilizing our overloaded operators:

```
int main()
{
    Person p1;        // default constructed Person
    Person p2("Gabby", "Doone", 'A', "Miss");
    Person p3("Renee", "Alexander", 'Z', "Dr.");
    p1.Print();
    p2.Print();
    p3.Print();
    p1 = p2;          // invoke overloaded assignment operator
    p1.Print();
    p2 = "Ms." + p2;  // invoke overloaded + operator
    p2.Print();       // then invoke overloaded = operator
    p1 = p2 = p3;     // overloaded = can handle cascaded =
    p2.Print();
    p1.Print();
    if (p2 == p2)     // overloaded comparison operator
        cout << "Same people" << endl;
    if (p1 == p3)
        cout << "Same people" << endl;
    return 0;
}
```

Finally, let us examine our `main()` function for the preceding program. We begin by instantiating three instances of `Person`, namely p1, p2, and p3; we then print their values using member function `Print()` for each instance.

Now, we invoke our overloaded assignment operator with the statement `p1 = p2;`. Under the hood, this translates to the following operator function invocation: `p1.operator=(p2);`. From this, we can clearly see that we are invoking the previously defined `operator=()` method of `Person`, which performs a deep copy from source object p2 to destination object p1. We apply `p1.Print();` to see our resulting copy.

Next, we invoke our overloaded `operator+` with `"Ms." + p2`. This portion of this line of code translates to the following operator function call: `operator+("Ms.", p2);`. Here, we simply invoke our previously described `operator+()` function, which is a non-member function and `friend` of the `Person` class. Because this function returns a `Person &`, we can cascade this function call to look more like the usual context of addition and additionally write `p2 = "Ms." + p2;`. In this full line of code, first, `operator+()` is invoked for `"Ms." + p2`. The return value of this invocation is `p2`, which is then used as the right-hand operand of the cascaded call to `operator=`. Notice that the left-hand operand to `operator=` also happens to be `p2`. Fortunately, the overloaded assignment operator checks for self-assignment.

Now, we see a cascaded assignment of `p1 = p2 = p3;`. Here, we are invoking the overloaded assignment operator twice. First, we invoke `operator=` with `p2` and `p3`. The translated call would be `p2.operator=(p3);`. Then, using the return value of the first function call, we would invoke `operator=` a second time. The nested, translated call for `p1 = p2 = p3;` would look like this: `p1.operator=(p2.operator=(p3));`.

Lastly in this program, we invoke the overloaded comparison operator twice. For example, each comparison of `if (p2 == p2)` or `if (p1 == p3)` merely calls the `operator==` member function we have defined previously. Recall that we've written this function to report `true` either if the objects are the same in memory or simply contain the same values, and return `false` otherwise.

Let's take a look at the output for this program:

```
No first name No last name
Miss Gabby A. Doone
Dr. Renee Z. Alexander
Miss Gabby A. Doone
Ms. Gabby A. Doone
Dr. Renee Z. Alexander
Dr. Renee Z. Alexander
Same people
Same people
```

We have now seen how to specify and utilize friend classes and friend functions, how to overload C++ operators, and have seen cases when these two concepts can complement each other. Let us now briefly recap the features we have learned in this chapter before moving forward to our next chapter.

Summary

In this chapter, we have furthered our C++ programming endeavors beyond OOP language features to include features that will enable us to write more extensible programs. We have learned how to utilize friend functions and friend classes and we have learned how to overload operators in C++.

We have seen that friend functions and classes should be used sparingly and with caution. They are not meant to provide a blatant means to circumvent access regions. Instead, they are meant to handle programming situations to allow access between two tightly coupled classes without providing the alternative of an *overly public* interface in either of those classes, which could be misused on a broader scale.

We have seen how to overload operators in C++ using operator functions, both as member and non-member functions. We have learned that overloading operators will allow us to extend the meaning of C++ operators to include user defined types in the same way they encompass standard types. We have also seen that, in some cases, friend functions or classes may come in handy to help implement operator functions so they may behave associatively.

We have added important features to our C++ repertoire through exploring friends and operator overloading, the latter of which will help us to ensure code we will soon write using templates can be used for nearly any data type, contributing to highly extensible and reusable code. We are now ready to move forward to *Chapter 13*, *Working with Templates*, so that we can continue expanding our C++ programming skills with essential language features that will make us better programmers. Let's move ahead!

Questions

1. Overload `operator=` in your `Shape` exercise from *Chapter 8*, *Mastering Abstract Classes*, or alternatively, overload `operator=` in your ongoing `LifeForm/Person/Student` classes as follows:

    Define `operator=` in `Shape` (or `LifeForm`) and override this method in all of its derived classes. Hint: the derived implementation of `operator=()` will do more work than its ancestor, yet could call its ancestor's implementation to perform the base class part of the work.

2. Overload `operator<<` in your `Shape` class (or `LifeForm` class) to print information about each `Shape` (or `LifeForm`). The arguments to this function should be an `ostream` `&` and a `Shape` `&` (or a `LifeForm` `&`). Note that `ostream` is from the C++ Standard Library (using namespace std;).

    You may either provide one function, `ostream &operator<<(ostream &, Shape &);`, and from it call a polymorphic `Print()`, which is defined in `Shape` and redefined in each derived class. Or, provide multiple `operator<<` methods to implement this functionality (one for each derived class). If using the `Lifeform` hierarchy, substitute `LifeForm` for `Shape`, in the aforementioned `operator<<` function signature.

3. Create an `ArrayInt` class to provide safe integer arrays with bounds checking. Overload `operator[]` to return an element if it exists in the array, or throw an exception if it is `OutOfBounds`. Add other methods to `ArrayInt`, such as `Resize()`, `RemoveElement()`,

and so on. Model the data comprising the array using a dynamically allocated array (that is, using `int *contents`) so that you can easily handle resizing. The code would begin as follows:

```
class ArrayInt // starting point for the class def.
{               // be sure to add: using std::to_string;
private:    // and also: using std::out_of_range;
    int numElements = 0;      // in-class init.
    int *contents = nullptr; // dynam. alloc. array
public:
    ArrayInt(int size); // set numElements and
                        // allocate contents
    // returns a referenceable memory location or
    // throws an exception
    int &operator[](int index)
    {
        if (index < numElements)
            return contents[index];
        else    // index is out of bounds
            throw std::out_of_range(
                        std::to_string(index));
    }
};


int main()
{
    ArrayInt a1(5); // Create ArrayInt of 5 elements
    try
    {
        a1[4] = 7;      // a1.operator[](4) = 7;
    }
    catch (const std::out_of_range &e)
    {
        cout << "Out of range: " << e.what() << endl;
    }
}
```

13
Working with Templates

This chapter will continue our pursuit of increasing your C++ programming repertoire beyond OOP concepts, with the continued goal of writing more extensible code. We will next explore creating generic code using C++ templates – both **template functions** and **template classes**. We will learn how template code, when written correctly, is the pinnacle in code reuse. In addition to exploring how to create both template functions and template classes, we will also understand how the appropriate use of operator overloading can make a template function reusable for nearly any type of data.

In this chapter, we will cover the following main topics:

- Exploring template basics to genericize code
- Understanding how to create and use template functions and template classes
- Understanding how operator overloading can make templates more extensible

Many object-oriented languages include the concept of programming with generics, allowing the types of classes and interfaces to be parameterized themselves. In some languages, generics are merely wrappers for casting objects to the desired type. In C++, the idea of generics is more comprehensive and is implemented using templates.

By the end of this chapter, you will be able to design more generic code by building both template functions and template classes. You will understand how operator overloading can ensure that a template function can become highly extensible for any data type. By pairing together well-designed template member functions with operator overloading, you will be able to create highly reusable and extensible template classes in C++.

Let's increase our understanding of C++ by expanding your programming repertoire by exploring templates.

Technical requirements

Online code for full program examples can be found in the following GitHub URL: `https://github.com/PacktPublishing/Deciphering-Object-Oriented-Programming-with-CPP/tree/main/Chapter13`. Each full program example can be found in the GitHub repository under the appropriate chapter heading (subdirectory) in a file that corresponds to the chapter number, followed by a dash, followed by the example number in the chapter at hand. For example, the first full program in this chapter can be found in the subdirectory `Chapter13` in a file named `Chp13-Ex1.cpp` under the aforementioned GitHub directory.

The CiA video for this chapter can be viewed at: `https://bit.ly/3A7lx0U`.

Exploring template basics to genericize code

Templates allow code to be generically specified in a manner that is abstracted from the data types primarily used within relevant functions or classes. The motivation for creating templates is to generically specify the definition of functions and classes that we repeatedly want to utilize, but with varying data types. The individualized versions of these components would otherwise differ only in the core data type utilized; these key data types can then be extracted and written generically.

When we then opt to utilize such a class or function with a specific type, rather than copying and pasting existing code from a similar class or function (with preset data types) and changing it slightly, the preprocessor instead would take the template code and *expand* it for our requested, bonafide type. This template *expansion* capability allows the programmer to write and maintain only one version of the genericized code, versus the many type-specific versions of code that would otherwise need to be written. The benefit is also that the preprocessor will do a more accurate expansion of the template code to a bonafide type than we might have done using a copy, paste, and slight modification method.

Let's take a moment to further investigate the motivation for using templates in our code.

Examining the motivation for templates

Imagine that we wish to create a class to safely handle dynamically allocated arrays for data type `int`, such as we have created in a solution for *Question 3* of *Chapter 12, Friends and Operator Overloading*. Our motivation may be to have an array type that can grow or shrink to any size (unlike native, fixed-sized arrays), yet have bounds checking for safe usage (unlike the raw manipulation of a dynamic array implemented using `int *`, which would unscrupulously allow us to access elements well beyond the length of our dynamic array allocation).

We may decide to create an `ArrayInt` class with the following beginning framework:

```
class ArrayInt
{
private:
```

```
    int numElements = 0;        // in-class initialization
    int *contents = nullptr; // dynamically allocated array
public:
    ArrayInt(int size): numElements(size)
    {
        contents = new int [size];
    }
    ~ArrayInt() { delete [] contents; }
    int &operator[](int index) // returns a referenceable
    {                        // memory location or throws exception
        if (index < numElements)
            return contents[index];
        else            // index selected is out of bounds
            throw std::out_of_range(std::to_string(index));
    }
};


int main()
{
    ArrayInt a1(5); // Create an ArrayInt of 5 elements
    try    // operator[] could throw an exception
    {
        a1[4] = 7;       // a1.operator[](4) = 7;
    }
    catch (const std::out_of_range &e)
    {
        cout << "Out of range: element " << e.what();
        cout << endl;
    }
}
```

In the previous code segment, notice that our `ArrayInt` class implements the data structure comprising the array using `int *contents`, which is dynamically allocated to the desired size in the constructor. We have overloaded `operator[]` to safely return only indexed values in the array that are within the proper range, and throw a `std::out_of_range` exception otherwise. We can add methods to `Resize()` an `ArrayInt` and so on. Overall, we love the safety and flexibility of this class.

Now, we may want to have an `ArrayFloat` class (or later, an `ArrayStudent` class). Rather than copying our baseline `ArrayInt` class and modifying it slightly to create an `ArrayFloat` class, for example, we may ask whether there is a more automated way to make this substitution. After all, what would we change in creating an `ArrayFloat` class using an `ArrayInt` class as a starting point? We would change the *type* of the data member `contents` – from an `int *` to a `float *`. We would change the *type* in the memory allocation in the constructor from `contents = new int [size];` to utilize `float` instead of `int` (and similarly so in any reallocation, such as in a `Resize()` method).

Rather than copying, pasting, and slightly modifying an `ArrayInt` class to create an `ArrayFloat` class, we can simply use a **template class** to genericize the *type* associated with the data manipulated within this class. Similarly, any functions relying on the specific data type will become **template functions**. We will examine the syntax for creating and utilizing templates shortly.

Using templates, we can instead create just one template class called `Array` where the type is genericized. At compile time, should the preprocessor detect we have utilized this class for type `int` or `float` in our code, the preprocessor will then provide the necessary template *expansions* for us. That is, by copying and pasting (behind the scenes) each template class (and its methods) and substituting in the data types that the preprocessor identifies we are using.

The resulting code, once expanded under the hood, is no smaller than if we had written the code for each individual class ourselves. But the point is that we did not have to tediously create, modify, test, and later maintain each minorly different class ourselves. This is done on our behalf by C++. This is the noteworthy purpose of template classes and template functions.

Templates are not restricted for use with primitive data types. For example, we may wish to create an `Array` of a user defined type, such as `Student`. We will need to ensure that all of our template member functions are meaningful for the data types that we actually expand the template class to utilize. We may need to overload selected operators so that our template member functions can work seamlessly with user defined types, just as they do with primitive types.

We will later see in this chapter an example illustrating how we may need to overload selected operators if we choose to expand a template class for user defined types so that the member functions of a class can work fluidly with any data type. Fortunately, we know how to overload operators!

Let's move forward to explore the mechanics of specifying and utilizing template functions and template classes.

Understanding template functions and classes

Templates provide the ability to create generic functions and classes by abstracting the data types associated with those functions and classes. Template functions and classes can both be carefully written in such a way as to genericize the relevant data types that underlie these functions and classes.

Let's begin by examining how to create and utilize template functions.

Creating and using template functions

Template functions parameterize the types of arguments in a function in addition to the arguments themselves. Template functions require the body of the function to be applicable to almost any data type. Template functions can be member or non-member functions. Operator overloading can help ensure that the bodies of template functions are applicable to user defined types – we'll see more of that shortly.

The keyword `template`, along with angle brackets, `< >`, and placeholders for the *type* names are used to specify a template function and its prototype.

Let's take a look at a template function that is not a member of a class (we will see examples of template member functions shortly). This example can be found, as a full working program, in our GitHub repository as follows:

https://github.com/PacktPublishing/Deciphering-Object-Oriented-Programming-with-CPP/blob/main/Chapter13/Chp13-Ex1.cpp

```cpp
// template function prototype
template <class Type1, class Type2>    // template preamble
Type2 ChooseFirst(Type1, Type2);

// template function definition
template <class Type1, class Type2>    // template preamble
Type2 ChooseFirst(Type1 x, Type2 y)
{
    if (x < y)
        return static_cast<Type2>(x);
    else
        return y;
}

int main()
{
    int value1 = 4, value2 = 7;
    float value3 = 5.67f;
    cout << "First: " << ChooseFirst(value1, value3);
    cout << endl;
    cout << "First: " << ChooseFirst(value2, value1);
```

```
        cout << endl;
    }
```

Looking at the previous function example, we first see a template function prototype. The preamble of `template <class Type1, class Type 2>` indicates that the prototype will be a template prototype and that placeholders `Type1` and `Type2` will be used instead of actual data types. The placeholders `Type1` and `Type2` may be (nearly) any name, following the rules of creating identifiers.

Then, to complete the prototype, we see `Type2 ChooseFirst(Type1, Type2);`, which indicates that the return type from this function will be of `Type2` and that the arguments of the `ChooseFirst()` function will be of `Type1` and `Type2` (which may certainly be expanded to be the same type).

Next, we see the function definition. It, too, begins with a preamble of `template <class Type1, class Type 2>`. Similar to the prototype, the function header `Type2 ChooseFirst(Type1 x, Type2 y)` indicates that formal parameters x and y are of types `Type1` and `Type2`, respectively. The body of this function is rather straightforward. We simply determine which of the two parameters should be ranked first in an ordering of the two values by using a simple comparison with the < operator.

Now, in `main()`, when the preprocessor portion of the compiler sees a call to `ChooseFirst()` with actual parameters, `int value1` and `float value3`, the preprocessor notices that `ChooseFirst()` is a template function. If no such version of `ChooseFirst()` yet exists to handle an `int` and a `float`, the preprocessor copies this template function and replaces `Type1` with `int` and `Type2` with `float` – creating on our behalf the appropriate version of this function to fit our needs. Notice that when `ChooseFirst(value2, value1)` is called and the types are both integers, the placeholder types of `Type1` and `Type2` will both be replaced with `int` when the template function is again expanded (under the hood) in our code by the preprocessor.

Though `ChooseFirst()` is a simple function, with it we can see the straightforward mechanics of creating a template function that genericizes key data types. We can also see how the preprocessor notices how the template function is used and takes on the effort on our behalf to expand this function, as needed, for our specific type usage.

Let's take a look at the output for this program:

```
First: 4
First: 4
```

Now that we have seen the basic mechanics of template functions, let us move forward to understand how we can expand this process to include template classes.

Creating and using template classes

Template classes parameterize the ultimate type of a class definition, and will additionally require template member functions for any methods that need to know the core data type being manipulated.

The keywords `template` and `class`, along with angle brackets, `< >`, and placeholders for the *type* names are used to specify a template class definition.

Let's take a look at a template class definition and its supporting template member functions. This example can be found as a complete program (with the necessary `#include` and `using` statements) in our GitHub repository as follows:

https://github.com/PacktPublishing/Deciphering-Object-Oriented-Programming-with-CPP/blob/main/Chapter13/Chp13-Ex2.cpp

```cpp
template <class Type>    // template class preamble
class Array
{
private:
    int numElements = 0;    // in-class initialization
    Type *contents = nullptr;// dynamically allocated array
public:.
    // Constructor and destructor will allocate, deallocate
    // heap memory to allow Array to be fluid in its size.
    // Later, you can use a smart pointer - or use the STL
    // vector class (we're building a similar class here!)
    Array(int size): numElements(size),
                     contents(new Type [size])
    { // note: allocation is handled in member init. list
    }
    ~Array() { delete [] contents; }
    void Print() const;
    Type &operator[](int index) // returns a referenceable
    {                // memory location or throws exception
        if (index < numElements)
            return contents[index];
        else    // index is out of bounds
            throw std::out_of_range
                        (std::to_string (index));
```

```cpp
    }
    void operator+(Type);    // prototype only
};


template <class Type>
void Array<Type>::operator+(Type item)
{
    // resize array as necessary, add new data element and
    // increment numElements
}


template <class Type>
void Array<Type>::Print() const
{
    for (int i = 0; i < numElements; i++)
        cout << contents[i] << " ";
    cout << endl;
}


int main()
{
    // Creation of int Array will trigger template
    // expansion by the preprocessor.
    Array<int> a1(3); // Create an int Array of 3 elements
    try    // operator[] could throw an exception
    {
        a1[2] = 12;
        a1[1] = 70;        // a1.operator[](1) = 70;
        a1[0] = 2;
        a1[100] = 10;// this assignment throws an exception
    }
    catch (const std::out_of_range &e)
    {
        cout << "Out of range: index " << e.what() << endl;
```

```
    }
    a1.Print();
}
```

In the preceding class definition, let's first notice the template class preamble of `template <class Type>`. This preamble specifies that the impending class definition will be that of a template class and that the placeholder `Type` will be used to genericize the data types primarily used within this class.

We then see the class definition for `Array`. Data member `contents` will be of the placeholder type of `Type`. Of course, not all data types will need to be genericized. Data member `int numElements` is perfectly reasonable as an integer. Next, we see an assortment of member functions prototyped and some defined inline, including overloaded `operator[]`. For the member functions defined inline, a template preamble is not necessary in front of the function definition. The only thing we need to do for inline functions is to genericize the data type using our placeholder, `Type`.

Let's now take a look at selected member functions. In the constructor, we now notice that the memory allocation of `contents = new Type [size];` merely uses the placeholder `Type` in lieu of an actual data type. Similarly, for overloaded `operator[]`, the return type of this method is `Type`.

However, looking at a member function that is not inline, we notice that the template preamble of `template <class Type>` must precede the member function definition. For example, let's consider the member function definition for `void Array<Type>::operator+(Type item);`. In addition to the preamble, the class name (preceding the member function name and scope resolution operator, `::`) in the function definition must be augmented to include the placeholder type `<Type>` in angle brackets. Also, any generic function parameters must use the placeholder type of `Type`.

Now, in our `main()` function, we merely use the data type of `Array<int>` to instantiate a safe, easily resizable array of integers. We could have alternatively used `Array<float>` had we instead wanted to instantiate an array of floating-point numbers. Under the hood, when we create an instance of a specific array type, the preprocessor notices whether we have previously expanded this class for that *type*. If not, the class definition and applicable template member functions are copied for us and the placeholder types are replaced with the type that we need. This is no fewer lines of code than if we had copied, pasted, and slightly modified the code ourselves; however, the point is that we only have one version to specify and maintain ourselves. This is less error-prone and easier for long-term maintenance.

Let's take a look at the output for this program:

```
 2  70  12
```

> **An interesting tangent – std::optional**
>
> In the previous example, `Array<Type>::operator[]` throws an `out_of_range` exception when the selected index is out of bounds. Sometimes, exception handling can be programmatically expensive. In such cases, using an optional return type may be a useful alternative. Remember, a valid return value for `operator[]` is a reference to the memory location for the array element in question. For the out-of-bounds index scenario, knowing we simply cannot return the corresponding memory location for an array element from this method (it would not make sense), an alternative to exception handling may be to use `std::optional<Type>` in the return value of the function.

Let's next take a look at a different full program example to pull together template functions and template classes.

Examining a full program example

It is useful to see an additional example that illustrates template functions and template classes. Let us expand on a `LinkList` program we reviewed most recently in *Chapter 12, Friends and Operator Overloading*; we will upgrade this program to utilize templates.

This complete program can be found in our GitHub repository as follows:

https://github.com/PacktPublishing/Deciphering-Object-Oriented-Programming-with-CPP/blob/main/Chapter13/Chp13-Ex3.cpp

```cpp
#include <iostream>
using std::cout;    // preferred to: using namespace std;
using std::endl;


// forward declaration with template preamble
template <class Type> class LinkList;


template <class Type>    // template preamble for class def.
class LinkListElement
{
private:
    Type *data = nullptr;
    LinkListElement *next = nullptr;
    // private access methods to be used in scope of friend
    Type *GetData() const { return data; }
    LinkListElement *GetNext() const { return next; }
```

```
        void SetNext(LinkListElement *e) { next = e; }
public:
    friend class LinkList<Type>;
    LinkListElement() = default;
    LinkListElement(Type *i): data(i), next(nullptr) { }
    ~LinkListElement(){ delete data; next = nullptr; }
};


// LinkList should only be extended as a protected/private
// base class; it does not contain a virtual destructor. It
// can be used as-is, or as implementation for another ADT.

template <class Type>
class LinkList
{
private:
    LinkListElement<Type> *head = nullptr, *tail = nullptr,
                                *current = nullptr;
public:
    LinkList() = default;
    LinkList(LinkListElement<Type> *e)
        { head = tail = current = e; }
    void InsertAtFront(Type *);
    LinkListElement<Type> *RemoveAtFront();
    void DeleteAtFront()  { delete RemoveAtFront(); }
    bool IsEmpty() const { return head == nullptr; }
    void Print() const;
    ~LinkList(){ while (!IsEmpty()) DeleteAtFront(); }
};
```

Let's examine the preceding template class definitions for `LinkListElement` and `LinkList`. Initially, we notice that the forward declaration of the `LinkList` class contains the necessary template preamble of `template class <Type>`. We also should notice that each class definition itself contains the same template preamble to dually specify that the class will be a template class and that the placeholder for the data type will be the identifier `Type`.

In the LinkListElement class, notice that the data type will be Type (the placeholder type). Also notice that the placeholder for the type will be necessary in the friend class specification of LinkList, that is, friend class LinkList<Type>;.

In the LinkList class, notice that any reference to the associated class of LinkListElement will include the type placeholder of <Type>. Notice, for example, this placeholder usage in the data member declaration of LinkListElement<Type> *head; or the return type of RemoveAtFront(), which is LinkListElement<Type>. Additionally, notice that the inline function definitions do not require a template preamble before each method; we are still covered by the preamble occurring before the class definition itself.

Now, let's move forward to take a look at the three non-inline member functions of the LinkList class:

```cpp
template <class Type>       // template preamble
void LinkList<Type>::InsertAtFront(Type *theItem)
{
    LinkListElement<Type> *newHead = nullptr;
    newHead = new LinkListElement<Type>(theItem);
    newHead->SetNext(head);   // newHead->next = head;
    head = newHead;
}


template <class Type>       // template preamble
LinkListElement<Type> *LinkList<Type>::RemoveAtFront()
{
    LinkListElement<Type> *remove = head;
    head = head->GetNext();   // head = head->next;
    current = head;     // reset current for usage elsewhere
    return remove;
}


template <class Type>       // template preamble
void LinkList<Type>::Print() const
{
    if (!head)
        cout << "<EMPTY>" << endl;
    LinkListElement<Type> *traverse = head;
    while (traverse)
    {
```

```
      Type output = *(traverse->GetData());
      cout << output << ' ';
      traverse = traverse->GetNext();
   }
   cout << endl;
}
```

As we examine the preceding code, we can see that in the non-inline methods of `LinkList`, the template preamble of `template <class Type>` appears before each member function definition. We also see that the class name tied with the scope resolution operator to the member function name is augmented with `<Type>`, for example, `void LinkList<Type>::Print()`.

We notice that the aforementioned template member functions require some part of their method to utilize the placeholder type, `Type`. For example, the `InsertAtFront(Type *theItem)` method uses both the placeholder `Type` as the data type of the formal parameter `theItem`, and to specify the associated class `LinkListElement<Type>` when declaring a local pointer variable `temp`. The `RemoveAtFront()` method similarly utilizes a local variable of type `LinkListElement<Type>`, hence necessitating its use as a template function. Similarly, `Print()` introduces a local variable of type `Type` to assist with output.

Let's now take a look at our `main()` function to see how we can utilize our template classes:

```
int main()
{
    LinkList<int> list1;   // create a LinkList of integers
    list1.InsertAtFront(new int (3000));
    list1.InsertAtFront(new int (600));
    list1.InsertAtFront(new int (475));
    cout << "List 1: ";
    list1.Print();
    // delete elements from list, one by one
    while (!(list1.IsEmpty()))
    {
        list1.DeleteAtFront();
        cout << "List 1 after removing an item: ";
        list1.Print();
    }
    LinkList<float> list2;   // create a LinkList of floats
    list2.InsertAtFront(new float(30.50));
    list2.InsertAtFront(new float (60.89));
```

```
        list2.InsertAtFront(new float (45.93));
        cout << "List 2: ";
        list2.Print();
}
```

In our preceding `main()` function, we utilize our template classes to create two types of linked lists, that is, a `LinkList` of integers with the declaration `LinkList<int> list1;` and a `LinkList` of floating-point numbers with the declaration `LinkList<float> list2;`.

In each case, we instantiate the various link lists, then add elements and print the respective lists. In the case of the first `LinkList` instance, we also demonstrate how elements can be successively removed from the list.

Let's take a look at the output for this program:

```
List 1: 475 600 3000
List 1 after removing an item: 600 3000
List 1 after removing an item: 3000
List 1 after removing an item: <EMPTY>


List 2: 45.93 60.89 30.5
```

Overall, we see that creating a `LinkList<int>` and a `LinkList<float>` is very easy. The template code is simply expanded behind the scenes to accommodate each data type we desire. We may then ask ourselves, how easy is it to create a linked list of `Student` instances? Very easy! We could simply instantiate `LinkList<Student> list3;` and call the appropriate `LinkList` methods, such as `list3.InsertAtFront(new Student("George", "Katz", 'C', "Mr.", 3.2, "C++", "123GWU"));`.

Perhaps we would like to include a means to order our elements in the template `LinkList` class, such as by adding an `OrderedInsert()` method (which typically relies on `operator<` or `operator>` for the comparison of elements). Would that work for all data types? That's a good question. It could, provided the code written in the method is generic to work for all data types. Can operator overloading help with this endeavor? Yes!

Now that we have seen the mechanics of template classes and functions in action, let's consider how we can ensure that our template classes and functions are fully extensible to work for any data type. To do this, let's consider how operator overloading can be of value.

Making templates more flexible and extensible

The addition of templates in C++ gives us the ability to make certain types of classes and functions generically specified a single time by the programmer, while behind the scenes, the preprocessor generates many versions of that code on our behalf. However, in order for a class to truly be extensible to expand for many different user defined types, code written within member functions must be universally applicable to any type of data. To help with this endeavor, operator overloading can be used to extend operations that may easily exist for standard types to include definitions for user defined types.

To recap, we know operator overloading can allow simple operators to work not only with standard types but also with user defined types. By overloading operators in our template code, we can ensure that our template code is highly reusable and extensible.

Let's consider how we can strengthen templates with the use of operator overloading.

Adding operator overloading to further genericize template code

Recall that when overloading an operator, it is important to promote the same meaning that the operator has for standard types. Imagine that we would like to add an `OrderedInsert()` method to our `LinkList` class. The body of this member function might rely on comparing two elements to see which one should go before the other. The easiest way to do this is using `operator<`. This operator is easily defined to work with standard types, but will it work with user defined types? It can, provided we overload the operator to work with the desired types.

Let's take a look at an example where we will need to overload an operator to make the member function code universally applicable:

```
template <class Type>
void LinkList<Type>::OrderedInsert(Type *theItem)
{
    current = head;
    if (*theItem < *(head->GetData()))
        InsertAtFront(theItem);  // add theItem before head
    else
        // Traverse list, add theItem in proper location
}
```

In the preceding template member function, we rely on `operator<` to be able to work with any data type in which we would like to utilize this template class. That is, when the preprocessor expands this code for a specific, user defined type, the `<` operator must work for whatever data type this method has been specifically expanded for.

Should we wish to create a `LinkList` of `Student` instances and apply an `OrderedInsert()` of one `Student` versus another, we then need to ensure that the comparison with `operator<` is defined for two `Student` instances. Of course, by default, `operator<` is only defined for standard types. But, if we simply overload `operator<` for `Student`, we can ensure that the `LinkList<Type>::OrderedInsert()` method will work for `Student` data types as well.

Let's take a look at how we can overload `operator<` for `Student` instances, both as a member function or as a non-member function:

```
// overload operator < As a member function of Student
bool Student::operator<(const Student &s)
{   // if this->gpa < s.gpa return true, else return false
    return this->gpa < s.gpa;
}
// OR, overload operator < as a non-member function
bool operator<(const Student &s1, const Student &s2)
{   // if s1.gpa < s2.gpa return true, else return false
    return s1.gpa < s2.gpa;
}
```

In the preceding code, we can recognize `operator<` implemented as either a member function of `Student` or as a non-member function. If you have access to the class definition for `Student`, the preferred approach would be to utilize the member function definition for this operator function. However, sometimes, we do not have access to modify a class. In such cases, we must utilize the non-member function approach. Nonetheless, in either implementation, we simply compare the `gpa` of the two `Student` instances, and return `true` if the first instance has a lower `gpa` than the second `Student` instance, and `false` otherwise.

Now that `operator<` has been defined for two `Student` instances, we can return to our prior template function of `LinkList<Type>::OrderedInsert(Type *)`, which utilizes operator `<` for comparison of two objects of type `Type` in the `LinkList`. When a `LinkList<Student>` is made somewhere in our code, the template code for `LinkList` and `LinkListElement` will be expanded by the preprocessor for `Student`; `Type` will be replaced with `Student`. When the expanded code is then compiled, the code in the expanded `LinkList<Student>::OrderedInsert()` will compile without error, as `operator<` has been defined for two `Student` objects.

What happens if we neglect to overload `operator<` for a given type, however, `OrderedInsert()` (or another method relying on `operator<`) is never called in our code on an object of that same expanded template type? Believe it or not, the code will compile and work without issue. In this case, we are not actually calling a function (that is, `OrderedInsert()`) that would require `operator<` to be implemented for that type. Because the function is never called, the template expansion for that member function is skipped. The compiler has no reason to discover that `operator<` should have been overloaded for the type in question (in order for the method to compile successfully). The uncalled method has simply not been expanded for the compiler to verify.

By using operator overloading to complement template classes and functions, we can make template code even further extensible by ensuring that typical operators used within method bodies can be made applicable to any type we would want to utilize in the template expansion. Our code becomes more widely applicable.

We have now seen how to utilize template functions and classes, and how operator overloading can enhance templates to create even more extensible code. Let us now briefly recap these concepts before moving forward to our next chapter.

Summary

In this chapter, we have furthered our C++ programming knowledge beyond OOP language features to include additional language features that will enable us to write more extensible code. We have learned how to utilize template functions and template classes, and how operator overloading nicely supports these endeavors.

We have seen that templates can allow us to generically specify a class or function with respect to the data type primarily used within that class or function. We have seen that template classes inevitably utilize template functions because those methods generally need to generically use the data upon which the class is built. We have seen that by taking advantage of operator overloading for user defined types, we can take advantage of method bodies written using simple operators to accommodate usage by more complex data types, making the template code much more useful and extensible.

The power of templates coupled with operator overloading (to make a method usable for nearly any type) makes C++'s implementation of generics much more powerful than simple type replacement.

We now understand that using templates can allow us to specify a class or function just one time more abstractly, and allow the preprocessor to generate many versions of that class or function for us, based upon specific data types that may be needed within the application.

By allowing the preprocessor to expand many versions of a template class or set of template functions for us based on types needed in an application, the work of creating many similar classes or functions (and maintaining those versions) is passed to C++, rather than the programmer. In addition to having less code for the user to maintain, changes made in the template classes or functions need only be made in one place – the preprocessor will re-expand the code without errors when needed.

We have added additional, useful features to our C++ repertoire through examining templates, which, combined with operator overloading, will ensure we can write highly extensible and reusable code for nearly any data type. We are now ready to continue forward with *Chapter 14, Understanding STL Basics*, so that we can continue extending our C++ programming skills with useful C++ library features that will make us better programmers. Let's move forward!

Questions

1. Convert your `ArrayInt` class from *Chapter 12, Friends and Operator Overloading,* to a template `Array` class to support a dynamically allocated array of any data type that can be easily resized and has built-in bounds checking.

    a. Consider what operators, if any, you will need to overload to allow generic code within each method to support any user defined types you may wish to store in the template `Array` type.

    b. Using your template `Array` class, create an array of `Student` instances. Utilize various member functions to demonstrate that various template functions operate correctly.

2. Using the template `LinkList` class, complete the implementation for `LinkList<Type>::OrderedInsert()`. Create a `LinkList` of `Student` instances in `main()`. After several `Student` instances have been inserted in the list using `OrderedInsert()`, verify that this method works correctly by displaying each `Student` and their gpa. The `Student` instances should be ordered from lowest to highest gpa. You may wish to use the online code as a starting point.

14
Understanding STL Basics

This chapter will continue our pursuit of increasing your C++ programming repertoire beyond OOP concepts by delving into a core C++ library that has become thoroughly integrated into the common usage of the language. We will explore the **Standard Template Library (STL)** in C++ by examining a subset of this library, representing common utilities that can both simplify our programming and make our code more easily understood by others who are undoubtedly familiar with the STL.

In this chapter, we will cover the following main topics:

- Surveying the contents and purpose of the STL in C++
- Understanding how to use essential STL containers – `list`, `iterator`, `vector`, `deque`, `stack`, `queue`, `priority_queue`, map, and map using a functor
- Customizing STL containers

By the end of this chapter, you will be able to utilize core STL classes to enhance your programming skills. Because you already understand the essential C++ language and OOP features in which libraries are built, you will see that you now have the ability to navigate and understand nearly any C++ class library, including the STL. By gaining familiarity with the STL, you will be able to enhance your programming repertoire significantly and become a more savvy and valuable programmer.

Let's increase our C++ toolkit by examining a very heavily utilized class library, the STL.

Technical requirements

Online code for full program examples can be found in the following GitHub URL: https://github.com/PacktPublishing/Deciphering-Object-Oriented-Programming-with-CPP/tree/main/Chapter14. Each full program example can be found in the GitHub repository under the appropriate chapter heading (subdirectory) in a file that corresponds to the chapter number, followed by a dash, followed by the example number in the chapter at hand. For example, the first full program in this chapter can be found in the subdirectory Chapter14 in a file named Chp14-Ex1.cpp under the aforementioned GitHub directory.

The CiA video for this chapter can be viewed at: `https://bit.ly/3PCL5IJ`.

Surveying the contents and purpose of the STL

The **Standard Template Library** in C++ is a library of standard classes and utilities that extend the C++ language. The use of the STL is so pervasive that it is as though the STL is a part of the language itself; it is an essential and integral part of C++. The STL in C++ has four key components comprising the library: **containers**, **iterators**, **functions**, and **algorithms**.

The STL has additionally influenced the C++ Standard Library in providing a set of programming standards; the two libraries actually share common features and components, most notably containers and iterators. We've already utilized components from the Standard Library, namely `<iostream>` for IOStreams, `<exception>` for exception handling, and `<new>` for operators `new()` and `delete()`. In this chapter, we will explore many overlapping components between the STL and the Standard Library in C++.

The STL has a full line of **container** classes. These classes encapsulate traditional data structures to allow similar items to be collected together and uniformly processed. There are several categories of container classes – sequential, associative, and unordered. Let's summarize these categories and provide a few examples of each:

- **Sequential containers**: Implement encapsulated data structures that can be accessed in a sequential manner, such as `list`, `queue`, or `stack`. It is interesting to note that `queue` and `stack` can be thought of as a customized or adaptive interface for a more basic container, such as a `list`. Nonetheless, a `queue` and `stack` still provide sequential access to their elements.

- **Associative containers**: Implement sorted, encapsulated data structures that can be searched quickly to retrieve an element such as `set` or `map`.

- **Unordered containers**: Implement unordered, encapsulated data structures that can be searched reasonably quickly, such as `unordered_set` or `unordered_map`.

In order for these container classes to be potentially used for any data type (and to preserve strong type checking), templates are utilized to abstract and genericize the data types of the collected items. In fact, we built our own container classes using templates in *Chapter 13, Working with Templates*, including `LinkList` and `Array`, so we already have a basic understanding of templatized container classes!

Additionally, the STL provides a full complement of **iterators**, which allow us to *walk through* or traverse containers. Iterators keep track of our current place without corrupting the content or ordering of the respective collections of objects. We will see how iterators allow us to process container classes more safely within the STL.

The STL also contains a plentiful supply of useful **algorithms**. Examples include sorting, counting the number of elements in a collection that may satisfy a condition, searching for particular elements or subsequences within elements, or copying elements in a variety of manners. Additional examples of algorithms include modifying a sequence of objects (replacing, swapping, and removing values), partitioning sets into ranges, or merging sets back together. Moreover, the STL contains many other useful algorithms and utilities.

Lastly, the STL includes functions. Actually, it would be more correct to say that the STL includes **functors**, or **function objects**. Functors are built around the ability to overload `operator()` (the function call operator), and by doing so, allow us to achieve parameterized flexibility through a function pointer. Though this is not an elementary feature of the STL we will immediately (or often) use, we will see one small, simple example of a functor in this chapter coupled with an STL container class, in the upcoming section *Examining STL map using a functor*.

In this chapter, we will focus on the container class section of the STL. Though we won't examine every STL container class in the STL, we will review a healthy assortment of these classes. We will notice that some of these container classes are similar to classes that we have built together in previous chapters of this book. Incidentally, during the incremental chapter progressions of this book, we have also built up our C++ language and OOP skills, which are necessary to decode a C++ class library such as STL.

Let's move forward to take a look at selective STL classes and test our C++ knowledge as we interpret each class.

Understanding how to use essential STL containers

In this section, we will put our C++ skills to the test by decoding various STL container classes. We will see that language features we have mastered, from core C++ syntax to OOP skills, have put us in a position to easily interpret the various components of STL we will now examine. Most notably, we will put our knowledge of templates to use! Our knowledge of encapsulation and inheritance, for example, will guide us to understand how to use various methods in STL classes. However, we will notice that virtual functions and abstract classes are extremely rare in the STL. The best way to gain competence with a new class within the STL will be to embrace the documentation detailing each class. With knowledge of C++, we can easily navigate through a given class to decode how to use it successfully.

The container classes in the C++ STL implement various **Abstract Data Types** (**ADTs**) by encapsulating the data structures that implement these higher-level concepts. We will examine core STL containers: `list`, `iterator`, `vector`, `deque`, `stack`, `queue`, `priority_queue`, and `map`.

Let's begin by examining how to utilize a very basic STL container, `list`.

Using STL list

The STL list class encapsulates the data structures necessary to implement a linked list. We can say that list implements the Abstract Data Type of a linked list. Recall that we have made our own linked list by creating LinkedListElement and LinkedList classes in *Chapter 6*, *Implementing Hierarchies with Inheritance*. STL list allows for easy insertion, deletion, and sorting of elements. Direct access to individual elements (known as *random access*) is not supported. Rather, you must iteratively traverse past a prior item in the linked list until you reach the desired item. STL list is a good example of a sequential container.

STL list actually supports bidirectional sequential access to its elements (it is implemented using a doubly-linked list). The STL additionally offers forward_list, allowing unidirectional sequential access to its elements with a smaller footprint than list; forward_list is implemented using a singly-linked list (much like our LinkedList class).

The STL list class has an assortment of member functions; we'll start by taking a look at a few popular methods in this example to get familiar with basic STL container class usage.

Now, let's take a look at how we can utilize the STL list class. This example can be found, as a full working program with necessary class definitions, in our GitHub as follows:

https://github.com/PacktPublishing/Deciphering-Object-Oriented-Programming-with-CPP/blob/main/Chapter14/Chp14-Ex1.cpp

```cpp
#include <list>
using std::list;

int main()
{
    list<Student> studentBody;    // create a list
    Student s1("Jul", "Li", 'M', "Ms.", 3.8, "C++",
            "117PSU");
    // Note: simple heap instance below, later you can opt
    // for a smart pointer to ease allocation/deallocation
    Student *s2 = new Student("Deb", "King", 'H', "Dr.",
                        3.8, "C++", "544UD");
    // Add Students to the studentBody list.
    studentBody.push_back(s1);
    studentBody.push_back(*s2);
    // The next 3 instances are anonymous objects in main()
    studentBody.push_back(Student("Hana", "Sato", 'U',
                    "Dr.", 3.8, "C++", "178PSU"));
```

```
studentBody.push_back(Student("Sara", "Kato", 'B',
                  "Dr.", 3.9, "C++", "272PSU"));
studentBody.push_back(Student("Giselle", "LeBrun", 'R',
                  "Ms.", 3.4, "C++", "299TU"));
while (!studentBody.empty())
{
    studentBody.front().Print();
    studentBody.pop_front();
}
delete s2;   // delete any heap instances
return 0;
}
```

Let's examine the aforementioned program segment, where we create and utilize an STL `list`. First, we `#include <list>` to include the appropriate STL header file. We also add `using std::list;` to include `list` from the standard namespace. Now, in `main()`, we can instantiate a list using `list<Student> studentBody;`. Our list will contain `Student` instances. Then, we create `Student s1` on the stack and `Student *s2` on the heap using an allocation with `new()`.

Next, we use `list::push_back()` to add both `s1` and `*s2` to the list. Notice that we are passing objects to `push_back()`. As we add `Student` instances to the `studentBody` list, the list will make copies of the objects internally and will properly clean up these objects when they are no longer members of the list. We need to keep in mind that if any of our instances have been allocated on the heap, such as `*s2`, we must delete our copy of that instance when we are done with it at the end of `main()`. Looking ahead to the end of `main()`, we can see that we appropriately `delete s2;`.

Next, we add three more students to the list. These `Student` instances do not have local identifiers. These students are instantiated within the call to `push_back()`, for example, `studentBody.push_back(Student("Hana", "Sato", 'U', "Dr.", 3.8, "C++", "178PSU"));`. Here, we are instantiating an *anonymous (stack) object* that will be properly popped off the stack and destructed once the call to `push_back()` concludes. Keep in mind, `push_back()` will also create its own local copy for these instances for their life expectancy within the `list`.

Now, in a while loop, we repeatedly check whether the list is `empty()` and if not, we examine the `front()` item and call our `Student::Print()` method. We then use `pop_front()` to remove that item from the list.

Let's take a look at the output for this program:

```
Ms. Jul M. Li with id: 117PSU GPA:  3.8 Course: C++
Dr. Deb H. King with id: 544UD GPA:  3.8 Course: C++
Dr. Hana U. Sato with id: 178PSU GPA:  3.8 Course: C++
```

```
Dr. Sara B. Kato with id: 272PSU GPA:  3.9 Course: C++
Ms. Giselle R. LeBrun with id: 299TU GPA:  3.4 Course: C++
```

Now that we have deciphered a simple STL `list` class, let us move forward to understand the idea of an `iterator` to complement a container such as `list`.

Using STL iterator

Quite often, we will need a non-destructive way to iterate through a collection of objects. For example, it is important to maintain the first, last, and current position in a given container, especially if the set may be accessed by more than one method, class, or thread. Using an **iterator**, the STL provides a common means to traverse any container class.

The use of iterators has definite benefits. A class can create an `iterator` that points to the first member in a collection. Iterators can then be moved to successive next members of the collection. Iterators can provide access to elements pointed to by the `iterator`.

Overall, the state information of a container can be maintained by an `iterator`. Iterators provide a safe means for interleaved access by abstracting the state information away from the container and instead into the iterator class.

We can think of an iterator as a bookmark within a book that two or more people are referencing. The first person reads the book sequentially, leaving the bookmark neatly where they expect to continue reading. While they step away, another person looks up an important item in the book and moves the bookmark to another location in the book to save their spot. When the first person returns, they find that they have lost their current location and are not where they expect to be. Each user should have had their own bookmark or iterator. The analogy is that an iterator (ideally) allows safe interleaved access to a resource that may be handled by multiple components within an application. Without an iterator, you may unintentionally modify a container without another user's knowledge. STL iterators mostly, but not always, live up to this ideal goal.

Let's take a look at how we can utilize an STL `iterator`. This example can be found as a complete program in our GitHub as follows:

```
https://github.com/PacktPublishing/Deciphering-Object-Orient-
ed-Programming-with-CPP/blob/main/Chapter14/Chp14-Ex2.cpp
```

```cpp
#include <list>
#include <iterator>
using std::list;
using std::iterator;
```

```cpp
bool operator<(const Student &s1, const Student &s2)
{   // overloaded operator< -- required to use list::sort()
    return s1.GetGpa() < s2.GetGpa();
}


int main()
{
    list<Student> studentBody;
    Student s1("Jul", "Li", 'M', "Ms.", 3.8, "C++",
                "117PSU");
    // Add Students to the studentBody list.
    studentBody.push_back(s1);
    // The next Student instances are anonymous objects
    studentBody.push_back(Student("Hana", "Sato", 'U',
                        "Dr.", 3.8, "C++", "178PSU"));
    studentBody.push_back(Student("Sara", "Kato", 'B',
                        "Dr.", 3.9, "C++", "272PSU"));
    studentBody.push_back(Student("Giselle", "LeBrun", 'R',
                        "Ms.", 3.4, "C++", "299TU"));
    studentBody.sort();  // sort() will rely on operator<
    // Though we'll generally prefer range-for loops, let's
    // understand and demo using an iterator for looping.
    // Create a list iterator; set to first item in list.
    // We'll next simplify iterator notation with 'auto'.
    list <Student>::iterator listIter =studentBody.begin();
    while (listIter != studentBody.end())
    {
        Student &temp = *listIter;
        temp.EarnPhD();
        ++listIter;    // prefer pre-inc (less expensive)
    }
    // Simplify iterator declaration using 'auto'
    auto autoIter = studentBody.begin();
    while (autoIter != studentBody.end())
    {
        (*autoIter).Print();
```

```
            ++autoIter;
    }
    return 0;
}
```

Let's take a look at our previously defined code segment. Here, we include both the `<list>` and `<iterator>` headers from the STL. We also add `using std::list;` and `using std::iterator;` to include `list` and `iterator` from the standard namespace. As in our previous `main()` function, we instantiate a `list` that can contain `Student` instances using `list<Student> studentbody;`. We then instantiate several `Student` instances and add them to the list using `push_back()`. Again, notice that several `Student` instances are *anonymous objects*, having no local identifier in `main()`. These instances will be popped off the stack when `push_back()` completes. This is no problem, as `push_back()` will create local copies for the list.

Now, we can sort the list using `studentBody.sort();`. It is important to note that this `list` method required us to overload `operator<` to provide a means of comparison between two `Student` instances. Luckily, we have! We have chosen to implement `operator<` by comparing `gpa`, but it could also have used `studentId` for comparison.

Now that we have a `list`, we can create an `iterator` and establish it to refer to the first item of the `list`. We do so by declaring `list <Student>::iterator listIter = studentBody.begin();`. With the iterator established, we can use it to safely loop through the `list` from start (as it is initialized) to `end()`. We assign a local reference variable `temp` to the loop iteration's current first element in the list with `Student &temp = *listIter;`. We then call a method on this instance with `temp.EarnPhD();`, and then we increment our iterator by one element using `++listIter;`.

In the subsequent loop, we simplify our declaration of the iterator using `auto`. The `auto` keyword allows the type of the iterator to be determined by its initial usage. Within this loop, we also eliminate the usage of `temp` – we simply deference the iterator first within parentheses and then invoke `Print()` by using `(*autoIter).Print()`. Using `++autoIter` simply advances to the next item in our list for processing.

Let's take a look at the sorted output for this program (sorted by `gpa`):

```
Dr. Giselle R. LeBrun with id: 299TU GPA:  3.4 Course: C++
Dr. Jul M. Li with id: 117PSU GPA:  3.8 Course: C++
Dr. Hana U. Sato with id: 178PSU GPA:  3.8 Course: C++
Dr. Sara B. Kato with id: 272PSU GPA:  3.9 Course: C++
```

Now that we have seen an `iterator` class in action, let's investigate a variety of additional STL container classes, starting with `vector`.

Using STL vector

The STL `vector` class implements the Abstract Data Type of a dynamic array. Recall that we have made our own dynamic array by creating an `Array` class in *Chapter 13, Working with Templates*. The STL version, however, will be far more extensive.

The `vector` (dynamic or resizable array) will expand as necessary to accommodate additional elements beyond its initial size. The `vector` class allows direct (that is, *random access*) to elements by overloading `operator []`. A `vector` allows elements to be accessed in constant time through direct access. It is not necessary to walk past all prior elements to access an element at a specific index.

However, adding elements in the middle of a `vector` is time-consuming. That is, adding to any location other than the end of the `vector` requires all elements past the insertion point to be internally shuffled; it may also require an internal resizing of the `vector`.

Clearly, `list` and `vector`, by comparison, have different strengths and weaknesses. Each is geared to different requirements of a collection of data. We can choose the one that best fits our needs.

Let's take a look at an assortment of common `vector` member functions. This is far from a complete list:

`size_type size() const;`	Returns number of elements
`void push_back (const value_type &element);`	Adds an element to the end
`bool empty() const;`	Returns TRUE if vector is empty
`void clear();`	Erases all elements in the vector
`iterator begin();`	Returns an iterator pointing to the first element
`iterator end();`	Returns an iterator pointing to the last element

The STL `vector` class additionally includes overloaded `operator=` (assignment replaces destination vector with source vector), `operator==` (comparison of vectors, element by element), and `operator []` (returns a reference to the requested location, that is, writable memory).

Let's take a look at how we can utilize the STL `vector` class with some of its basic operations. This example can be found, as a full working program, in our GitHub as follows:

https://github.com/PacktPublishing/Deciphering-Object-Orient-ed-Programming-with-CPP/blob/main/Chapter14/Chp14-Ex3.cpp

```
#include <vector>
using std::vector;

int main()
```

```cpp
{   // instantiate two vectors
    vector<Student> studentBody1, studentBody2;
    // add 3 Students, which are anonymous objects
    studentBody1.push_back(Student("Hana", "Sato", 'U',
                            "Dr.", 3.8, "C++", "178PSU"));
    studentBody1.push_back(Student("Sara", "Kato", 'B',
                            "Dr.", 3.9, "C++", "272PSU"));
    studentBody1.push_back(Student("Giselle", "LeBrun",
                            'R', "Ms.", 3.4, "C++", "299TU"));

    // Compare this loop to next loop using an iterator and
    // also to the preferred range-for loop further beyond
    for (int i = 0; i < studentBody1.size(); i++)
        studentBody1[i].Print();   // print first vector

    studentBody2 = studentBody1;   // assign one to another
    if (studentBody1 == studentBody2)
        cout << "Vectors are the same" << endl;

    // Notice: auto keyword simplifies iterator declaration
    for (auto iter = studentBody2.begin();
            iter != studentBody2.end(); iter++)
        (*iter).EarnPhD();

    // Preferred range-for loop (and auto to simplify type)
    for (const auto &student : studentBody2)
        student.Print();

    if (!studentBody1.empty())   // clear first vector
        studentBody1.clear();
    return 0;
}
```

In the previously listed code segment, we #include <vector> to include the appropriate STL header file. We also add using std::vector; to include vector from the standard namespace. Now, in main(), we can instantiate two vectors using vector<Student> studentBody1, studentBody2;. We can then use the vector::push_back() method to add several Student instances in succession to our first vector. Again, notice that the Student instances are *anonymous objects* in main(). That is, there is no local identifier that references them – they are created only to be placed into our vector, which makes a local copy of each instance upon insertion. Once we have elements in our vector, we then loop through our first vector, printing each Student using studentBody1[i].Print();.

Next, we demonstrate the overloaded assignment operator for vector by assigning one vector to another using studentBody1 = studentBody2;. Here, we make a deep copy from right to left in the assignment. We can then test whether the two vectors are equal using the overloaded comparison operator within a conditional statement, that is, if (studentBody1 == studentBody2).

We then apply EarnPhD() to the contents of the second vector in a for loop using an iterator specified with auto iter = studentBody2.begin();. The auto keyword allows the type of the iterator to be determined by its initial usage. We then print out the contents of our second vector using a preferred range-for loop (as well as using auto to simplify the variable type in the range-for loop). Finally, we look through our first vector, testing whether it is empty(), and then clear elements one by one using studentBody1.clear();. We have now seen a sampling of the vector methods and their capabilities.

Let's take a look at the output for this program:

```
Dr. Hana U. Sato with id: 178PSU GPA:  3.8 Course: C++
Dr. Sara B. Kato with id: 272PSU GPA:  3.9 Course: C++
Ms. Giselle R. LeBrun with id: 299TU GPA:  3.4 Course: C++
Vectors are the same
Everyone to earn a PhD
Dr. Hana U. Sato with id: 178PSU GPA:  3.8 Course: C++
Dr. Sara B. Kato with id: 272PSU GPA:  3.9 Course: C++
Dr. Giselle R. LeBrun with id: 299TU GPA:  3.4 Course: C++
```

Next, let's investigate the STL deque class to further our knowledge of STL containers.

Using STL deque

The STL deque class (pronounced *deck*) implements the Abstract Data Type of a double-ended queue. This ADT extends the notion that a queue is first in, first out. Instead, the deque class allows greater flexibility. Adding elements at either end of a deque is quick. Adding elements in the middle of a deque is time-consuming. A deque is a sequential container, though more flexible than a list.

You might imagine that a deque is a specialization of a queue; it is not. Instead, the flexible deque class will serve as a basis to implement other container classes, which we will see shortly. In these cases, private inheritance will allow us to conceal deque as an underlying implementation (with vast functionality) for more restrictive, specialized classes.

Let's take a look at an assortment of common deque member functions. This is far from a complete list:

iterator begin();	Returns an iterator pointing to the first element
iterator end();	Returns an iterator pointing to the last element
size_type size() const;	Returns number of element
void push_back (const value_type &element);	Adds an element to end
void push_front (const value_type &element);	Adds an element to the front
void pop_back();	Removes an element from the end
void pop_front();	Removes an element from the front
bool empty() const;	Returns TRUE if empty
void clear();	Erases all elements

The STL deque class additionally includes overloaded operator= (assignment of the source to destination deque) and operator[] (returns a reference to requested location – writable memory).

Let's take a look at how we can utilize the STL deque class. This example can be found, as a full working program, in our GitHub as follows:

https://github.com/PacktPublishing/Deciphering-Object-Orient-ed-Programming-with-CPP/blob/main/Chapter14/Chp14-Ex4.cpp

```cpp
#include <deque>
using std::deque;

int main()
{
    deque<Student> studentBody;    // create a deque
    Student s1("Tim", "Lim", 'O', "Mr.", 3.2, "C++",
            "111UD");
    // the remainder of the Students are anonymous objects
    studentBody.push_back(Student("Hana", "Sato", 'U',
                    "Dr.",3.8, "C++", "178PSU"));
```

```
    studentBody.push_back(Student("Sara", "Kato", 'B',
                           "Dr.", 3.9, "C++", "272PSU"));
    studentBody.push_front(Student("Giselle", "LeBrun",
                           'R',"Ms.", 3.4, "C++", "299TU"));
    // insert one past the beginning
    studentBody.insert(std::next(studentBody.begin()),
    Student("Anne", "Brennan", 'B', "Ms.", 3.9, "C++",
            "299CU"));

    studentBody[0] = s1;   // replace 0th element;
                           // no bounds checking!
    while (!studentBody.empty())
    {
        studentBody.front().Print();
        studentBody.pop_front();
    }
    return 0;
}
```

In the previously listed code segment, we #include <deque> to include the appropriate STL header file. We also add using std::deque; to include deque from the standard namespace. Now, in main(), we can instantiate a deque to contain Student instances using deque<Student> studentBody;. We then call either deque::push_back() or deque::push_front() to add several Student instances (some anonymous objects) to our deque. We are getting the hang of this! Now, we insert a Student one position past the front of our deque using studentBody. insert(std::next(studentBody.begin()), Student("Anne", "Brennan", 'B', "Ms.", 3.9, "C++", "299CU"));.

Next, we take advantage of overloaded operator[] to insert a Student into our deque using studentBody[0] = s1;. Please be warned that operator[] does not do any bounds checking on our deque! In this statement, we insert Student s1 into the 0th position in the deque, instead of the Student that once occupied that position. A safer bet is to use the deque::at() method, which will incorporate bounds checking. Regarding the aforementioned assignment, we also want to ensure that operator= has been overloaded for both Person and Student, as each class has dynamically allocated data members.

Now, we loop through until our deque is empty(), extracting and printing the front element of the deque using studentBody.front().Print();. With each iteration, we also pop the front item from our deque using studentBody.pop_front();.

Let's take a look at the output for this program:

```
Mr. Tim O. Lim with id: 111UD GPA:  3.2 Course: C++
Ms. Anne B. Brennan with id: 299CU GPA:  3.9 Course: C++
Dr. Hana U. Sato with id: 178PSU GPA:  3.8 Course: C++
Dr. Sara B. Kato with id: 272PSU GPA:  3.9 Course: C++
```

Now that we have a feel for a `deque`, let's next investigate the STL `stack` class.

Using STL stack

The STL `stack` class implements the Abstract Data Type of a stack. The stack ADT supports the **last in, first out** (**LIFO**) order for the insertion and removal of members. True to form for an ADT, the STL `stack` includes a public interface that does not advertise its underlying implementation. After all, a stack might change its implementation; the ADTs usage should not depend in any manner on its underlying implementation. The STL `stack` is considered an adaptive interface of a basic sequential container.

Recall that we have made our own `Stack` class in *Chapter 6, Implementing Hierarchies with Inheritance*, using a private base class of `LinkedList`. The STL version will be more extensive; interestingly, it is implemented using `deque` as its underlying private base class. With `deque` as a private base class of the STL `stack`, the more versatile underlying capabilities of `deque` are hidden; only the applicable methods are used to implement the stack's public interface. Also, because the means of implementation is hidden, a `stack` may be implemented using another container class at a later date without impacting its usage.

Let's take a look at an assortment of common `stack` member functions. This is far from a complete list. It is important to note that the public interface for `stack` is far smaller than that of its private base class, `deque`:

`size_type size() const;`	Returns number of elements
`void push (const value_type &element);`	Adds an element to top
`void pop();`	Removes an element from top
`bool empty() const;`	Returns TRUE if empty
`value_type &top();`	Returns top element
`const value_type &top() const;`	Returns top element (read only)

The STL `stack` class additionally includes overloaded `operator=` (assignment of source to destination stack), `operator==` and `operator!=` (equality/inequality of two stacks), and `operator<`, `operator>`, `operator<=`, and `operator>=` (comparison of stacks).

Let's take a look at how we can utilize the STL stack class. This example can be found, as a full working program, in our GitHub as follows:

https://github.com/PacktPublishing/Deciphering-Object-Oriented-Programming-with-CPP/blob/main/Chapter14/Chp14-Ex5.cpp

```
#include <stack>     // template class preamble
using std::stack;

int main()
{
    stack<Student> studentBody;    // create a stack
    // add Students to the stack (anonymous objects)
    studentBody.push(Student("Hana", "Sato", 'U', "Dr.",
                          3.8, "C++", "178PSU"));
    studentBody.push(Student("Sara", "Kato", 'B', "Dr.",
                          3.9, "C++", "272PSU"));
    studentBody.push(Student("Giselle", "LeBrun", 'R',
                          "Ms.", 3.4, "C++", "299TU"));

    while (!studentBody.empty())
    {
        studentBody.top().Print();
        studentBody.pop();
    }
    return 0;
}
```

In the aforementioned code segment, we #include <stack> to include the appropriate STL header file. We also add using std::stack; to include stack from the standard namespace. Now, in main(), we can instantiate a stack to contain Student instances using stack<Student> studentBody;. We then call stack::push() to add several Student instances to our stack. Notice that we are using the traditional push() method, which contributes to the ADT of a stack.

We then loop through our stack while it is not empty(). Our goal is to access and print the top element using studentBody.top().Print();. We then neatly pop our top element off the stack using studentBody.pop();.

Let's take a look at the output for this program:

```
Ms. Giselle R. LeBrun with id: 299TU GPA:   3.4 Course: C++
Dr. Sara B. Kato with id: 272PSU GPA:   3.9 Course: C++
Dr. Hana U. Sato with id: 178PSU GPA:   3.8 Course: C++
```

Next, let's investigate the STL queue class to further increase our STL container repertoire.

Using STL queue

The STL queue class implements the ADT of a queue. As the stereotypical queue class, STL's queue class supports the **first in, first out** (**FIFO**) order of insertion and removal of members.

Recall that we made our own Queue class in *Chapter 6, Implementing Hierarchies with Inheritance*; we derived our Queue from our LinkedList class using private inheritance. The STL version will be more extensive; the STL queue class is implemented using deque as its underlying implementation (also using private inheritance). Remember, because the means of implementation are hidden with private inheritance, a queue may be implemented using another data type at a later date without impacting its public interface. The STL queue class is another example of an adaptive interface for a basic sequential container.

Let's take a look at an assortment of common queue member functions. This is far from a complete list. It is important to note that the public interface of queue is far smaller than that of its private base class, deque:

`size_type size size() const;`	Returns number of elements
`bool empty() const;`	Returns TRUE if empty
`value_type &front();`	Returns front element
`const value_type &front() const;`	Returns front element (read only)
`value_type &back();`	Returns back element
`const value_type &back() const;`	Returns back element (read only)
`void push (const value_type &element);`	Adds element to back of queue
`void pop();`	Removes element from front of queue

The STL queue class additionally includes overloaded operator= (assignment of source to destination queue), operator== and operator!= (equality/inequality of two queues), and operator< , operator>, operator<=, and operator>= (comparison of queues).

Let's take a look at how we can utilize the STL queue class. This example can be found, as a full working program, in our GitHub as follows:

https://github.com/PacktPublishing/Deciphering-Object-Orient-ed-Programming-with-CPP/blob/main/Chapter14/Chp14-Ex6.cpp

```cpp
#include <queue>
using std::queue;

int main()
{
    queue<Student> studentBody;   // create a queue
    // add Students to the queue (anonymous objects)
    studentBody.push(Student("Hana", "Sato", 'U', "Dr.",
                            3.8, "C++", "178PSU"));
    studentBody.push(Student("Sara", "Kato", 'B' "Dr.",
                            3.9, "C++", "272PSU"));
    studentBody.push(Student("Giselle", "LeBrun", 'R',
                            "Ms.", 3.4, "C++", "299TU"));
    while (!studentBody.empty())
    {
        studentBody.front().Print();
        studentBody.pop();
    }
    return 0;
}
```

In the previous code segment, we first #include <queue> to include the appropriate STL header file. We also add using std::queue; to include queue from the standard namespace. Now, in main(), we can instantiate a queue to contain Student instances using queue<Student> studentBody;. We then call queue::push() to add several Student instances to our queue. Recall that with the queue ADT, push() implies that we are adding an element at the end of the queue. Some programmers prefer the term *enqueue* to describe this operation; however, the STL has selected to name this operation push(). With the queue ADT, pop() will remove an item from the front of the queue; a better term is *dequeue*, however, that is not what the STL has chosen. We can adapt.

We then loop through our queue while it is not empty(). Our goal is to access and print the front element using studentBody.front().Print();. We then neatly pop our front element off the queue using studentBody.pop();. Our work is complete.

Let's take a look at the output for this program:

```
Dr. Hana U. Sato with id: 178PSU GPA:   3.8 Course: C++
Dr. Sara B. Kato with id: 272PSU GPA:   3.9 Course: C++
Ms. Giselle R. LeBrun with id: 299TU GPA:   3.4 Course: C++
```

Now that we have tried a `queue`, let's investigate the STL `priority_queue` class.

Using STL priority queue

The STL `priority_queue` class implements the Abstract Data Type of a priority queue. The priority queue ADT supports a modified FIFO order of insertion and removal of members; the elements are *weighted*. The front element is of the largest value (determined by overloaded `operator<`) and the rest of the elements follow in sequence from the next greatest to the least. The STL `priority_queue` class is considered an adaptive interface for a sequential container.

Recall that we implemented our own `PriorityQueue` class in *Chapter 6, Implementing Hierarchies with Inheritance*. We used public inheritance to allow our `PriorityQueue` to specialize our `Queue` class, adding additional methods to support the priority (weighted) enqueuing scheme. The underlying implementation of `Queue` (with private base class `LinkedList`) was hidden. By using public inheritance, we allowed our `PriorityQueue` to be able to be generalized as a `Queue` through upcasting (which we understood once we learned about polymorphism and virtual functions in *Chapter 7, Utilizing Dynamic Binding through Polymorphism*). We made an acceptable design choice: *PriorityQueue Is-A* (specialization of) *Queue* and at times may be treated in its more general form. We also recall that neither a `Queue` nor a `PriorityQueue` could be upcast to their underlying implementation of a `LinkedList`, as `Queue` was derived privately from `LinkedList`; we cannot upcast past a non-public inheritance boundary.

Contrastingly, the STL version of `priority_queue` is implemented using the STL `vector` as its underlying implementation. Recall that because the means of implementation is hidden, a `priority_queue` may be implemented using another data type at a later date without impacting its public interface.

An STL `priority_queue` allows an inspection, but not a modification, of the top element. The STL `priority_queue` does not allow insertion through its elements. That is, elements may only be added resulting in an order from greatest to least. Accordingly, the top element may be inspected, and the top element may be removed.

Let's take a look at an assortment of common `priority_queue` member functions. This is not a complete list. It is important to note that the public interface of `priority_queue` is far smaller than that of its private base class, `vector`:

`size_type size() const;`	Returns number of elements
`bool empty() const;`	Returns TRUE if empty
`const value_type &top() const;`	Returns top element (read only)
`void push (const value_type &element);`	Adds an element to the back
`void pop();`	Removes an element from the front
`void emplace(args);`	A new element is constructed using args and then added

Unlike the previously examined container classes, the STL `priority_queue` does not overload operators, including `operator=`, `operator==`, and `operator<`.

The most interesting method of `priority_queue` is that of `void emplace(args);`. This is the member function that allows the priority enqueuing mechanism to add items to this ADT. We also notice that `top()` must be used to return the top element (versus `front()`, which a `queue` utilizes). But then again, an STL `priority_queue` is not implemented using a `queue`). To utilize `priority_queue`, we `#include <queue>`, just as we would for a `queue`.

Because the usage of `priority_queue` is so similar to `queue`, we will instead explore it further, programming-wise, in our question set at the end of this chapter.

Now that we have seen many examples of sequential container types in STL (including adaptive interfaces), let's next investigate the STL `map` class, an associative container.

Examining STL map

The STL `map` class implements the Abstract Data Type of a hash table. The class `map` allows for elements in the hash table or map to be stored and retrieved quickly using a **key** or index. The key can be numerical or any other data type. Only one key may be associated with a single element of value. However, the STL container `multimap` can be used instead should there be more than one piece of data that needs to be associated with a single key.

Hash tables (maps) are fast for storage and lookup of data. The performance is a guaranteed $O(log(n))$. The STL `map` is considered an associative container, as it associates a key to a value to quickly retrieve a value.

Let's take a look at an assortment of common map member functions. This is not a complete list:

`bool empty() const;`	Returns TRUE if empty, FALSE otherwise
`size_type size() const;`	Returns number of element
`void clear();`	Empties out the map
`size_type erase(const key_type &k);`	Erases element with key of k
`iterator erase(const_iterator pos);`	Erases element at iterator
`void swap(map &);`	Swaps contents of 2 maps
`iterator insert(iterator pos, const value_type &val);`	Insert val at pos
`iterator begin();`	Returns iterator pointing to the first element
`iterator end();`	Returns iterator pointing past the end
`iterator find(const key_type &k);`	Find the element whose key is k

The STL class map additionally includes overloaded operator `operator==` (comparison of maps, element by element) implemented as a global function. STL map also includes overloaded `operator[]` (returns a reference to the map element associated with a key that is used as an index; this is writable memory).

Let's take a look at how we can utilize the STL map class. This example can be found, as a full working program, in our GitHub as follows:

https://github.com/PacktPublishing/Deciphering-Object-Oriented-Programming-with-CPP/blob/main/Chapter14/Chp14-Ex7.cpp

```cpp
#include <map>
using std::map;
using std::pair;

bool operator<(const Student &s1, const Student &s2)
{   // We need to overload operator< to compare Students
    return s1.GetGpa() < s2.GetGpa();
}

int main()
{
    Student s1("Hana", "Lo", 'U', "Dr.", 3.8, "C++",
               "178UD");
```

```
Student s2("Ali", "Li", 'B', "Dr.", 3.9, "C++",
          "272UD");
Student s3("Rui", "Qi", 'R', "Ms.", 3.4, "C++",
          "299TU");
Student s4("Jiang", "Wu", 'C', "Ms.", 3.8, "C++",
          "887TU");

// create three pairings of ids to Students
pair<string, Student> studentPair1
                        (s1.GetStudentId(), s1);
pair<string, Student> studentPair2
                        (s2.GetStudentId(), s2);
pair<string, Student> studentPair3
                        (s3.GetStudentId(), s3);

// Create map of Students w string keys
map<string, Student> studentBody;
studentBody.insert(studentPair1);   // insert 3 pairs
studentBody.insert(studentPair2);
studentBody.insert(studentPair3);
// insert using virtual indices per map
studentBody[s4.GetStudentId()] = s4;

// Iterate through set with map iterator - let's
// compare to range-for and auto usage just below
map<string, Student>::iterator mapIter;
mapIter = studentBody.begin();
while (mapIter != studentBody.end())
{
    // set temp to current item in map iterator
    pair<string, Student> temp = *mapIter;
    Student &tempS = temp.second;   // get 2nd element
    // access using mapIter
    cout << temp.first << " ";
    cout << temp.second.GetFirstName();
```

```
            // or access using temporary Student, tempS
            cout << " " << tempS.GetLastName() << endl;
            ++mapIter;
    }
    // Now, let's iterate through our map using a range-for
    // loop and using 'auto' to simplify the declaration
    // (this decomposes the pair to 'id' and 'student')
    for (auto &[id, student] : studentBody)
        cout << id << " " << student.GetFirstName() << " "
                << student.GetLastName() << endl;
    return 0;
}
```

Let's examine the preceding code segments. Again, we include the applicable header file with #include <map>. We also add using std::map; and using std::pair; to include map and pair from the standard namespace. Next, we instantiate four Student instances. Next, we create three pair instances to associate a grouping between each Student and its key (that is, with their respective studentId) using the declaration pair<string, Student> studentPair1 (s1. GetStudentId(), s1);. This may seem confusing to read, but let's break this declaration down into its components. Here, the instance's data type is pair<string, Student>, the variable name is studentPair1, and (s1.GetStudentId(), s1) are the arguments passed to the specific pair instance's constructor.

We will be making a hash table (map) of Student instances to be indexed by a key (which is their studentId). Next, we declare a map to hold the collection of Student instances with map<string, Student> studentBody;. Here, we indicate that the association between the key and element will be between a string and a Student. We then declare a map iterator with map<string, Student>::iterator mapIter; using the same data types.

Now, we simply insert the three pair instances into the map. An example of this insertion is studentBody.insert(studentPair1);. We then insert a fourth Student, s4, into the map using the map's overloaded operator[] with the following statement: studentBody[s4.GetStudentId()] = s4;. Notice that the studentId is used as the index value in operator[]; this value will become the key value for the Student in the hash table.

Next, we declare and establish the map iterator to the beginning of the map and then process the map while it is not at the end(). Within the loop, we set a variable, temp, to the pair at the front of the map, indicated by the map iterator. We also set tempS as a temporary reference to a Student in the map, which is indicated by temp.second (the second value in the current pair managed by the map iterator). We now can print out each Student instance's studentId (the key, which is a string) using temp.first (the first item in the current pair). In the same statement, we can then print out each Student instance's firstName using temp.second.GetFirstName()

(since the `Student` corresponding to the key is the second item in the current `pair`). Similarly, we could also use `tempS.GetLastName()` to print a student's `lastName`, as `tempS` was previously initialized to the second element in the current `pair` at the beginning of each loop iteration.

Finally, as an alternative to the more tedious approach demonstrated previously used to iterate through the `map` (taking apart the `pair` manually), let's examine the final loop in our program. Here, we utilize a range-for loop to process the map. The use of `auto` with `&[id, student]` will specify the type of data that we will iterate. The brackets (`[]`) will decompose the `pair`, binding the iterative elements to `id` and `student`, respectively, as identifiers. Notice the ease at which we can now iterate over the `studentBody` map.

Let's take a look at the output for this program:

```
178UD  Hana Lo
272UD  Ali Li
299TU  Rui Qi
887TU  Jiang Wu
178UD  Hana Lo
272UD  Ali Li
299TU  Rui Qi
887TU  Jiang Wu
```

Next, let's take a look at an alternative with an STL map, which will introduce us to the STL `functor` concept.

Examining STL map using a functor

The STL map class has great flexibility, like many STL classes. In our past map example, we assumed that a means for comparison was present in our `Student` class. We had, after all, overloaded `operator<` for two `Student` instances. What happens, however, if we cannot revise a class that has not provided this overloaded operator and we also choose not to overload `operator<` as an external function?

Fortunately, we may specify a third data type for the template type expansion when instantiating a map or map iterator. This additional data type will be a specific type of class, known as a functor. A **functor** is an object that can be treated as though it is a function or function pointer. We will create a class (or struct) to represent our functor type, and within that class (or struct), we must overload `operator()`. It is within overloaded `operator()` that we will provide a means of comparison for the objects in question. A functor essentially simulates encapsulating a function pointer by overloading `operator()`.

Let's take a look at how we might revise our map example to utilize a simple functor. This example can be found, as a full working program, in our GitHub as follows:

https://github.com/PacktPublishing/Deciphering-Object-Orient-
ed-Programming-with-CPP/blob/main/Chapter14/Chp14-Ex8.cpp

```cpp
#include <map>
using std::map;
using std::pair;

struct comparison     // This struct represents a 'functor'
{                     // that is, a 'function object'
    bool operator() (const string &key1,
                     const string &key2) const
    {
        int ans = key1.compare(key2);
        if (ans >= 0) return true;  // return a boolean
        else return false;
    }
    // default constructor and destructor are adequate
};

int main()
{
    Student s1("Hana", "Sato", 'U', "Dr.", 3.8, "C++",
            "178PSU");
    Student s2("Sara", "Kato", 'B', "Dr.", 3.9, "C++",
            "272PSU");
    Student s3("Jill", "Long", 'R', "Dr.", 3.7, "C++",
            "234PSU");
    // Now, map is maintained in sorted (decreasing) order
    // per <comparison> functor using operator()
    map<string, Student, comparison> studentBody;
    map<string, Student, comparison>::iterator mapIter;
    // The remainder of the program is similar to prior
}   // map program. See online code for complete example.
```

In the previously mentioned code fragment, we first introduce a user defined type of `comparison`. This can be a `class` or a `struct`. Within this structure definition, we have overloaded the function call operator (`operator()`) and provided a means of comparison between two `string` keys for `Student` instances. This comparison will allow `Student` instances to be inserted in an order determined by the comparison functor.

Now, when we instantiate our `map` and map iterators, we specify as the third parameter for the template type expansion our `comparison` type (the functor). And, neatly embedded within this type is the overloaded function call operator, `operator()`, which will provide our needed comparison. The remaining code will be similar to our original `map` program.

Certainly, functors may be used in additional, more advanced ways beyond what we have seen here with the container class `map`. Nonetheless, you now have a flavor for how a functor can apply to the STL.

Now that we have seen how to utilize a variety of STL container classes, let's consider why we may want to customize an STL class, and how to do so.

Customizing STL containers

Most classes in C++ can be customized in some fashion, including classes in the STL. However, we must be aware of design decisions made within the STL that will limit how we may customize these components. Because the STL container classes purposely do not include virtual destructors or other virtual functions, we should not use specialization via public inheritance to extend these classes. Note that C++ will not stop us, but we know from *Chapter 7, Using Dynamic Binding through Polymorphism*, that we should never override non-virtual functions. STL's choice to not include virtual destructors and other virtual functions to allow further specialization of these classes was a solid design choice made long ago when STL containers were crafted.

We could, however, use private or protected inheritance, or the concepts of containment or association to use an STL container class as a building block. That is, to hide the underlying implementation of a new class, where the STL provides a solid, yet hidden implementation for the new class. We would simply provide our own public interface for the new class and, under the hood, delegate the work to our underlying implementation (whether that be a private or protected base class, or a contained or associated object).

Extreme care and caution must be taken when extending any template class, including those in the STL using private or protected base classes. This caution will also apply to containing or associating to other template classes. Template classes are generally not compiled (or syntax checked) until an instance of the template class with a specific type is created. This means that any derived or wrapper classes that are created can only be fully tested when instances of specific types are created.

Appropriate overloaded operators will need to be put in place for new classes so that these operators will work automatically with customized types. Keep in mind that some operator functions, such as operator=, are not explicitly inherited from base to derived class and need to be written with each new class. This is appropriate since derived classes will likely have more work to accomplish than found in the generalized versions of operator=. Remember, if you cannot modify the class definition of a class requiring a selected overloaded operator, you must implement that operator function as an external function.

In addition to customizing containers, we may also choose to augment an algorithm based on an existing algorithm within the STL. In this case, we would use one of the many STL functions as part of a new algorithm's underlying implementation.

Customizing classes from existing libraries comes up routinely in programming. For example, consider how we extended the Standard Library exception class to create customized exceptions in *Chapter 11, Handling Exceptions* (though that scenario utilized public inheritance, which will not apply to customizing STL classes). Keep in mind that the STL offers a very full complement of container classes. Rarely will you find the need to augment STL classes – perhaps only with a very domain-specific class need. Nonetheless, you now know the caveats involved in customizing STL classes. Remember, care and caution must always be used when augmenting a class. We can now see the need to employ proper OO component testing for any classes we create.

We have now considered how to potentially customize STL container classes and algorithms within our programs. We have also seen quite a few STL container class examples in action. Let us now briefly recap these concepts before moving forward to our next chapter.

Summary

In this chapter, we have furthered our C++ knowledge beyond OOP language features to gain familiarity with the C++ Standard Template Library. As this library is used so commonly in C++, it is essential that we understand both the scope and breadth of the classes it contains. We are now prepared to utilize these useful, well-tested classes in our code.

We have looked at quite a few STL examples; by examining selective STL classes, we should feel empowered to understand the remainder of the STL (or any C++ library) on our own.

We have seen how to use common and essential STL classes such as list, iterator, vector, deque, stack, queue, priority_queue, and map. We have also seen how to utilize a functor in conjunction with a container class. We have been reminded that we now have the tools to potentially customize any class, even those from class libraries such as STL through private or protected inheritance, or with containment or association.

We have additionally seen through examining selected STL classes that we have the skills to understand the remaining depth and breadth of the STL, as well as decode many additional class libraries that are available to us. As we navigate the prototypes of each member function, we notice key language concepts, such as the use of `const`, or that a method returns a reference to an object representing writable memory. Each prototype reveals the mechanics for the usage of the new class. It is very exciting to have come this far with our programming endeavors!

We have now added additional, useful features to our C++ repertoire through browsing the STL in C++. Usage of the STL (to encapsulate traditional data structures) will ensure that our code can easily be understood by other programmers who are also undoubtedly using the STL. Relying on the well-tested STL for these common containers and utilities ensures that our code remains more bug-free.

We are now ready to continue forward with *Chapter 15, Testing Classes and Components*. We want to complement our C++ programming skills with useful OO component testing skills. Testing skills will help us understand whether we have created, extended, or augmented classes in a robust fashion. These skills will make us better programmers. Let's continue onward!

Questions

1. Replace your template `Array` class from your exercise from *Chapter 13, Working with Templates*, with an STL `vector`. Create a `vector` of `Student` instances. Use `vector` operations to insert, retrieve, print, compare, and remove objects from the vector. Alternatively, utilize an STL `list`. Use this opportunity to utilize the STL documentation to navigate the full set of operations available for these classes.

    a. Consider what operators, if any, you will need to overload. Consider whether you will need an `iterator` to provide safe interleaved access to your collection.

    b. Create a second `vector` of `Student` instances. Assign one to another. Print both vectors.

2. Modify the `map` from this chapter to index the hash table (map) of `Student` instances based on `lastName` rather than `studentId`.

3. Modify the `queue` example from this chapter to instead utilize `priority_queue`. Be sure to make use of the priority enqueueing mechanism `priority_queue::emplace()` to add elements into the `priority_queue`. You will also need to utilize `top()` instead of `front()`. Note that `priority_queue` can be found in the `<queue>` header file.

4. Try out an STL algorithm using `sort()`. Be sure to `#include <algorithm>`. Sort an array of integers. Keep in mind that many containers have sorting mechanisms built in, but native collection types, such as a language-supplied array, will not (which is why you should use a basic array of integers).

15

Testing Classes and Components

This chapter will continue our pursuit of increasing your C++ programming repertoire beyond OOP concepts through exploring means to test the classes and components that comprise our OO programs. We will explore various strategies to help ensure that the code we write will be well-tested and robust.

This chapter shows how to test your OO programs through testing individual classes, as well as testing the various components that work together.

In this chapter, we will cover the following main topics:

- Understanding the canonical class form and creating robust classes
- Creating drivers to test classes
- Testing classes related by inheritance, association, or aggregation
- Testing exception handling mechanisms

By the end of this chapter, you will have various techniques in your programming arsenal to ensure that your code is well-tested before it goes into production. Having the skills to consistently produce robust code will help you become a more beneficial programmer.

Let's increase our C++ skills set by examining various techniques for OO testing.

Technical requirements

Online code for full program examples can be found in the following GitHub URL: `https://github.com/PacktPublishing/Deciphering-Object-Oriented-Programming-with-CPP/tree/main/Chapter15`. Each full program example can be found in the GitHub repository under the appropriate chapter heading (subdirectory) in a file that corresponds to the chapter number, followed by a dash, followed by the example number in the chapter at hand. For example, the first full program in this chapter can be found in the subdirectory `Chapter15` in a file named `Chp15-Ex1.cpp` under the aforementioned GitHub directory.

The CiA video for this chapter can be viewed at: `https://bit.ly/3AxyLFH`.

Contemplating OO testing

Software testing is immensely important prior to any code deployment. Testing object-oriented software will require different techniques than other types of software. Because OO software contains relationships between classes, we must understand how to test dependencies and relationships that may exist between classes. Additionally, each object may progress through different states based on the order that operations are applied to each instance, as well as through specific interactions with related objects (for example, via association). The overall flow of control through an OO application is much more complex than with procedural applications, as the combinations and order of operations applied to a given object and influences from associated objects are numerous.

Nonetheless, there are metrics and processes we can apply to test OO software. These range from understanding idioms and patterns we can apply for class specification, to creating drivers to test classes both independently and as they relate to other classes. These processes can further include creating scenarios to provide likely sequences of events or states that objects may progress through. Relationships between objects, such as inheritance, association, and aggregation become very important in testing; related objects can influence the state of an existing object.

Let's begin our quest in testing OO software by understanding a simple pattern that we can often apply to classes we develop. This idiom will ensure that a class is potentially complete, with no unexpected behavior. We will start with the canonical class form.

Understanding the canonical class form

For many classes in C++, it is reasonable to follow a pattern for class specification to ensure that a new class contains a full set of desired components. The **canonical class form** is a robust specification of a class that enables class instances to provide uniform behavior (analogous to standard data types) in areas such as initialization, assignment, argument passing, and usage in return values from functions. The canonical class form will apply to most classes that are intended for either instantiation or that will serve as public base classes for new derived classes. Classes that are intended to serve as private or protected base classes (even if they may be instantiated themselves) may not follow all parts of this idiom.

A class following **orthodox** canonical form will include the following:

- A default constructor (or an =default prototype to explicitly allow this interface)
- A copy constructor
- An overloaded assignment operator
- A virtual destructor

Though any of the aforementioned components may be prototyped with =default to explicitly utilize the default, system-supplied implementations, modern preferences are moving away from such practices (as these prototypes are generally redundant). The exception is the default constructor whose interface you will not otherwise get without using =default when other constructors are present.

A class following the **extended** canonical form will additionally include the following:

- A *move* copy constructor
- A *move* assignment operator

Let's look at each component of the canonical class form in the next subsections.

Default constructor

A **default constructor** is necessary for simple instantiation. Though a default (empty) constructor will be provided if a class contains no constructors, it is important to recall that a default constructor will not be provided if a class contains constructors with other signatures. It is best to provide a default constructor with reasonable, basic initialization, or alternatively, add =default to the default constructor prototype; this is especially useful when in-class initialization is utilized.

Additionally, a default constructor for a given class' base class will be called in the absence of an alternate base class constructor specification in the member initialization list. If a base class has no such default constructor (and one hasn't been provided because a constructor with another signature exists), the implicit call to the base class constructor will be flagged as an error.

Let's also consider multiple inheritance situations in which a diamond-shaped hierarchy occurs, and virtual base classes are used to eliminate duplication of the most base class subobjects within instances of the most derived class. In this scenario, the default constructor for the now *shared* base class subobject is called unless otherwise specified in the member initialization list of the derived class responsible for creating the diamond shape. This occurs even if non-default constructors are specified in the member initialization list at the middle level; remember these specifications are ignored when the mid-levels specify a potentially shared virtual base class.

Copy constructor

A **copy constructor** is often crucial for all objects containing pointer data members. Unless a copy constructor is supplied by the programmer, a system-supplied copy constructor will be linked in when necessary in the application. The system-supplied copy constructor performs a member-wise (shallow) copy of all data members. This means that multiple instances of a class may contain pointers to *shared* pieces of memory representing the data that should have been individualized. Unless resource sharing is intended, raw pointer data members in the newly instantiated object will want to allocate their own memory and copy the data values from the source object into this memory. Also, remember to use the member initialization list in a derived class copy constructor to specify the base class' copy constructor to copy the base class data members. Certainly, copying the base class subobject in a deep fashion is crucial; additionally, the base class data members are inevitably private, so selecting the base class copy constructor in the derived class' member initialization list is very important.

By specifying a copy constructor, we also help provide an expected manner for the creation of objects passed (or returned) by value from a function. Ensuring deep copies in these scenarios is crucial. The user may think these copies are *by value*, yet if their pointer data members are actually shared with the source instance, it's not truly passing (or returning) an object by value.

Overloaded assignment operator

An **overloaded assignment operator**, much like the copy constructor, is often also crucial for all objects containing pointer data members. The default behavior for the system-supplied assignment operator is a shallow assignment of data from source to destination object. Again, when data members are raw pointers, unless the two objects want to share the resources for heap data members, it is highly recommended that the assignment operator should be overloaded. Allocated space in the destination object should be equal to the source data member sizes for any such pointer data members. The contents (data) should then be copied from source to destination object for each pointer data member.

Also, remember that an overloaded assignment operator is not *inherited*; each class is responsible for writing its own version. This makes sense, as the derived class inevitably has more data members to copy than the assignment operator function in its base class. However, when overloading an assignment operator in a derived class, remember to call the base class' assignment operator to perform a deep assignment of inherited base class members (which may be private and otherwise inaccessible).

Virtual destructor

A **virtual destructor** is required when using public inheritance. Often, derived class instances are collected in a group and generalized by a set of base class pointers. Recall that upcasting in this fashion is only possible to public base classes (not to protected or private base classes). When pointers to objects are generalized in this fashion, a virtual destructor is crucial to allow the correct starting point in the destructor sequence to be determined through dynamic (that is, runtime) binding versus static binding. Recall that static binding would choose the starting destructor based on the pointer's type, not what type the object actually is. A good rule of thumb is if a class has one or more virtual functions, be sure to ensure that you also have a virtual destructor (even if it is only a virtual destructor prototype utilizing =default).

Move copy constructor

A **move copy constructor** is much like a copy constructor and often crucial for all objects containing pointer data members; however, the goal is to conserve memory as well as optimize performance (such as by eliminating unnecessary copies of objects). Rather than the newly constructed object becoming a deep copy of the source object, the motivation is to instead *move* the inner allocated memory resources from the source instance to the newly allocated instance. With this goal of moving resource ownership, we simply perform assignments for the pointer data members from the source object to the data members pointed to by this. We then must null the source object's pointers to those data members so that both instances do not *share* the dynamically allocated data members. We have, in essence, moved (the memory for) the pointer data members.

What about the non-pointer data members? The memory for these data members will be copied as usual. The memory for the non-pointer data members and the memory for the pointers themselves (not the memory pointed to by those pointers), still reside in the source instance. As such, the best we can do is designate a null value (nullptr) for the source object's pointers and place a 0 (or similar) value in the non-pointer data members to indicate that these members are no longer relevant.

We will use the move() function, found in the C++ Standard Library, to indicate a move copy constructor as follows:

```
Person p1("Alexa", "Gutierrez", 'R', "Ms.");
Person p2(move(p1));  // move copy constructor
Person p3 = move(p2); // also the move copy constructor
```

Additionally, with classes related by inheritance, we will also use move() in the member initialization list of the derived class move copy constructor. This will specify the base class move copy constructor to help initialize the subobject.

Move assignment operator

A **move assignment operator** is much like an overloaded assignment operator and is often crucial for all objects containing pointer data members. However, the goal is to again conserve memory by *moving* the dynamically allocated data of the source object to the destination object (versus performing a deep assignment). As with the overloaded assignment operator, we will test for self-assignment and then delete any previously dynamically allocated data members from the (pre-existing) destination object. However, we will then simply copy the pointer data members from the source object to those in the destination object. We will also null out the pointers in the source object so that the two instances do not share these dynamically allocated data members.

Also, much like the move copy constructor, non-pointer data members will be simply copied from source to destination object and replaced with a `nullptr` value in the source object to indicate non-usage.

We will again use the `move()` function as follows:

```
Person p3("Alexa", "Gutierrez", 'R', "Ms.");
Person p5("Xander", "LeBrun", 'R', "Dr.");
p5 = move(p3);   // move assignment; replaces p5
```

Additionally, with classes related by inheritance, we can again specify that the move assignment operator of the derived class will call the base class move assignment operator to help complete the task.

Bringing the components of canonical class form together

Let's see an example of a pair of classes that embrace the canonical class form. We will start with our `Person` class. This example can be found, as a complete program, in our GitHub repository:

```
https://github.com/PacktPublishing/Deciphering-Object-Orient-
ed-Programming-with-CPP/blob/main/Chapter15/Chp15-Ex1.cpp
```

```
class Person
{
private:     // Note slightly modified data members
    string firstName, lastName;
    char middleInitial = '\0';   // in-class initialization
    // pointer data member to demo deep copy and operator =
    char *title = nullptr;       // in-class initialization
protected: // Assume usual protected member functions exist
public:
    Person() = default;       // default constructor
    // Assume other usual constructors exist
    Person(const Person &);   // copy constructor
    Person(Person &&);        // move copy constructor
    virtual ~Person() { delete [] title }; // virtual dest.
    // Assume usual access functions and virtual fns. exist
    Person &operator=(const Person &);   // assignment op.
    Person &operator=(Person &&);   // move assignment op.
};
```

```
// copy constructor
Person::Person(const Person &p): firstName(p.firstName),
    lastName(p.lastName), middleInitial(p.middleInitial)
{
    // Perform a deep copy for the pointer data member
    // That is, allocate memory, then copy contents
    title = new char [strlen(p.title) + 1];
    strcpy(title, p.title);
}


// overloaded assignment operator
Person &Person::operator=(const Person &p)
{
    if (this != &p)  // check for self-assignment
    {
        // delete existing Person ptr data mbrs. for 'this'
        delete [] title;
        // Now re-allocate correct size and copy from source
        // Non-pointer data members are simply copied from
        // source to destination object
        firstName = p.firstName; // assignment btwn. strings
        lastName = p.lastName;
        middleInitial = p.middleInitial;
        title = new char [strlen(p.title) + 1]; // mem alloc
        strcpy(title, p.title);
    }
    return *this;   // allow for cascaded assignments
}
```

In the previous class definition, we notice that Person contains a default constructor, copy constructor, overloaded assignment operator, and a virtual destructor. Here, we have embraced the orthodox canonical class form as a pattern applicable for a class that might one day serve as a public base class. Also notice that we have added the prototypes for the move copy constructor and move assignment operator to additionally embrace the extended canonical class form.

The prototypes of the move copy constructor Person(Person &&); and the move assignment operator Person &operator=(Person &&); contain parameters of type Person &&. These are examples of **r-value references**. Arguments that can serve **l-value references**, such as Person &, will bind to the original copy constructor and overloaded assignment operator, whereas r-value reference parameters will bind to the applicable move methods instead.

Let's now look at the definitions for the methods contributing to the extended canonical class form – the move copy constructor and the move assignment operator for Person:

```cpp
// move copy constructor
Person::Person(Person &&p): firstName(p.firstName),
    lastName(p.lastName), middleInitial(p.middleInitial),
    title(p.title)  // dest ptr takes over src ptr's memory
{
    // Overtake source object's dynamically alloc. memory
    // or use simple assignments (non-ptr data members)
    // to copy source object's members in member init. list
    // Then null-out source object's ptrs to that memory
    // Clear source obj's string mbrs, or set w null char
    p.firstName.clear(); // set src object to empty string
    p.lastName.clear();
    p.middleInitial = '\0'; // null char indicates non-use
    p.title = nullptr; // null out src ptr; don't share mem
}


// move overloaded assignment operator
Person &Person::operator=(Person &&p)
{
    if (this != &p)          // check for self-assignment
    {
        // delete destination object's ptr data members
        delete [] title;
        // for ptr mbrs: overtake src obj's dynam alloc mem
        // and null source object's pointers to that memory
        // for non-ptr mbrs, a simple assignment suffices
        // followed by clearing source data member
        firstName = p.firstName;  // string assignment
        p.firstName.clear();   // clear source data member
```

```
        lastName = p.lastName;
        p.lastName.clear();
        middleInitial = p.middleInitial; // simple =
        p.middleInitial = '\0'; // null char shows non-use
        title = p.title; // ptr assignment to take over mem
        p.title = nullptr;   // null out src pointer
    }
    return *this;   // allow for cascaded assignments
}
```

Notice, in the preceding move copy constructor for data members that are pointers, we overtake the source object's dynamically allocated memory by using simple pointer assignments in the member initialization list (versus memory allocation such as we would employ in a deep copy constructor). We then place a `nullptr` value in the source object's pointer data members in the body of the constructor. For non-pointer data members, we simply copy the values from the source to the destination object and place a zeroed or empty value (such as `'\0'` for `p.middleInitial` or using `clear()` for `p.firstName`) in the source object to indicate its further non-use.

In the move assignment operator, we check for self-assignment and then employ the same scheme to merely move the dynamically allocated memory from the source object to the destination object with a simple pointer assignment. We copy simple data members as well, and of course, replace source object data values with either null pointers (`nullptr`) or zeroed values to indicate further non-use. The return value of `*this` allows for cascaded assignments.

Now, let's see how a derived class, `Student`, employs both the orthodox and extended canonical class form while utilizing its base class components to aid in the implementation of selected idiom methods:

```
class Student: public Person
{
private:
    float gpa = 0.0;           // in-class initialization
    string currentCourse;
    // one pointer data member to demo deep copy and op=
    const char *studentId = nullptr; // in-class init.
    static int numStudents;
public:
    Student();                      // default constructor
    // Assume other usual constructors exist
    Student(const Student &);  // copy constructor
    Student(Student &&);        // move copy constructor
```

```cpp
    ~Student() override;        // virtual destructor
    // Assume usual access functions exist
    // as well as virtual overrides and additional methods
    Student &operator=(const Student &);  // assignment op.
    Student &operator=(Student &&);  // move assignment op.
};
// See online code for default constructor implementation
// as well as implementation for other usual member fns.


// copy constructor
Student::Student(const Student &s): Person(s),
                gpa(s.gpa), currentCourse(s.currentCourse)
{   // Use member init. list to specify base copy
    // constructor to initialize base sub-object
    // Also use mbr init list to set most derived data mbrs
    // Perform deep copy for Student ptr data members
    // use temp - const data can't be directly modified
    char *temp = new char [strlen(s.studentId) + 1];
    strcpy (temp, s.studentId);
    studentId = temp;
    numStudents++;
}


// Overloaded assignment operator
Student &Student::operator=(const Student &s)
{
    if (this != &s)    // check for self-assignment
    {   // call base class assignment operator
        Person::operator=(s);
        // delete existing Student ptr data mbrs for 'this'
        delete [] studentId;
        // for ptr members, reallocate correct size and copy
        // from source; for non-ptr members, just use =
        gpa = s.gpa;   // simple assignment
        currentCourse = s.currentCourse;
```

```
        // deep copy of pointer data mbr (use a temp since
        // data is const and can't be directly modified)
        char *temp = new char [strlen(s.studentId) + 1];
        strcpy (temp, s.studentId);
        studentId = temp;
    }
    return *this;
}
```

In the preceding class definition, we again see that Student contains a default constructor, a copy constructor, an overloaded assignment operator, and a virtual destructor to complete the orthodox canonical class form.

Notice, however, that in the Student copy constructor, we specify the use of the Person copy constructor through the member initialization list. Similarly, in the Student overloaded assignment operator, once we check for self-assignment, we call the overloaded assignment operator in Person to help us complete the task using Person::operator=(s);.

Let's now look at the method definitions contributing to the extended canonical class form of Student – the move copy constructor and the move assignment operator:

```
// move copy constructor
Student::Student(Student &&s): Person(move(s)), gpa(s.gpa),
    currentCourse(s.currentCourse),
    studentId(s.studentId) // take over src obj's resource
{
    // First, use mbr. init. list to specify base move copy
    // constructor to initialize base sub-object. Then
    // overtake source object's dynamically allocated mem.
    // or use simple assignments (non-ptr data members)
    // to copy source object's members in mbr. init. list.
    // Then null-out source object's ptrs to that memory or
    // clear out source obj's string mbrs. in method body
    s.gpa = 0.0;      // then zero-out source object member
    s.currentCourse.clear();  // clear out source member
    s.studentId = nullptr; // null out src ptr data member
    numStudents++;   // it is a design choice whether or not
    // to inc. counter; src obj is empty but still exists
}
```

```
// move assignment operator
Student &Student::operator=(Student &&s)
{
    // make sure we're not assigning an object to itself
    if (this != &s)
    {
        Person::operator=(move(s));  // call base move oper=
        delete [] studentId;  // delete existing ptr data mbr
        // for ptr data members, take over src objects memory
        // for non-ptr data members, simple assignment is ok
        gpa = s.gpa; // assignment of source to dest data mbr
        s.gpa = 0.0; // zero out source object data member
        currentCourse = s.currentCourse; // string assignment
        s.currentCourse.clear(); // set src to empty string
        studentId = s.studentId; // pointer assignment
        s.studentId = nullptr;   // null out src ptr data mbr
    }
    return *this;  // allow for cascaded assignments
}
```

Notice, in the previously listed Student move copy constructor, we specify the utilization of the base class move copy constructor in the member initialization list. The remainder of the Student move copy constructor is similar to that found in the Person base class.

Likewise, let's notice in the Student move assignment operator, the call to the base class move operator= with Person::operator=(move(s));. The remainder of this method is similar to that found in the base class.

A good rule of thumb is that most non-trivial classes should minimally utilize the orthodox canonical class form. Of course, there are exceptions. For example, a class that will only serve as a protected or private base class need not have a virtual destructor because derived class instances cannot be upcast past a non-public inheritance boundary. Similarly, if we have a good reason to not want copies or to disallow an assignment, we can prohibit copies or assignments using the = delete specification in the extended signature of either of these methods.

Nonetheless, the canonical class form will add robustness to classes that embrace this idiom. The uniformity among classes utilizing this idiom with respect to their implementation of initialization, assignment, and argument passing will be valued by programmers.

Let's move forward to take a look at a complementary idea to the canonical class form, that of robustness.

Ensuring a class is robust

An important feature of C++ is the ability to build libraries of classes for widespread reuse. Whether we wish to achieve this goal, or simply wish to provide reliable code for our own organization's use, our code must be robust. A **robust class** will be well-tested, should follow the canonical class form (except for requiring a virtual destructor in protected and private base classes), and be portable (or included in a platform-specific library). Any class that is a candidate for reuse, or which is to be used in any professional capacity, must absolutely be robust.

A robust class must ensure that all instances of a given class are fully constructed. A **fully constructed object** is one in which all data members are appropriately initialized. All constructors for a given class (including copy constructors) must be verified to initialize all data members. The values with which data members are loaded should be checked for range suitability. Remember, an uninitialized data member is a potential disaster! Precautions should be made in the event that a given constructor does not complete properly or if the initial values of data members are inappropriate.

Fully constructed objects may be validated using a variety of techniques. A rudimentary (and not advised) technique is to embed a status data member into each class (or derive or embed a status ancestor/member). Set the status member to 0 in the member initialization list and to 1 as the last line of the constructor. Probe this value after instantiation. The huge downfall of this approach is that users will certainly forget to probe the *fully constructed* success flag.

An alternative to the simple, aforementioned scheme is to utilize in-class initialization for all simple data types, resetting these members in the member initialization list of each alternate constructor to the desired values. After instantiation, the values may again be probed to determine whether an alternate constructor completed successfully. This is still far from an ideal implementation.

A much better technique is to utilize exception handling. Embedding exception handling inside each constructor is ideal. If data members are not initialized within a suitable range, first try to re-enter their values, or open an alternate database for input, for example. As a last resort, you can throw an exception to report the *not fully constructed object*. We will more closely examine exception handling with respect to testing later in this chapter.

Meanwhile, let us move forward with a technique to rigorously test our classes and components – creating drivers to test classes.

Creating drivers to test classes

In *Chapter 5*, *Exploring Classes in Detail*, we briefly talked about breaking our code into source and header files. Let us briefly recap. Typically, the header file will be named after the class (such as `Student.h`) and will contain the class definition, plus any inline member function definitions. By placing inline functions in a header file, they will be properly re-expanded should their implementations change (as the header is subsequently included in each source file, creating a dependency with that header).

The implementation for the methods of each class will be placed in a corresponding source code file (such as `Student.cpp`), which will include the header on which it is based (that is, `#include "Student.h"`). Note that the double quotes imply that this header is in our current working directory; we could also specify a path as to where to find the header. By comparison, the angle brackets used with C++ libraries tell the preprocessor to look in predesignated directories by the compiler. Also, note that each derived class header file will include the header file for its base class (so that it may see member function prototypes).

Note that any static data member or method definitions will appear in their corresponding source code files (so that only one definition per application will exist).

With this header and source code file structure in mind, we can now create a driver to test each individual class or each grouping of closely related classes (such as those related through association or aggregation). Classes related through inheritance can be tested in their own, individual driver files. Each driver file can be named to reflect the class that is being tested, such as `StudentDriver.cpp`. The driver file will include the relevant header files for the class(es) being tested. Of course, the source files from the classes in question would be compiled and linked to the driver file as part of the compilation process.

The driver file can simply contain a `main()` function as a test bed to instantiate the class(es) in question and serve as a scope to test each member function. The driver will test default instantiation, typical instantiation, copy construction, assignment between objects, and each of the additional methods in the class(es). Should virtual destructors or other virtual functions exist, we should instantiate derived class instances (in the derived class' driver), upcasting these instances to be stored using base class pointers, and then invoke the virtual functions to verify that the correct behaviors occur. In the case of a virtual destructor, we can trace which destructor is the entry point in the destruction sequence by deleting a dynamically allocated instance (or waiting for a stack instance to go out of scope) and single-stepping through our debugger to verify all is as expected.

We can also test that objects are fully constructed; we will see more on this topic shortly.

Assuming we have our usual `Person` and `Student` class hierarchy, here is a simple driver (the file containing `main()`) to test the `Student` class. This driver can be found in our GitHub repository. To make a complete program, you will also need to compile and link together the `Student.cpp` and `Person.cpp` files found in this same directory. Here is the GitHub repository URL for the driver:

```
https://github.com/PacktPublishing/Deciphering-Object-Orient-
ed-Programming-with-CPP/blob/main/Chapter15/Chp15-Ex2.cpp
```

```cpp
#include "Person.h"   // include relevant class header files
#include "Student.h"
using std::cout;       // preferred to: using namespace std;
using std::endl;
constexpr int MAX = 3;
```

```
int main() // Driver to test Student class, stored in above
{           // filename for chapter example consistency
    // Test all instantiation means, even copy constructor
    Student s0; // Default construction
    // alternate constructor
    Student s1("Jo", "Li", 'H', "Ms.", 3.7, "C++",
            "UD1234");
    Student s2("Sam", "Lo", 'A', "Mr.", 3.5, "C++",
            "UD2245");
    // These initializations implicitly invoke copy const.
    Student s3(s1);
    Student s4 = s2;    // This is also initialization
    // Test the assignment operator
    Student s5("Ren", "Ze", 'A', "Dr.", 3.8, "C++",
            "BU5563");
    Student s6;
    s6 = s5;  // this is an assignment, not initialization
    // Test each public method. A sample is shown here
    s1.Print();  // Be sure to test each method!

    // Generalize derived instances as base types
    // Do the polymorphic operations work as expected?
    Person *people[MAX] = { }; // initialized with nullptrs
    // base instance for comparison
    people[0] = new Person("Juliet", "Martinez", 'M',
                        "Ms.");
    // derived instances, generalized with base class ptrs.
    people[1] = new Student("Zack", "Moon", 'R', "Dr.",
                        3.8, "C++", "UMD1234");
    people[2] = new Student("Gabby", "Doone", 'A', "Dr.",
                        3.9, "C++", "GWU4321");
    for (auto *item : people)  // loop through all elements
    {
        item->IsA();
        cout << "  ";
```

```
        item->Print();
    }
    // Test destruction sequence (dynam. alloc. instances)
    for (auto *item : people)  // loop thru all elements
        delete item;    // engage virtual dest. sequence
    return 0;
}
```

Briefly reviewing the preceding program fragment, we can see that we have tested each means for instantiation, including the copy constructor. We've also tested the assignment operator, verified each member function works (an example method is shown), and verified that the virtual functions (including the virtual destructor), work as intended.

Now that we have seen a basic driver test our classes, let's consider some additional metrics we can use when testing classes related via inheritance, association, or aggregation.

Testing related classes

With OO programs, it is not sufficient to simply test an individual class for completeness and robustness, though these are good starting points. Completeness entails not only following the canonical class form but also ensuring that data members have a safe means for access using appropriate access methods (labeled as const when not modifying the instance). Completeness also verifies that the required interface as specified by the OO design has been implemented.

Robustness leads us to verify that all of the aforementioned methods had been tested within an appropriate driver, evaluated for platform independence, and verified that each means for instantiation leads to a fully constructed object. We can augment this type of testing with threshold testing of data members, for instance, noting when exceptions are thrown. Completeness and robustness, though seemingly comprehensive, are actually the most straightforward means for OO component testing.

The more challenging means for testing is to test the interaction between related classes.

Testing classes related through inheritance, association, or aggregation

Classes related through various object relationships require various additional means for component testing. Objects with various relationships with one another can impact the state progression a given instance may have during its life expectancy within the application. This type of testing will require the most detailed effort. We will find that scenarios will be useful to help us capture the usual interactions between related objects, leading to more comprehensive ways to test classes that interact with one another.

Let's begin by considering how we can test classes related to inheritance.

Adding strategies to test inheritance

Classes related through public inheritance need to have virtual functions verified. For example, have all intended derived class methods been overridden? Remember, a derived class does not need to override all virtual functions specified in its base class if base class behaviors are still deemed appropriate at the derived class level. It will be necessary to compare the implementation to the design to ensure that we have overridden all required polymorphic operations with suitable methods.

Certainly, the binding of virtual functions is done at runtime (that is, dynamic binding). It will be important to create derived class instances and store them using base class pointers so that the polymorphic operations can be applied. We then need to verify that the derived class behavior shines through. If not, perhaps we may find ourselves in an unintended function hiding situation, or perhaps the base class operation wasn't marked `virtual` as intended (keeping in mind that the keywords `virtual` and `override` at the derived class level, though nice and recommended, are optional and do not affect the dynamic behavior).

Though classes related through inheritance have unique testing strategies, remember that instantiation will create a single object, that is, of a base class or of a derived class type. When we instantiate one such type, we have one such instance, not a pair of instances working together. A derived class merely has a base class subobject, which is part of itself. Let's consider how this compares with associated objects or aggregates, which can be separate objects (association), potentially interacting with their companions.

Adding strategies to test aggregation and association

Classes related through association or aggregation may be multiple instances communicating with one another and causing state changes with one another. This is certainly more complex than the object relationship of inheritance.

Classes related via aggregation are generally easier to test than those related via association. Thinking of the most common form of aggregation (composition), the embedded (inner) object is part of the outer (whole) object. When the outer object is instantiated, we get the memory for the inner object embedded within the *whole*. The memory layout is not tremendously different (other than the potential ordering) when compared to the memory layout of a derived class instance, which contains a base class subobject. In each case, we are still dealing with a single instance (even though it has embedded *parts*). The point of comparison with testing, however, is that operations applied to the *whole* are often delegated to the *parts* or components. We will rigorously need to test the operations, on the whole, to ensure that they delegate necessary information to each of the parts.

Classes related via the lesser-used form of a general aggregation (where the whole contains pointers to the parts versus the typical embedded object implementation of composition) have similar issues to an association, as the implementation is similar. With that in mind, let's take a look at testing issues relating to associated objects.

Classes related via an association are often independently existing objects, which at some point in the application have created a link to one another. There may or may not be a predetermined point in the application when the two objects create a link to one another. Operations applied to one object may cause a change in the associated object. For example, let us consider a `Student` and a `Course`. Both may exist independently, then at some point in the application, a `Student` may add a `Course` with `Student::AddCourse()`. By doing so, not only does a particular `Student` instance now contain a link to a specific `Course` instance, but the `Student::AddCourse()` operation has caused a change in the `Course` class. That particular `Student` instance is now part of a particular `Course` instance's roster. At any point, the `Course` may be canceled, rippling a change in all `Student` instances who are enrolled in that `Course`. These changes reflect states in which each associated object may exist. For example, a `Student` may be in a state of *currently enrolled*, or *dropping* a `Course`. There are many possibilities. How do we test all of them?

Adding scenarios to aid in testing object relationships

The notion of a scenario comes up in object-oriented analysis as a means to both create OO designs and test them. A **scenario** is a descriptive walkthrough of a likely series of events that will occur in an application. A scenario will feature classes and how they may interact with one another for a specific situation. Many related scenarios can be collected into the OO concept of a **use case**. In the OO analysis and design phases, scenarios help determine which classes may exist in the application as well as operations and relationships each may have. In testing, scenarios can be reused to form the basis for driver creation to test various object relationships. With this in mind, a series of drivers can be developed to test numerous scenarios (that is, use cases). This type of modeling will more thoroughly be able to provide a test bed for related objects than the initial, simple means of testing for completeness and robustness.

Another area of concern between any type of related classes is that of version control. What happens, for example, if a base class definition or default behavior changes? How will that impact a derived class? How will that impact associated objects? With each change, we inevitably will need to revisit component testing for all related classes.

Next, let's consider how exception handling mechanisms factor into OO component testing.

Testing exception handling mechanisms

Now that we can create drivers to test each class (or a grouping of related classes), we will want to understand which methods in our code may throw exceptions. For these scenarios, we will want to add try blocks within the driver to ensure we know how to handle each potential exception thrown. Before doing so, we should ask ourselves, did we include adequate exception handling in our code during the development process? For example, considering instantiation, do our constructors check whether an object is fully constructed? Do they throw exceptions if not? If the answer is no, our classes may not be as robust as we had anticipated.

Let's consider embedding exception handling into a constructor, and how we may construct a driver to test all potential means for instantiation.

Embedding exception handling in constructors to create robust classes

We may recall from our recent *Chapter 11, Handling Exceptions*, that we can create our own exception classes, derived from the C++ Standard Library `exception` class. Let's assume that we have created such a class, namely `ConstructionException`. If at any point in a constructor we are not able to properly initialize a given instance to provide a fully constructed object, we can throw a `ConstructionException` from any constructor. The implication of potentially throwing a `ConstructionException` is that we now should enclose instantiation within try blocks and add matching catch blocks to anticipate a `ConstructionException` that may be thrown. Keep in mind, however, that instances declared within the scope of a try block have scope only within the `try-catch` pairing.

The good news is that if an object does not complete construction (that is, if an exception is thrown before the constructor completes), the object will technically not exist. If an object does not technically exist, there will be no necessary clean up of a partially instantiated object. We will, however, need to think about what this means to our application if an instance we anticipate does not fully construct. How will that alter the progression of our code? Part of testing is to ensure that we have considered all ways in which our code may be used and bulletproof our code accordingly!

It is important to note that the introduction of `try` and `catch` blocks may alter our program flow, and it is crucial to include this type of testing in our drivers. We may seek scenarios that account for the `try` and `catch` blocks as we conduct our testing.

We have now seen how we can augment our test drivers to accommodate classes that may throw exceptions. We have also discussed in this chapter adding scenarios in our drivers to help track the states between objects with relationships and, of course, simple class idioms we can follow to set us up for success. Let us now briefly recap these concepts before moving forward to our next chapter.

Summary

In this chapter, we have increased our ability to become better C++ programmers by examining various OO class and component testing practices and strategies. Our primary goal is to ensure that our code is robust, well-tested, and can be deployed error-free to our various organizations.

We have considered programming idioms, such as following the canonical class form to ensure that our classes are complete and have expected behavior for construction/destruction, assignment, and usage in argument passing and as return values from functions. We have talked about what it means to create a robust class – one that follows the canonical class form that is also well-tested, platform-independent, and tested for fully constructed objects.

We have also explored how to create drivers to test individual classes or sets of related classes. We have established a checklist of items to test individual classes within a driver. We have looked more thoroughly at object relationships to understand that objects that interact with one another require more sophisticated testing. That is, as objects move from state to state, they may be impacted by associated objects, which can further alter their course of progression. We've added utilizing scenarios as test cases for our drivers to better capture the dynamic states in which instances may move within an application.

Finally, we have taken a look at how exception handling mechanisms can impact how we test our code. We have augmented our drivers to account for the flow of control that try and catch blocks may redirect our applications from their anticipated, typical progression.

We are now ready to continue forward with the next part of our book, design patterns and idioms in C++. We will start with *Chapter 16*, *Using the Observer Pattern*. In the remaining chapters, we will understand how to apply popular design patterns and employ them in our coding. These skills will make us better programmers. Let's move forward!

Questions

1. Consider a pair of classes from one of your previous exercises containing an object relationship (hint – public inheritance will be easier to consider than association).

    a. Do your classes follow the canonical class form? Orthodox or extended? Why, or why not? If they do not and should, revise the classes to follow this idiom.

    b. Would you consider your classes robust? Why, or why not?

2. Create a driver (or two) to test your pair of classes:

    a. Be sure to test for the usual checklist of items (construction, assignment, destruction, the public interface, upcasting (if applicable), and use of a virtual function).

    b. (Optional) If you selected two classes related using association, create a separate driver to follow a typical scenario detailing the interaction of the two classes.

    c. Be sure to include testing of exception handling in one of your test drivers.

3. Create a `ConstructionException` class (derived from the C++ Standard Library `exception`). Embed checks within your constructors in a sample class to throw a `ConstructionException` when necessary. Be sure to enclose all forms of instantiation of this class within the appropriate `try` and `catch` block pairings.

Part 4:
Design Patterns and Idioms in C++

The goal of this part is to expand your C++ repertoire, beyond OOP and other necessary skills, to include knowledge of core design patterns. Design patterns provide well-proven techniques and strategies to solve recurring types of OO problems. This section introduces common design patterns and demonstrates in depth how to apply these patterns by building on previous examples within the book in creative ways. Each chapter contains detailed code examples to exemplify each pattern.

The initial chapter in this section introduces the idea of design patterns and discusses the advantages of utilizing such patterns within coding solutions. The initial chapter also introduces the Observer pattern and provides an in-depth program to appreciate the various components of this pattern.

The next chapter explains the Factory Method pattern and likewise provides detailed programs, showing how to implement the Factory Method pattern with and without an Object Factory. This chapter additionally compares an Object Factory to an Abstract Factory.

The following chapter introduces the Adapter pattern and provides implementation strategies and program examples using inheritance versus association to implement the Adapter class. Additionally, an adapter as a simple wrapper class is illustrated.

The Singleton pattern is examined in the following chapter. Following two simple examples, a paired-class implementation is demonstrated with a detailed example. Registries to accommodate Singletons are also introduced.

The final chapter in this section and book introduces the pImpl pattern to reduce compile-time dependencies within your code. A basic implementation is provided and then expanded upon using unique pointers. Performance issues are additionally explored relating to this pattern.

This part comprises the following chapters:

- *Chapter 16, Using the Observer Pattern*
- *Chapter 17, Applying the Factory Pattern*
- *Chapter 18, Applying the Adapter Pattern*
- *Chapter 19, Using the Singleton Pattern*
- *Chapter 20, Removing Implementation Details Using the pImpl Pattern*

Using the Observer Pattern

This chapter will begin our quest to expand your C++ programming repertoire beyond OOP concepts, with the goal of enabling you to solve recurring types of coding problems by utilizing common design patterns. Design patterns will also enhance code maintenance and provide avenues for potential code reuse.

The goal of the fourth section of the book, beginning with this chapter, is to demonstrate and explain popular design patterns and idioms and learn how to implement them effectively in C++.

In this chapter, we will cover the following main topics:

- Understanding the advantage of utilizing design patterns
- Understanding the Observer pattern and how it contributes to OOP
- Understanding how to implement the Observer pattern in C++

By the end of this chapter, you will understand the utility of employing design patterns in your code, as well as understand the popular **Observer pattern**. We will see an example implementation of this pattern in C++. Utilizing common design patterns will help you become a more beneficial and valuable programmer by enabling you to embrace more sophisticated programming techniques.

Let's increase our programming skillset by examining various design patterns, starting in this chapter with the Observer pattern.

Technical requirements

Online code for full program examples can be found in the following GitHub URL: https://github.com/PacktPublishing/Deciphering-Object-Oriented-Programming-with-CPP/tree/main/Chapter16. Each full program example can be found in the GitHub repository under the appropriate chapter heading (subdirectory) in a file that corresponds to the chapter number, followed by a dash, followed by the example number in the chapter at hand. For example, the first full program in this chapter can be found in the subdirectory Chapter16 in a file named Chp16-Ex1.cpp under the aforementioned GitHub directory.

The CiA video for this chapter can be viewed at: `https://bit.ly/3A8ZWoy`.

Utilizing design patterns

Design patterns represent a grouping of well-tested programming solutions for recurring types of programming conundrums. Design patterns represent the high-level concept of a design issue and how a generalized collaboration between classes can provide a solution that can be implemented in a variety of ways.

There are many well-identified design patterns that have been recognized and described in the past 25+ years of software development. We will look at some popular patterns in the remaining chapters of this book to give you a feel of how we can incorporate popular software design solutions into our coding arsenal of techniques.

Why might we choose to utilize a design pattern? To start, once we have identified a type of programming problem, we can make use of a *tried and true* solution that other programmers have tested comprehensively. Additionally, once we employ a design pattern, other programmers immersing themselves in our code (for maintenance or future enhancements) will have a basic understanding of the techniques we have chosen, as core design patterns have become an industry standard.

Some of the earliest design patterns came about nearly 50 years ago, with the advent of the **Model-View-Controller** paradigm, later simplified at times to **Subject-View**. For example, Subject-View is a rudimentary pattern in which an object of interest (the **Subject**) will be loosely coupled with its method of display (its **View**). The Subject and its View communicate with a one-to-one association. Sometimes Subjects can have multiple Views, in which case the Subject is associated with many View objects. If one View changes, a state update can be sent to the Subject, who can then send necessary messages to the other Views so that they, too, can be updated to reflect how the new state may have modified their particular View.

The original **Model-View-Controller** (**MVC**) pattern, emanating from early OOP languages such as Smalltalk, has a similar premise, except that a Controller object delegates events between the Model (that is, the Subject) and its View (or Views). These preliminary paradigms influenced early design patterns; the elements of Subject-View or MVC can be seen conceptually as a rudimentary basis for core design patterns today.

Many of the design patterns we will review in the remainder of this book will be adaptations of patterns originally described by the *Gang of Four* (Erich Gamma, Richard Helm, Ralph Johnson, and John Vlissides) in *Design Patterns, Elements of Reusable Object-Oriented Software*. We will apply and adapt these patterns to solve problems stemming from applications we have introduced in earlier chapters of this book.

Let's begin our pursuit of understanding and utilizing popular design patterns by investigating a pattern in action. We will start with a behavioral pattern known as the **Observer pattern**.

Understanding the Observer pattern

In the **Observer pattern**, an object of interest will maintain a list of observers who are interested in state updates of the main object. The observers will maintain a link to their object of interest. We will refer to the main object of interest as the **Subject**. The list of interested objects is known collectively as the **Observers**. The Subject will inform any Observer of relevant state changes. The Observers, once notified of any state changes of the Subject, will take any appropriate next action themselves (usually through a virtual function invoked on each Observer by the Subject).

Already, we can imagine how an Observer pattern may be implemented using associations. In fact, the Observer represents a one-to-many association. The Subject, for example, may use an STL `list` (or `vector`) to collect a set of Observers. Each Observer will contain an association to the Subject. We can imagine an important operation on the Subject, corresponding to a state change in the Subject, issuing an update to its list of Observers to *notify* them of the state change. The `Notify()` method is, in fact, invoked when a Subject's state changes and uniformly applies polymorphic Observer `Update()` methods on each of the Subject's list of Observers. Before we get swept up in implementation, let's consider the key components comprising the Observer pattern.

The Observer pattern will include the following:

- A Subject, or object of interest. The Subject will maintain a list of Observer objects (a many-sided association).

- A Subject will provide an interface to `Register()` or `Remove()` an Observer.

- A Subject will include a `Notify()` interface, which will update its Observers when the Subject's state has changed. The Subject will `Notify()` Observers by calling a polymorphic `Update()` method on each Observer in its collection.

- An Observer class will be modeled as an abstract class (or interface).

- An Observer interface will provide an abstract, polymorphic `Update()` method to be called when its associated Subject has changed its state.

- An association between each Observer to its Subject will be maintained in a concrete class, derived from Observer. Doing so will alleviate awkward casting (compared to maintaining the Subject link in the abstract Observer class).

- Both classes will be able to maintain their current state.

The aforementioned `Subject` and `Observer` classes are specified generically so that they may be combined with a variety of concrete classes (mostly through inheritance) that desire to use the Observer pattern. A generic Subject and Observer provide a great opportunity for reuse. With a design pattern, many core elements of a pattern can often be set up more generically to allow for greater reuse of the code itself, not only the reuse in the concept of the solution (pattern).

Let's move forward to see a sample implementation of the Observer pattern.

Implementing the Observer pattern

To implement the Observer pattern, we will first need to define our Subject and Observer classes. We will then need to derive concrete classes from these classes to incorporate our application specifics and to put our pattern in motion. Let's get started!

Creating an Observer, Subject, and domain-specific derived classes

In our example, we will create Subject and Observer classes to establish the framework for *registering* an Observer with a Subject and for the Subject to Notify() its set of observers of a state change it may have. We will then derive from these base classes descendent classes we are accustomed to seeing – Course and Student, where Course will be our concrete Subject and Student will become our concrete Observer.

The application we will model will involve a course registration system and the concept of a waitlist. As we have seen before in *Question 2* of *Chapter 10, Implementing Association, Aggregation, and Composition*, we will model a Student having an association to many Course instances, and a Course having an association to many Student instances. The Observer pattern will come into play when we model our waitlist.

Our Course class will be derived from Subject. The list of observers that our Course will inherit will represent the Student instances on this Course's waitlist. The Course will also have a list of Student instances, representing students who have been successfully enrolled in the course at hand.

Our Student class will be derived from both Person and Observer. The Student will include a list of Course instances in which that Student is currently enrolled. The Student will also have a data member, waitListedCourse, which will correspond to an association to a Course that the Student is waiting to add. This *waitlisted* Course represents the Subject from which we will receive notifications. A notification will correspond to a state change indicating that the Course now has room for a Student to add the Course.

It is from Observer that Student will inherit the polymorphic operation Update(), which will correspond to the Student being notified that a spot is now open in the Course. Here, in Student::Update(), we will include the mechanics to add a student's waitListedCourse (provided the course is open and has available seats). If the addition is successful, we will release the Student from the course's waitlist (the list of observers inherited by Course from Subject). Naturally, the Student will be added to the current student list in the Course and the Course will appear in that student's current course list.

Specifying the Observer and the Subject

Let's break down our example into components, starting with the pair of classes to specify our `Observer` and `Subject`. This complete program can be found in our GitHub:

https://github.com/PacktPublishing/Deciphering-Object-Oriented-Programming-with-CPP/blob/main/Chapter16/Chp16-Ex1.cpp

```cpp
#include <list>    // partial list of #includes
#include <iterator>
using std::cout;    // prefered to: using namespace std;
using std::endl;
using std::setprecision;
using std::string;
using std::to_string;
using std::list;

constexpr int MAXCOURSES = 5, MAXSTUDENTS = 5;


// Simple enums for states; we could have also made a
// hierarchy of states, but let's keep it simple
enum State { Initial = 0, Success = 1, Failure = 2 };
// More specific states for readability in subsequent code
enum StudentState { AddSuccess = State::Success,
                    AddFailure = State::Failure };
enum CourseState { OpenForEnrollment = State::Success,
                   NewSpaceAvailable = State::Success,
                   Full = State::Failure };
class Subject;   // forward declarations
class Student;

class Observer   // Observer is an abstract class
{
private:
    // Represent a state as an int, to eliminate type
    // conversions between specific and basic states
    int observerState = State::Initial;   // in-class init.
protected:
```

```
        Observer() = default;
        Observer(int s): observerState(s) { }
        void SetState(int s) { observerState = s; }
    public:
        int GetState() const { return observerState; }
        virtual ~Observer() = default;
        virtual void Update() = 0;
    };
```

In the previous class definition, we introduce our abstract Observer class. Here, we include an observerState and protected constructors to initialize this state. We include a protected SetState() method to update this state from the scope of a derived class. We also include a public GetState() method. The addition of GetState() will facilitate implementation within our Subject's Notify() method by allowing us to easily check whether the state of our Observer has changed. Though state information has historically been added to derived classes of both Observer and Subject, we will instead generalize state information in these base classes. This will allow our derived classes to remain more pattern-independent and instead focused on the essence of the application.

Notice that our destructor is virtual, and we introduce an abstract method virtual void Update() = 0; to specify the interface our Subject will invoke on its list of observers to delegate updates to these Observer instances.

Now, let's take a look at our Subject base class:

```
class Subject     // Treated as an abstract class, due to
{                 // protected constructors. However, there's
private:          // no pure virtual function
    list<class Observer *> observers;
    int numObservers = 0;
    // Represent a state as an int, to eliminate
    // type conversions between specific and basic states
    int subjectState = State::Initial;
    list<Observer *>::iterator newIter;
protected:
    Subject() = default;
    Subject(int s): subjectState(s) { } // note in-class
                                        // init. above
    void SetState(int s) { subjectState = s; }
```

```
public:
    int GetState() const { return subjectState; }
    int GetNumObservers() const { return numObservers; }
    virtual ~Subject() = default;
    virtual void Register(Observer *);
    virtual void Release(Observer *);
    virtual void Notify();
};
```

In the aforementioned `Subject` class definition, we see that our `Subject` includes an STL `list` to collect its `Observer` instances. It also includes the `subjectState` and a counter to reflect the number of observers. Also, we include a data member to keep track of an uncorrupted iterator. We'll see this will be handy once we erase an element (`list::erase()` is an operation that will invalidate a current iterator).

Our `Subject` class will also have protected constructors and a `SetState()` method, which initializes or sets the `Subject`'s state. Though this class is not technically abstract (it does not contain a pure virtual function), its constructors are protected to simulate an abstract class; this class is only intended to be constructed as a subobject within a derived class instance.

In the public interface, we have some access functions to get the current state or number of observers. We also have a virtual destructor, and virtual functions for `Register()`, `Release()`, and `Notify()`. We will provide implementations for the latter three methods at this base class level.

Let's next take a look at the default implementations of `Register()`, `Release()`, and `Notify()` in our `Subject` base class:

```
void Subject::Register(Observer *ob)
{
    observers.push_back(ob); // Add an Observer to the list
    numObservers++;
}


void Subject::Release(Observer *ob) // Remove an Observer
{                                   // from the list
    bool found = false;
    // loop until we find the desired Observer
    // Note auto iter will be: list<Observer *>::iterator
    for (auto iter = observers.begin();
        iter != observers.end() && !found; ++iter)
    {
```

```
            if (*iter == ob)// if we find observer that we seek
            {
                // erase() element, iterator is now corrupt.
                // Save returned (good) iterator;
                // we'll need it later
                newIter = observers.erase(iter);
                found = true;   // exit loop after found
                numObservers--;
            }
        }
    }
}


void Subject::Notify()
{   // Notify all Observers
    // Note auto iter will be: list<Observer *>::iterator
    for (auto iter = observers.begin();
         iter != observers.end(); ++iter)
    {
        (*iter)->Update(); // AddCourse, then Release
        // Observer. State 'Success' is represented
        // generally for Observer (at this level we have
        // no knowledge of how Subject and Observer have
        // been specialized). In our application, this
        // means a Student (observer) added a course,
        // got off waitlist (so waitlist had a Release),
        // so we update the iterator
        if ((*iter)->GetState() == State::Success)
            iter = newIter; // update the iterator since
                            // erase() invalidated this one
    }
    if (!observers.empty())
    {   // Update last item on waitlist
        Observer *last = *newIter;
        last->Update();
    }
}
```

In the aforementioned `Subject` member functions, let's begin by examining the `void Subject::Register(Observer *)` method. Here, we simply add the `Observer *` specified as a parameter to our STL `list` of observers (and increase the counter for the number of observers).

Next, let's consider the inverse of `Register()` by reviewing `void Subject::Release(Observer *)`. Here, we iterate through our list of observers until we find the one we are seeking. We then call `list::erase()` on that current item, set our `found` flag to `true` (to leave the loop), and decrement the number of observers. Also, notice that we save the return value of `list::erase()`, which is an updated (and valid) iterator to the list of observers. The iterator `iter` in the loop has been invalidated with our call to `list::erase()`. We save this revised iterator in a data member `newIter` so that we can access it shortly.

Finally, let's take a look at the `Notify()` method in `Subject`. This method will be called once there is a state change in the `Subject`. The goal will be to `Update()` all observers on the `Subject`'s observer list. To do just that, we look through our list. One by one, we grab an `Observer` with the list iterator `iter`. We call `Update()` on the current `Observer` with `(*iter)->Update();`. We can tell whether the update has been a success for a given `Observer` by checking the observer's state using `if ((*iter)->GetState() == State::Success)`. With a state of *Success*, we know the observer's actions will have caused the `Release()` function we just reviewed to be called on itself. Because the `list::erase()` used in `Release()` has invalidated the iterator, we now get the correct and revised iterator using `iter = newIter;`. Finally, outside of the loop, we call `Update()` on the last item in the list of observers.

Deriving concrete classes from Subject and Observer

Let's continue moving forward with this example by taking a look at our concrete classes derived from `Subject` or `Observer`. Let's start with `Course`, derived from `Subject`:

```
class Course: public Subject
{   // inherits Observer list;
    // Observer list represents Students on waitlist
private:
    string title;
    int number = 0;   // course num, total num students set
    int totalStudents = 0; // using in-class initialization
    Student *students[MAXSTUDENTS] = { }; // initialize to
                                          // nullptrs
public:
    Course(const string &title, int num): number(num)
    {
        this->title = title;   // or rename parameter
```

```cpp
            // Note: in-class init. is in-lieu of below:
            // for (int i = 0; i < MAXSTUDENTS; i++)
                // students[i] = nullptr;
    }
    // destructor body shown as place holder to add more
    // work that will be necessary
    ~Course() override
    {       /* There's more work to add here! */     }
    int GetCourseNum() const { return number; }
    const string &GetTitle() const { return title; }
    const AddStudent(Student *);
    void Open()
    {   SetState(CourseState::OpenForEnrollment);
        Notify();
    }
    void PrintStudents() const;
};


bool Course::AddStudent(Student *s)
{   // Should also check Student hasn't been added to Course
    if (totalStudents < MAXSTUDENTS)  // course not full
    {
        students[totalStudents++] = s;
        return true;
    }
    else return false;
}


void Course::PrintStudents() const
{
    cout << "Course: (" << GetTitle() <<
            ") has the following students: " << endl;
    for (int i = 0; i < MAXSTUDENTS &&
                        students[i] != nullptr; i++)
    {
        cout << "\t" << students[i]->GetFirstName() << " ";
```

```
            cout << students[i]->GetLastName() << endl;
    }
}
```

In our aforementioned `Course` class, we include data members for the course title and number as well as for the total number of students currently enrolled. We also have our list of students currently enrolled, indicated by `Student *students[MAXNUMBERSTUDENTS];`. Additionally, keep in mind that we inherit the STL `list` of observers from our `Subject` base class. This list of `Observer` instances will represent the `Student` instances comprising our waitlist (of students) for the `Course`.

The `Course` class additionally includes a constructor, a virtual destructor, and simple access functions. Note that the virtual destructor has more work to do than shown – if a `Course` destructs, we must remember to first remove (but not delete) `Student` instances from the `Course`. Our `bool Course::AddStudent(Student *)` interface will allow us to add a `Student` to a `Course`. Of course, we should ensure that the `Student` has not already added the `Course` in the body of this method.

Our `void Course::Open();` method will be invoked on a `Course` to indicate that the course is now available to add students. Here, we will first set the state to `Course::OpenForEnrollment` (clearly indicating *Open for Enrollment* with the enumerated type) and then call `Notify()`. Our `Notify()` method in base class `Subject` loops through each `Observer`, calling polymorphic `Update()` on each observer. Each `Observer` is a `Student`; `Student::Update()` will allow each `Student` on the waitlist to try to add the `Course`, which now is open to receive students. With a successful addition to the course's current student list, a `Student` will then request `Release()` of its position on the waitlist (as an `Observer`).

Next, let's take a look at our class definition for `Student`, our concrete class derived from both `Person` and `Observer`:

```
class Person { }; // Assume our typical Person class here

class Student: public Person, public Observer
{
private:
    float gpa = 0.0;       // in-class initialization
    const string studentId;
    int currentNumCourses = 0;
    Course *courses[MAXCOURSES] = { }; // set to nullptrs
    // Course we'd like to take - we're on the waitlist.
    Course *waitListedCourse = nullptr;   // Our Subject
                                  // (in specialized form)
```

```
        static int numStudents;
public:
    Student();  // default constructor
    Student(const string &, const string &, char,
            const string &, float, const string &, Course *);
    Student(const string &, const string &, char,
            const string &, float, const string &);
    Student(const Student &) = delete; // Copies disallowed
    ~Student() override;    // virtual destructor
    void EarnPhD();
    float GetGpa() const { return gpa; }
    const string &GetStudentId() const
        { return studentId; }
    void Print() const override;  // from Person
    void IsA() const override;   // from Person
    void Update() override;        // from Observer
    virtual void Graduate(); // newly introduced virtual fn
    bool AddCourse(Course *);
    void PrintCourses() const;
    static int GetNumberStudents() { return numStudents; }
};
```

Briefly reviewing the aforementioned class definition for Student, we can see that this class is derived from both Person and Observer using multiple inheritance. Let's assume our Person class is as we have used in the past many times.

In addition to the usual components of our Student class, we add the data member Course *waitListedCourse;, which will model the association to our Subject. This data member will model the idea of a Course that we would very much like to add, yet currently cannot, that is, a *waitlisted* course. Here, we are implementing the concept of a single waitlisted course, but we could easily expand the example to include a list supporting multiple waitlisted courses. Notice that this link (data member) is declared in the form of the derived type, Course, not the base type, Subject. This is typical in the Observer pattern and will help us avoid dreaded down-casting as we override our Update() method in Student. It is through this link that we will conduct our interaction with our Subject and the means by which we will receive updates from our Subject as it changes states.

We also notice that we have virtual void Update() override; prototyped in Student. This method will allow us to override the pure virtual Update() method specified by Observer.

Next, let's review a selection of the various new member functions for Student:

```cpp
// Assume most Student member functions are as we are
// accustomed to seeing. All are available online.
// Let's look at ONLY those that may differ:

// Note that the default constructor for Observer() will be
// invoked implicitly, thus it is not needed in init list
// below (it is shown in comment as a reminder it's called)
Student::Student(const string &fn, const string &ln,
    char mi, const string &t, float avg, const string &id,
    Course *c): Person(fn, ln, mi, t), // Observer(),
    gpa(avg), studentId(id), currentNumCourses(0)
{
    // Below nullptr assignment is no longer needed with
    // above in-class initialization; otherwise, add here:
    // for (int i = 0; i < MAXCOURSES; i++)
        // courses[i] = nullptr;
    waitListedCourse = c;  // set initial waitlisted Course
                           // (Subject)
    c->Register(this); // Add the Student (Observer) to
                       // the Subject's list of Observers
    numStudents++;
}


bool Student::AddCourse(Course *c)
{
    // Should also check Student isn't already in Course
    if (currentNumCourses < MAXCOURSES)
    {
        courses[currentNumCourses++] = c;  // set assoc.
        c->AddStudent(this);               // set back-link
        return true;
    }
    else  // if we can't add the course,
    {   // add Student (Observer) to the Course's Waitlist,
        c->Register(this);  // stored in Subject base class
```

```
        waitListedCourse = c; // set Student (Observer)
                              // link to Subject
      return false;
    }
  }
```

Let's review the previously listed member functions. Since we are accustomed to most of the necessary components and mechanics in the `Student` class, we will focus on the newly added `Student` methods, starting with an alternate constructor. In this constructor, let us assume that we set most of the data members as usual. The key additional lines of code here are `waitListedCourse = c;` to set our waitlist entry to the desired `Course` (`Subject`), as well as `c->Register(this);`, where we add the `Student` (`Observer`) to the `Subject`'s list (the formal waitlist for the course).

Next, in our `bool Student::AddCourse(Course *)` method, we first check that we haven't exceeded our maximum allowed courses. If not, we go through the mechanics to add the association to link a `Student` and `Course` in both directions. That is, `courses[currentNumCourses++] = c;` to have the student's current course list contain an association to the new `Course`, as well as `c->AddStudent(this);` to ask the current `Course` to add the `Student` (namely, `this`) to its enrolled student list.

Let's continue by reviewing the remainder of the new member functions for `Student`:

```
void Student::Update()
{   // Course state changed to 'Open For Enrollment', etc.
    // so we can now add it.
    if ((waitListedCourse->GetState() ==
        CourseState::OpenForEnrollment) ||
        (waitListedCourse->GetState() ==
        CourseState::NewSpaceAvailable))
    {
        if (AddCourse(waitListedCourse)) // success Adding
        {
            cout << GetFirstName() << " " << GetLastName();
            cout << " removed from waitlist and added to ";
            cout << waitListedCourse->GetTitle() << endl;
            // Set observer's state to AddSuccess
            SetState(StudentState::AddSuccess);
            // Remove Student from Course's waitlist
            waitListedCourse->Release(this); // Remove Obs.
                                             // from Subject
```

```
            waitListedCourse = nullptr; // Set Subject link
        }                                // to null
    }
}

void Student::PrintCourses() const
{
    cout << "Student: (" << GetFirstName() << " ";
    cout << GetLastName() << ") enrolled in: " << endl;
    for (int i = 0; i < MAXCOURSES &&
                    courses[i] != nullptr; i++)
        cout << "\t" << courses[i]->GetTitle() << endl;
}
```

Continuing with the remainder of our previously mentioned `Student` member functions, next, in our polymorphic `void Student::Update()` method, we conduct the desired adding of a waitlisted course. Recall, `Notify()` will be called when there is a state change on our `Subject` (`Course`). One such state change may be when a `Course` is *Open for Enrollment*, or perhaps a state of *New Space Available* now exists following a `Student` dropping the `Course`. `Notify()` then calls `Update()` on each `Observer`. Our `Update()` has been overridden in `Student` to get the state of the `Course` (`Subject`). If the state indicates the `Course` is now *Open for Enrollment* or has a *New Space Available*, we try `AddCourse(waitListedCourse);`. If this is a success, we set the state of the `Student` (`Observer`) to `StudentState::AddSuccess` (*Add Success*) to indicate that we have been successful in our `Update()`, which means we've added the `Course`. Next, since we have added the desired course to our current course list, we can now remove ourselves from the `Course`'s waitlist. That is, we will want to remove ourselves (`Student`) as an `Observer` from the `Subject` (the `Course`'s waitlist) using `waitListedCourse->Release(this);`. Now that we have added our desired waitlisted course, we can also remove our link to the `Subject` using `waitListedCourse = nullptr;`.

Lastly, our aforementioned `Student` code includes a method to print the currently enrolled courses of the `Student` with `void Student::PrintCourses();`. This method is pretty straightforward.

Bringing the pattern components together

Let us now bring all of our various components together by taking a look at our `main()` function to see how our Observer pattern is orchestrated:

```
int main()
{   // Instantiate several courses
    Course *c1 = new Course("C++", 230);
```

```cpp
Course *c2 = new Course("Advanced C++", 430);
Course *c3 = new Course("C++ Design Patterns", 550);
// Instantiate Students, select a course to be on the
// waitlist for -- to be added when registration starts
Student s1("Anne", "Chu", 'M', "Ms.", 3.9, "66CU", c1);
Student s2("Joley", "Putt", 'I', "Ms.", 3.1,
        "585UD", c1);
Student s3("Geoff", "Curt", 'K', "Mr.", 3.1,
        "667UD", c1);
Student s4("Ling", "Mau", 'I', "Ms.", 3.1, "55TU", c1);
Student s5("Jiang", "Wu", 'Q', "Dr.", 3.8, "88TU", c1);
cout << "Registration is Open" << "\n";
cout << "Waitlist Students to be added to Courses";
cout << endl;
// Sends a message to Students that Course is Open.
c1->Open(); // Students on waitlist will automatically
c2->Open(); // be Added (as room allows)
c3->Open();


// Now that registration is open, add more courses
cout << "During open registration, Students now adding
        additional courses" << endl;
s1.AddCourse(c2);  // Try to add more courses
s2.AddCourse(c2);  // If full, we'll be added to
s4.AddCourse(c2);  // a waitlist
s5.AddCourse(c2);
s1.AddCourse(c3);
s3.AddCourse(c3);
s5.AddCourse(c3);
cout << "Registration complete\n" << endl;
c1->PrintStudents();   // print each Course's roster
c2->PrintStudents();
c3->PrintStudents();
s1.PrintCourses();  // print each Student's course list
s2.PrintCourses();
s3.PrintCourses();
```

```
        s4.PrintCourses();
        s5.PrintCourses();
        return 0;
}
```

Reviewing our aforementioned `main()` function, we first instantiate three `Course` instances. We next instantiate five `Student` instances, utilizing a constructor that allows us to provide an initial `Course` that each `Student` would like to add when course registration commences. Note that these `Students` (`Observers`) will be added to the waitlist for their desired courses (`Subject`). Here, a `Subject` (`Course`) will have a list of `Observers` (`Students`) who wish to add the course when registration opens.

Next, we see that a `Course` that many `Student` instances desire becomes *Open for Enrollment* for registration with `c1->Open();`. `Course::Open()` sets the state of the `Subject` to `CourseState::OpenForEnrollment`, easily indicating the course is *Open for Enrollment*, and then calls `Notify()`. As we know, `Subject::Notify()` will call `Update()` on the `Subject`'s list of observers. It is here that an initial waitlisted `Course` instance will be added to a student's schedule and be subsequently removed as an `Observer` from the `Subject`'s waitlist.

Now that registration is open, each `Student` will try to add more courses in the usual manner using `bool Student::AddCourse(Course *)`, such as with `s1.AddCourse(c2);`. Should a `Course` be full, the `Student` will be added to the `Course`'s waitlist (modeled as the inherited `Subject`'s list of observers, which are in fact, derived `Student` types). Recall, `Course` inherits from `Subject`, which keeps a list of students interested in adding a particular course (the waitlist of observers). When the `Course` state changes to *New Space Available*, students on the waitlist (via data member `observers`) will be notified, and the `Update()` method on each `Student` will subsequently call `AddCourse()` for that `Student`.

Once we have added various courses, we will then see each `Course` print its roster of students, such as `c2->PrintStudents()`. Likewise, we will then see each `Student` print the respective courses in which they are enrolled, such as with `s5.PrintCourses();`.

Let's take a look at the output for this program:

```
Registration is Open
Waitlist Students to be added to Courses
Anne Chu removed from waitlist and added to C++
Goeff Curt removed from waitlist and added to C++
Jiang Wu removed from waitlist and added to C++
Joley Putt removed from waitlist and added to C++
Ling Mau removed from waitlist and added to C++
During open registration, Students now adding more courses
```

```
Registration complete
Course: (C++) has the following students:
        Anne Chu
        Goeff Curt
        Jiang Wu
        Joley Putt
        Ling Mau
Course: (Advanced C++) has the following students:
        Anne Chu
        Joley Putt
        Ling Mau
        Jiang Wu
Course: (C++ Design Patterns) has the following students:
        Anne Chu
        Goeff Curt
        Jiang Wu
Student: (Anne Chu) enrolled in:
        C++
        Advanced C++
        C++ Design Patterns
Student: (Joley Putt) enrolled in:
        C++
        Advanced C++
Student: (Goeff Curt) enrolled in:
        C++
        C++ Design Patterns
Student: (Ling Mau) enrolled in:
        C++
        Advanced C++
Student: (Jiang Wu) enrolled in:
        C++
        Advanced C++
        C++ Design Patterns
```

We have now seen an implementation of the Observer pattern. We have folded the more generic `Subject` and `Observer` classes into the framework of classes we are accustomed to seeing, namely `Course`, `Person`, and `Student`. Let us now briefly recap what we have learned relating to patterns before moving forward to our next chapter.

Summary

In this chapter, we have begun our pursuit to become better C++ programmers by expanding our repertoire beyond OOP concepts to include the utilization of design patterns. Our primary goal is to enable you to solve recurring types of coding problems using *tried and true* solutions by applying common design patterns.

We have first understood the purpose of design patterns and the advantage of employing them in our code. We have then specifically understood the premise behind the Observer pattern and how it contributes to OOP. Finally, we have taken a look at how we may implement the Observer pattern in C++.

Utilizing common design patterns, such as the Observer pattern, will help you more easily solve recurring types of programming problems in a manner understood by other programmers. A key tenant in OOP is to strive for the reuse of components whenever possible. By utilizing design patterns, you will be contributing to reusable solutions with more sophisticated programming techniques.

We are now ready to continue forward with our next design pattern in *Chapter 17, Implementing the Factory Pattern*. Adding more patterns to our collection of skills makes us more versatile and valued programmers. Let's continue forward!

Questions

1. Using the online code for the example in this chapter as a starting point and the solution from a previous exercise (*Question 3, Chapter 10, Implementing Association, Aggregation, and Composition*):

   a. Implement (or modify your previous) `Student::DropCourse()`. When a `Student` drops a `Course`, this event will cause the `Course` state to become state 2, *New Space Available*. With the state change, `Notify()` will then be called on the `Course` (`Subject`), which will then `Update()` the list of observers (students on the waitlist). `Update()` will indirectly allow waitlisted `Student` instances, if any, to now add the `Course`.

   b. Lastly, in `DropCourse()`, remember to remove the dropped course from the student's current course list.

2. What other examples can you imagine that might easily incorporate the Observer pattern?

Applying the Factory Pattern

This chapter will continue our pursuit to expand your C++ programming repertoire beyond core OOP concepts, with the goal of enabling you to solve recurring types of coding problems utilizing common design patterns. We know that incorporating design patterns can enhance code maintenance and provide avenues for potential code reuse.

Continuing to demonstrate and explain popular design patterns and idioms and learning how to implement them effectively in C++, we continue our quest with the Factory pattern, more precisely known as the **Factory Method pattern**.

In this chapter, we will cover the following main topics:

- Understanding the Factory Method pattern and how it contributes to OOP
- Understanding how to implement the Factory Method pattern with and without an Object Factory, and comparing an Object Factory to an Abstract Factory

By the end of this chapter, you will understand the popular Factory Method pattern. We will see two example implementations of this pattern in C++. Adding additional core design patterns to your programming repertoire will enable you to become a more sophisticated and valuable programmer.

Let's increase our programming skillset by examining another common design pattern, the Factory Method pattern.

Technical requirements

Online code for full program examples can be found in the following GitHub URL: `https://github.com/PacktPublishing/Deciphering-Object-Oriented-Programming-with-CPP/tree/main/Chapter17`. Each full program example can be found in the GitHub repository under the appropriate chapter heading (subdirectory) in a file that corresponds to the chapter number, followed by a dash, followed by the example number in the chapter at hand. For example, the first full program in this chapter can be found in the subdirectory `Chapter17` in a file named `Chp17-Ex1.cpp` under the aforementioned GitHub directory.

The CiA video for this chapter can be viewed at: `https://bit.ly/3QOmCC1`.

Understanding the Factory Method pattern

The **Factory pattern**, or **Factory Method pattern**, is a creational design pattern that allows the creation of objects without needing to specify the exact (derived) class that will be instantiated. A Factory Method pattern provides an interface for creating an object, yet allows details within the creation method to decide which (derived) class to instantiate.

A Factory Method pattern is also known as a **virtual constructor**. Much as a virtual destructor has the specific destructor (which is the entry point of the destruction sequence) determined at runtime through dynamic binding, the concept of a virtual constructor is such that the desired object to instantiate is uniformly determined at runtime.

We cannot always anticipate the specific mix of related derived class objects needed in an application. A Factory Method (or virtual constructor) can create, upon request, an instance of one of many related derived class types, based on the input provided. A derived class object will be returned as its base class type by the Factory Method, allowing objects to be both created and stored more generically. Polymorphic operations can be applied to the newly created (upcasted) instances, allowing relevant derived class behaviors to shine through. A Factory Method promotes loose coupling with client code by removing the need to bind specific derived class types in the client code itself. The client merely utilizes the Factory Method to create and provide appropriate instances.

With a Factory Method pattern, we will specify an abstract class (or interface) for collecting and specifying the general behaviors of derived classes we wish to create. The abstract class or interface in this pattern is known as **Product**. We then create the derived classes that we may want to instantiate, overriding any necessary abstract methods. The various concrete derived classes are known as **Concrete Products**.

We then specify a Factory Method whose purpose is to host an interface for uniformly creating instances of Concrete Products. The Factory Method can either be placed in the abstract Product class or in a separate Object Factory class; an **Object Factory** represents a class with the task of creating Concrete Products. This Factory (creation) Method will be static if placed within the abstract Product class and optionally static if instead placed within an Object Factory class. The Factory Method will decide which specific Concrete Product to manufacture, based on a consistent list of input parameters. The Factory Method will return a generalized Product pointer to the Concrete Product. Polymorphic methods can be applied to the newly created object to elicit its specific behavior.

The Factory Method pattern will include the following:

- An abstract **Product** class (or interface).
- Multiple **Concrete Product** derived classes.

- A **Factory Method** in either the abstract Product class or in a separate **Object Factory** class. The Factory Method will have a uniform interface to create an instance of any of the Concrete Product types.

- Concrete Products will be returned by the Factory Method as generalized Product instances.

Keep in mind that a Factory Method (regardless of whether it is in an Object Factory) produces Products. A Factory Method provides a uniform manner for producing many related Product types. Multiple Factory Methods can exist to produce unique Product lines; each Factory Method can be distinguished by a meaningful name, even if their signatures happen to be the same.

Let's move forward to see two sample implementations of the Factory Method pattern.

Implementing the Factory Method pattern

We will explore two common implementations of the Factory Method pattern. Each will have design trade-offs, certainly worthy of discussion!

Let's start with the technique in which the Factory Method is placed in the abstract Product class.

Including the Factory Method in the Product class

To implement the Factory Method pattern, we will first need to create our abstract Product class as well as our Concrete Product classes. These class definitions will begin the foundation on which to build our pattern.

In our example, we will create our Product using a class we are accustomed to seeing – Student. We will then create Concrete Product classes, namely GradStudent, UnderGradStudent, and NonDegreeStudent. We will include a Factory Method in our Product (Student) class with a consistent interface to create any of the derived Product types.

The components we will model complement our framework for our existing Student application by adding classes to differentiate students based on their educational degree goals. The new components provide the basis for a university matriculation (new Student admission) system.

Let us assume that rather than instantiating a Student, our application will instantiate various types of Student – GradStudent, UnderGradStudent, or NonDegreeStudent – based on their learning goals. The Student class will include an abstract polymorphic Graduate() operation; each derived class will override this method with varying implementations. For example, a GradStudent seeking a Ph.D. may have more degree-related criteria to satisfy in the GradStudent::Graduate() method than other specializations of Student. They may require credit hours to be verified, a passing grade point average to be verified, and verification that their dissertation has been accepted. In contrast, an UnderGradStudent might only have their credit hours and overall grade point average to be corroborated.

The abstract Product class will include a static method, `MatriculateStudent()`, as the Factory Method to create various types of students (the Concrete Product types).

Defining the abstract Product class

Let's first take a look at the mechanics for the implementation of our Factory Method, beginning by examining the definition for our abstract Product class, `Student`. This example can be found, as a complete program, in our GitHub repository:

https://github.com/PacktPublishing/Deciphering-Object-Oriented-Programming-with-CPP/blob/main/Chapter17/Chp17-Ex1.cpp

```
// Assume Person class exists with its usual implementation

class Student: public Person    // Notice that Student is now
{                               // an abstract class
private:
    float gpa = 0.0;  // in-class initialization
    string currentCourse;
    const string studentId;
    static int numStudents;
public:
    Student();  // default constructor
    Student(const string &, const string &, char,
       const string &, float, const string &,
       const string &);
    Student(const Student &);  // copy constructor
    ~Student() override;  // virtual destructor
    float GetGpa() const { return gpa; }
    const string &GetCurrentCourse() const
        { return currentCourse; }
    const string &GetStudentId() const
        { return studentId; }
    void SetCurrentCourse(const string &); // proto. only
    void Print() const override;
    string IsA() const override { return "Student"; }
    virtual void Graduate() = 0;   // Student is abstract
```

```
    // Create a derived Student type based on degree sought
    static Student *MatriculateStudent(const string &,
        const string &, const string &, char,
        const string &, float, const string &,
        const string &);
    static int GetNumStudents() { return numStudents; }
};
// Assume all the usual Student member functions exist
```

In the previous class definition, we introduce our abstract Student class, which is derived from Person (a concrete and hence instantiable class). This has been accomplished with the introduction of the abstract method virtual void Graduate() = 0;. In our student matriculation example, we will be following the design decision that only specific types of students should be instantiated, that is, derived class types GradStudent, UnderGradStudent, or NonDegreeStudent.

In the preceding class definition, notice our Factory Method, with the prototype static Student *MatriculateStudent();. This method will use a uniform interface and will provide the means for the creation of various derived class types of Student. We will examine this method in detail once we have seen the class definitions for the derived classes.

Defining the Concrete Product classes

Now, let's take a look at our Concrete Product classes, starting with GradStudent:

```
class GradStudent: public Student
{
private:
    string degree;    // PhD, MS, MA, etc.
public:
    GradStudent() = default;// default constructor
    GradStudent(const string &, const string &,
        const string &, char, const string &, float,
        const string &, const string &);
    // Prototyping default copy constructor isn't necessary
    // GradStudent(const GradStudent &) = default;
    // Since the most base class has virt dtor prototyped,
    // it is not necessary to prototype default destructor
    // ~GradStudent() override = default; // virtual dtor
    void EarnPhD();
    string IsA() const override { return "GradStudent"; }
    void Graduate() override;
```

```
};

// Assume alternate constructor is implemented
// as expected. See online code for full implementation.

void GradStudent::EarnPhD()
{
    if (!degree.compare("PhD")) // only PhD candidates can
        ModifyTitle("Dr.");      // EarnPhd(), not MA and MS
}                                // candidates

void GradStudent::Graduate()
{   // Here, we can check that the required num of credits
    // have been met with a passing gpa, and that their
    // doctoral or master's thesis has been completed.
    EarnPhD(); // Will change title only if a PhD candidate
    cout << "GradStudent::Graduate()" << endl;

}
```

In the aforementioned GradStudent class definition, we add a degree data member to indicate a degree of "PhD", "MS", or "MA", and adjust the constructors and destructor, as necessary. We have moved EarnPhD() to GradStudent, as this method is not applicable to all Student instances. Instead, EarnPhD() is applicable to a subset of GradStudent instances; we will award the title of "Dr." only to the Ph.D. candidates.

In this class, we have overridden IsA() to return "GradStudent". We have also overridden Graduate() to go through the graduation checklist that is applicable for graduate students, calling EarnPhD() if those checklist items have been met.

Now, let's take a look at our next Concrete Product class, UnderGradStudent:

```
class UnderGradStudent: public Student
{
private:
    string degree;  // BS, BA, etc
public:
    UnderGradStudent() = default;// default constructor
    UnderGradStudent(const string &, const string &,
        const string &, char, const string &, float,
        const string &, const string &);
```

```
    // Prototyping default copy constructor isn't necessary
    // UnderGradStudent(const UnderGradStudent &) =default;
    // Since the most base class has virt dtor prototyped,
    // it is not necessary to prototype default destructor
    // ~UnderGradStudent() override = default; // virt dtor
    string IsA() const override
        { return "UnderGradStudent"; }
    void Graduate() override;
};


// Assume alternate constructor is implemented
// as expected. See online code for full implementation.

void UnderGradStudent::Graduate()
{   // Verify that num of credits and gpa requirements have
    // been met for major and any minors or concentrations.
    // Have all applicable university fees been paid?
    cout << "UnderGradStudent::Graduate()" << endl;
}
```

Quickly taking a look at the previously defined `UnderGradStudent` class, we notice that it is very similar to `GradStudent`. This class even includes a `degree` data member. Keep in mind that not all `Student` instances will receive degrees, so we don't want to generalize this attribute by defining it in `Student`. Though we could have introduced a shared base class of `DegreeSeekingStudent` for `UnderGradStudent` and `GradStudent` to collect this commonality, that fine level of granularization would add an additional layer almost unnecessarily. The duplication here is a design trade-off.

The key difference between these two sibling classes is the overridden `Graduate()` method. We can imagine that the checklist for an undergraduate student for graduation may be quite different from that of a graduate student. For this reason, we can reasonably differentiate the two classes. Otherwise, they are very much the same.

Now, let's take a look at our next Concrete Product class, `NonDegreeStudent`:

```
class NonDegreeStudent: public Student
{
public:
    NonDegreeStudent() = default;  // default constructor
```

```
    NonDegreeStudent(const string &, const string &, char,
        const string &, float, const string &,
        const string &);
    // Prototyping default copy constructor isn't necessary
    // NonDegreeStudent(const NonDegreeStudent &s)
    //     =default;
    // Since the most base class has virt dtor prototyped,
    // it is not necessary to prototype default destructor
    // ~NonDegreeStudent() override = default; // virt dtor
    string IsA() const override
        { return "NonDegreeStudent"; }
    void Graduate() override;
};


// Assume alternate constructor is implemented as expected.
// See online code for full implementation.

void NonDegreeStudent::Graduate()
{   // Check if applicable tuition has been paid.
    // There is no credit or gpa requirement.
    cout << "NonDegreeStudent::Graduate()" << endl;
}
```

Taking a comparably quick look at the aforementioned NonDegreeStudent class, we notice that this Concrete Product is similar to its sibling classes. However, there is no degree data member within this class. Also, the overridden Graduate() method has less verification to undertake than in the overridden versions of this method for either the GradStudent or UnderGradStudent classes.

Examining the Factory Method definition

Next, let's take a look at our Factory Method, a static method in our Product (Student) class:

```
// Creates a Student based on the degree they seek
// This is a static Student method (keyword in prototype)
Student *Student::MatriculateStudent(const string &degree,
    const string &fn, const string &ln, char mi,
    const string &t, float avg, const string &course,
    const string &id)
```

```
{
    if (!degree.compare("PhD") || !degree.compare("MS")
        || !degree.compare("MA"))
        return new GradStudent(degree, fn, ln, mi, t, avg,
                                    course, id);
    else if (!degree.compare("BS") ||
            !degree.compare("BA"))
        return new UnderGradStudent(degree, fn, ln, mi, t,
                                    avg, course, id);
    else if (!degree.compare("None"))
        return new NonDegreeStudent(fn, ln, mi, t, avg,
                                    course, id);
}
```

The aforementioned static method of Student, MatriculateStudent(), represents the Factory Method to create various Products (concrete Student instances). Here, based on the degree type that the Student seeks, one of GradStudent, UnderGradStudent, and NonDegreeStudent will be instantiated. Notice that the signature of MatriculateStudent() can handle the parameter requirements for any of the derived class constructors. Also notice that any of these specialized instance types will be returned as a base class pointer of the abstract Product type (Student).

An interesting option within the Factory Method, MatriculateStudent(), is that this method is not obligated to instantiate a new derived class instance. Instead, it may recycle a previous instance that may still be available. For example, imagine a Student is temporarily unregistered in the university (due to late payment), yet has been kept available on a list of *pending students*. The MatriculateStudent() method may instead choose to return a pointer to such an existing Student. *Recycling* is an alternative within a Factory Method!

Bringing the pattern components together

Finally, let's now bring all of our various components together by taking a look at our main() function to see how our Factory Method pattern is orchestrated:

```
int main()
{
    Student *scholars[MAX] = { }; // init. to nullptrs
    // Student is now abstract; cannot instantiate directly
    // Use Factory Method to make derived types uniformly
    scholars[0] = Student::MatriculateStudent("PhD",
        "Sara", "Kato", 'B', "Ms.", 3.9, "C++", "272PSU");
```

```
    scholars[1] = Student::MatriculateStudent("BS",
        "Ana", "Sato", 'U', "Ms.", 3.8, "C++", "178PSU");
    scholars[2] = Student::MatriculateStudent("None",
        "Elle", "LeBrun", 'R', "Miss", 3.5, "C++", "111BU");
    for (auto *oneStudent : scholars)
    {
        oneStudent->Graduate();
        oneStudent->Print();
    }
    for (auto *oneStudent : scholars)
        delete oneStudent;    // engage virt dtor sequence
    return 0;
}
```

Reviewing our aforementioned `main()` function, we first create an array of pointers for potentially specialized `Student` instances in their generalized `Student` form. Next, we invoke the static Factory Method `Student::MatriculateStudent()`, within the abstract Product class, to create the appropriate Concrete Product (derived `Student` class type). We create one of each of the derived `Student` types – `GradStudent`, `UnderGradStudent`, and `NonDegreeStudent`.

We then loop through our generalized collection, calling `Graduate()` and then `Print()` for each instance. For students earning a Ph.D. (`GradStudent` instances), their title will be changed to `"Dr."` by the `GradStudent::Graduate()` method. Finally, we iterate through another loop to deallocate each instance's memory. Thankfully, `Student` has included a virtual destructor so that the destruction sequence starts at the proper level.

Let's take a look at the output for this program:

```
GradStudent::Graduate()
  Dr. Sara B. Kato with id: 272PSU GPA:  3.9 Course: C++
UnderGradStudent::Graduate()
  Ms. Ana U. Sato with id: 178PSU GPA:  3.8 Course: C++
NonDegreeStudent::Graduate()
  Miss Elle R. LeBrun with id: 111BU GPA:  3.5 Course: C++
```

An advantage of the preceding implementation is that it is very straightforward. However, we can see a close coupling exists between the abstract Product, containing the Factory Method (which constructs the derived class types), and the derived Concrete Products. Yet in OOP, a base class will ideally have no knowledge of any descendent types.

A disadvantage to this closely coupled implementation is that the abstract Product class must include a means for instantiation for each of its descendants in its static creation method, `MatriculateStudent()`. Adding new derived classes now affects the abstract base class definition – it needs to be recompiled. What if we don't have access to the source code for this base class? Is there a way to decouple the dependencies that exist between the Factory Method and the Products that the Factory Method will create? Yes, there is an alternate implementation.

Let us now take a look at an alternate implementation of the Factory Method pattern. We will instead use an Object Factory class to encapsulate our Factory Method of `MatriculateStudent()`, rather than including this method in the abstract Product class.

Creating an Object Factory class to encapsulate the Factory Method

For our alternative implementation of the Factory Method pattern, we will create our abstract Product class with a slight deviation from its previous definition. We will, however, create our concrete Product classes as before. These class definitions, collectively, will again begin the framework on which to base our pattern.

In our revised example, we will define our Product again as the `Student` class. We will also again derive Concrete Product classes of `GradStudent`, `UnderGradStudent`, and `NonDegreeStudent`. This time, however, we will not include a Factory Method in our Product (`Student`) class. Instead, we will create a separate Object Factory class that will include the Factory Method. As before, the Factory Method will have a uniform interface to create any of the derived Product types. The Factory Method need not be static, as it was in our last implementation.

Our Object Factory class will include `MatriculateStudent()` as the Factory Method to create various types of `Student` instances (the Concrete Product types).

Defining the abstract Product class without the Factory Method

Let's take a look at the mechanics for our alternate implementation of the Factory Method pattern, beginning by examining the definition for our abstract Product class, `Student`. This example can be found, as a complete program, in our GitHub repository at the following URL:

https://github.com/PacktPublishing/Deciphering-Object-Oriented-Programming-with-CPP/blob/main/Chapter17/Chp17-Ex2.cpp

```
// Assume Person class exists with its usual implementation

class Student: public Person    // Notice Student is
{                               // an abstract class
private:
```

```
        float gpa = 0.0;    // in-class initialization
        string currentCourse;
        const string studentId;
        static int numStudents; // Remember, static data mbrs
                    // are also shared by all derived instances
    public:
        Student();   // default constructor
        Student(const string &, const string &, char,
            const string &, float, const string &,
            const string &);
        Student(const Student &);   // copy constructor
        ~Student() override;   // destructor
        float GetGpa() const { return gpa; }
        const string &GetCurrentCourse() const
            { return currentCourse; }
        const string &GetStudentId() const
            { return studentId; }
        void SetCurrentCourse(const string &); // proto. only
        void Print() const override;
        string IsA() const override { return "Student"; }
        virtual void Graduate() = 0;   // Student is abstract
        static int GetNumStudents() { return numStudents; }
    };
```

In our aforementioned class definition for Student, the key difference from our prior implementation is that this class no longer contains a static MatriculateStudent() method to serve as the Factory Method. Student is merely an abstract base class. Remember, all graduate students, undergraduate students, and non-degree students are all specializations of Student, therefore static int numStudents is a shared, collective count of all types of Student.

Defining the Concrete Product classes

With that in mind, let's take a look at the derived (Concrete Product) classes:

```
class GradStudent: public Student
{    // Implemented as in our last example
};


class UnderGradStudent: public Student
```

```
{    // Implemented as in our last example
};


class NonDegreeStudent: public Student
{    // Implemented as in our last example
};
```

In our previously listed class definitions, we can see that our Concrete derived Product classes are identical to our implementation for these classes as in our first example.

Adding the Object Factory class with the Factory Method

Next, let us introduce an Object Factory class that includes our Factory Method:

```
class StudentFactory      // Object Factory class
{
public:
    // Factory Method creates Student based on degree sought
    Student *MatriculateStudent(const string &degree,
        const string &fn, const string &ln, char mi,
        const string &t, float avg, const string &course,
        const string &id)
    {
        if (!degree.compare("PhD") || !degree.compare("MS")
            || !degree.compare("MA"))
            return new GradStudent(degree, fn, ln, mi, t,
                                        avg, course, id);
        else if (!degree.compare("BS") ||
                !degree.compare("BA"))
            return new UnderGradStudent(degree, fn, ln, mi,
                                        t, avg, course, id);
        else if (!degree.compare("None"))
            return new NonDegreeStudent(fn, ln, mi, t, avg,
                                        course, id);
    }
};
```

In the aforementioned Object Factory class definition (the `StudentFactory` class), we minimally include the Factory Method specification, namely, `MatriculateStudent()`. The method is very similar to that in our prior example. However, by capturing the creation of Concrete Products in an Object Factory, we have decoupled the relationship between the abstract Product and the Factory Method.

Bringing the pattern components together

Next, let's compare our `main()` function to that of our original example to visualize how our revised components implement the Factory Method pattern:

```cpp
int main()
{
    Student *scholars[MAX] = { }; // init. to nullptrs
    // Create an Object Factory for Students
    StudentFactory *UofD = new StudentFactory();
    // Student is now abstract, cannot instantiate directly
    // Ask the Object Factory to create a Student
    scholars[0] = UofD->MatriculateStudent("PhD", "Sara",
                "Kato", 'B', "Ms.", 3.9, "C++", "272PSU");
    scholars[1] = UofD->MatriculateStudent("BS", "Ana",
                "Sato", 'U', "Dr.", 3.8, "C++", "178PSU");
    scholars[2] = UofD->MatriculateStudent("None", "Elle",
                "LeBrun", 'R', "Miss", 3.5, "C++", "111BU");
    for (auto *oneStudent : scholars)
    {
        oneStudent->Graduate();
        oneStudent->Print();
    }
    for (auto *oneStudent : scholars)
        delete oneStudent;   // engage virt dtor sequence
    delete UofD; // delete factory that created various
    return 0;    // types of students
}
```

Considering our previously listed `main()` function, we see that we have again created an array of pointers to the abstract Product type (`Student`). We have then instantiated an Object Factory that can create various `Student` instances of Concrete Product types with `StudentFactory *UofD = new StudentFactory();`. As with the previous example, one instance of each derived type `GradStudent`, `UnderGradStudent`, and `NonDegreeStudent` is created by the Object Factory based upon the degree type sought by each student. The remainder of the code in `main()` is as found in our prior example.

Our output will be the same as our last example.

The advantage of the Object Factory class over our prior means of implementation is that we have removed the dependency of object creation from our abstract Product class (in the Factory Method) with knowledge of what the derived class types are. That is, should we expand our hierarchy to include new Concrete Product types, we do not have to modify the abstract Product class. Of course, we will need to have access to modify our Object Factory class, `StudentFactory`, to augment our `MatriculateStudent()` Factory Method.

A pattern related to this implementation, an **Abstract Factory**, is an additional pattern that allows individual factories with a similar purpose to be grouped together. An Abstract Factory can be specified to provide a means to unify similar Object Factories; it is a factory that will create factories, adding yet another level of abstraction to our original pattern.

We have now seen two implementations of the Factory Method pattern. We have folded the concepts of Product and Factory Method into the framework of classes we are accustomed to seeing, namely `Student`, and descendants of `Student`. Let's now briefly recap what we have learned relating to patterns before moving forward to our next chapter.

Summary

In this chapter, we have continued our pursuit to become better C++ programmers by expanding our knowledge of design patterns. In particular, we have explored the Factory Method pattern, both conceptually and with two common implementations. Our first implementation included placing the Factory Method in our abstract Product class. Our second implementation removed the dependency between our Abstract Product and our Factory Method by instead adding an Object Factory class to contain our Factory Method. We also very briefly discussed the notion of an Abstract Factory.

Utilizing common design patterns, such as the Factory Method pattern, will help you more easily solve recurring types of programming problems in a manner understood by other programmers. By utilizing core design patterns, you will be contributing to well-understood and reusable solutions with more sophisticated programming techniques.

We are now ready to continue forward with our next design pattern, in *Chapter 18, Implementing the Adapter Pattern*. Adding more patterns to our collection of skills makes us more versatile and valued programmers. Let's continue onward!

Questions

1. Using the solution from a previous exercise (*Question 1*, *Chapter 8*, *Mastering Abstract Classes*), augment your code as follows:

   a. Implement the Factory Method pattern to create various shapes. You will have already created an abstract base class of `Shape` and derived classes such as `Rectangle`, `Circle`, `Triangle`, and possibly `Square`.

   b. Choose whether to implement your Factory Method as a static method in `Shape` or as a method in a `ShapeFactory` class (introducing the latter class if necessary).

2. What other examples can you imagine that might easily incorporate the Factory Method pattern?

18

Applying the Adapter Pattern

This chapter will extend our quest to expand your C++ programming skills beyond core OOP concepts, with the goal of enabling you to solve recurring types of coding problems utilizing common design patterns. Incorporating design patterns in coding solutions can not only provide elegant solutions but also enhance code maintenance and provide potential opportunities for code reuse.

The next core design pattern that we will learn how to implement effectively in C++ is the **Adapter pattern**.

In this chapter, we will cover the following main topics:

- Understanding the Adapter pattern and how it contributes to OOP
- Understanding how to implement the Adapter pattern in C++

By the end of this chapter, you will understand the essential Adapter pattern and how it can be used to either allow two incompatible classes to communicate or to upgrade unseemly code to well-designed OO code. Adding another key design pattern to your knowledge set will refine your programming skills to help make you a more valuable programmer.

Let's increase our programming skillset by examining another common design pattern, the Adapter pattern.

Technical requirements

Online code for full program examples can be found in the following GitHub URL: `https://github.com/PacktPublishing/Deciphering-Object-Oriented-Programming-with-CPP/tree/main/Chapter18`. Each full program example can be found in the GitHub repository under the appropriate chapter heading (subdirectory) in a file that corresponds to the chapter number, followed by a dash, followed by the example number in the chapter at hand. For example, the first full program in this chapter can be found in the subdirectory `Chapter18` in a file named `Chp18-Ex1.cpp` under the aforementioned GitHub directory.

The CiA video for this chapter can be viewed at: `https://bit.ly/3Kaxckc`.

Understanding the Adapter pattern

The **Adapter pattern** is a structural design pattern that provides a means for converting an existing, undesirable interface of a class to an interface that another class expects. An **Adapter class** will be the link for communication between two existing components, adapting the interfaces so that the two may share and exchange information. An Adapter allows two or more classes to work together that otherwise could not do so.

Ideally, an Adapter will not add functionality but will add the preferred interface for usage (or conversion) to either allow one class to be used in an expected manner or for two otherwise incompatible classes to communicate with one another. In its most simple form, an Adapter simply converts an existing class to support an expected interface as may be specified in an OO design.

An Adapter can be either associated with or derived from the class for which it is providing an adaptive interface. If inheritance is used, a private or protected base class is appropriate to hide the underlying implementation. If instead, the Adapter class is associated with the class having the undesirable interface, the methods in the Adapter class (with the new interfaces) will merely delegate the work to its associated class.

The Adapter pattern can also be used to add an OO interface to (that is, to *wrap an OO interface around*) a series of functions or other classes, allowing assorted existing components to be utilized more naturally in an OO system. This specific type of Adapter is known as a **wrapper class**. The originating functions or utilities may even be written in another language, such as C (which would then require special language tags, such as `extern C`, to allow the linker to resolve linkage conventions between the two languages).

Utilizing the Adapter pattern has benefits. The Adapter allows the reuse of existing code by providing a shared interface to allow otherwise unrelated classes to communicate. The OO programmer will now use the Adapter class directly, allowing for easier maintenance of the application. That is, most programmer interaction will be with a well-designed Adapter class, rather than with two or more odd components. A small drawback of using an Adapter is a slightly decreased performance from the added layer of code. However, most often, reusing existing components through providing a clean interface to support their interaction is a winning proposition, despite a (hopefully small) performance trade-off.

The Adapter pattern will include the following:

- An **Adaptee** class, which represents the class with desirable utilities, yet which exists in a form that is not suitable or as expected.

- An **Adapter** class, which adapts the interface of the Adaptee class to meet the needs of the desired interface.

- A **Target** class, which represents the specific, desired interface of the application at hand. A class may be both a Target and an Adapter.

- Optional **Client** classes, which will interact with the Target class to fully define the application at hand.

An Adapter pattern allows the reuse of qualified, existing components that do not meet the interface needs of current application designs.

Let's move forward to see two common applications of the Adapter pattern; one will have two potential means for implementation.

Implementing the Adapter pattern

Let's explore two common uses of the Adapter pattern. That is, creating an Adapter to bridge the gap between two incompatible class interfaces or building an Adapter to simply wrap an existing set of functions with an OO interface.

We will start with the usage of an *Adapter* providing a connector between two (or more) incompatible classes. The *Adaptee* will be a well-tested class that we would like to reuse (but which has an undesirable interface), and the *Target* classes will be those specified in our OO design for an application in the making. Let's now specify an Adapter to allow our Adaptee to work with our Target classes.

Using an Adapter to provide a necessary interface to an existing class

To implement the Adapter pattern, we will first need to identify our Adaptee class. We will then create an Adapter class to modify the interface of the Adaptee. We will also identify our Target class, representing the class we need to model per our OO design. At times, our Adapter and Target may be rolled into a single class. In an actual application, we will additionally have Client classes, representing the full complement of classes found in the final application at hand. Let's start with the Adaptee and Adapter classes, as these class definitions will begin the foundation on which to build our pattern.

In our example, we will specify our Adaptee class as one we are accustomed to seeing – Person. We will imagine that our planet has recently become aware of many other exoplanets capable of supporting life and that we have benevolently made allies with each such civilization. Further imagining that the various software systems on Earth would like to welcome and include our new friends, including Romulans and Orkans, we would like to adapt some of our existing software to easily accommodate the new demographics of our exoplanet neighbors. With that in mind, we will transform our Person class to include more interplanetary terminology by creating an Adapter class, Humanoid.

In our forthcoming implementation, we will use private inheritance to inherit Humanoid (Adapter) from Person (Adaptee), therefore hiding the underlying implementation of the Adaptee. We could have alternatively associated a Humanoid with a Person (an implementation we will also review in this section). We can then flesh out some derived classes of Humanoid within our hierarchy, such as Orkan, Romulan, and Earthling, to accommodate the intergalactic application at hand. The Orkan, Romulan, and Earthling classes can be considered our Target classes, or those that our application will instantiate. We will choose to make our Adapter class, Humanoid, abstract so that it is not directly instantiable. Because our specific derived classes (Target classes) can be generalized by their abstract base class type (Humanoid) in our application (Client), we can also consider Humanoid a Target class. That is, Humanoid can be viewed primarily as an Adapter, but secondarily as a generalized Target class.

Our various Client classes can utilize derived classes of Humanoid, making instances of each of its concrete descendants. These instances may be stored in their own specialized type or genericized using Humanoid pointers. Our implementation is a modern take on the well-used Adapter design pattern.

Specifying the Adaptee and Adapter (private inheritance technique)

Let's take a look at the mechanics of the first usage of our Adapter pattern, beginning by reviewing the definition for our Adaptee class, Person. This example can be found, as a complete program, in our GitHub repository:

https://github.com/PacktPublishing/Deciphering-Object-Orient-ed-Programming-with-CPP/blob/main/Chapter18/Chp18-Ex1.cpp

```cpp
// Person is the Adaptee class (class requiring adaptation)
class Person
{
private:
    string firstName, lastName, title, greeting;
    char middleInitial = '\0';  // in-class initialization
protected:
    void ModifyTitle(const string &);
public:
    Person() = default;   // default constructor
    Person(const string &, const string &, char,
          const string &);
    // default copy constructor prototype is not necessary
    // Person(const Person &) = default;  // copy ctor
    // Default op= suffices, so we'll comment out proto.
    // (see online code for review of implementation)
    // Person &operator=(const Person &); // assignment op.
    virtual ~Person()= default;  // virtual destructor
```

```
        const string &GetFirstName() const
            { return firstName; }
        const string &GetLastName() const
            { return lastName; }
        const string &GetTitle() const { return title; }
        char GetMiddleInitial() const { return middleInitial; }
        void SetGreeting(const string &);
        virtual const string &Speak() { return greeting; }
        virtual void Print() const;
};


    // Assume constructors, destructor, and non-inline methods
    // are implemented as expected (see online code)
```

In the previous class definition, we notice that our `Person` class definition is as we have been accustomed to seeing it in many other examples throughout this book. This class is instantiable; however, `Person` is not an appropriate class to instantiate in our intergalactic application. Instead, the expected interface would be to utilize that found in `Humanoid`.

With that in mind, let's take a look at our Adapter class, `Humanoid`:

```
class Humanoid: private Person    // Humanoid is abstract
{
protected:
    void SetTitle(const string &t) { ModifyTitle(t); }
public:
    Humanoid() = default;
    Humanoid(const string &, const string &,
            const string &, const string &);
    // default copy constructor prototype not required
    // Humanoid(const Humanoid &h) = default;
    // default op= suffices, so commented out below, but
    // let's review how we'd write op= if needed
    // note explicit Humanoid downcast after calling base
    // class Person::op= to match return type needed here
    // Humanoid &operator=(const Humanoid &h)
    //       { return dynamic_cast<Humanoid &>
    //                (Person::operator=(h)); }
```

```
        // dtor proto. not required since base dtor is virt.
        // ~Humanoid() override = default; // virt destructor
        // Added interfaces for the Adapter class
        const string &GetSecondaryName() const
            { return GetFirstName(); }
        const string &GetPrimaryName() const
            { return GetLastName(); }
        // scope resolution needed in method to avoid recursion
        const string &GetTitle() const
            { return Person::GetTitle();}
        void SetSalutation(const string &m) { SetGreeting(m); }
        virtual void GetInfo() const { Print(); }
        virtual const string &Converse() = 0; // abstract class
};

Humanoid::Humanoid(const string &n2, const string &n1,
    const string &planetNation, const string &greeting):
    Person(n2, n1, ' ', planetNation)
{
    SetGreeting(greeting);
}


const string &Humanoid::Converse()   // default definition
{                       // for pure virtual function - unusual
    return Speak();
}
```

In the aforementioned Humanoid class, our goal is to provide an Adapter to contribute to the expected interface that our intergalactic application requires. We simply derive Humanoid from Person using private inheritance, hiding the public interfaces found in Person from use outside the scope of Humanoid. We understand that the target application (Client) would not wish for the public interfaces found in Person to be utilized by various subtypes of Humanoid instances. Notice that we are not adding functionality, only adapting the interface.

We then notice the public methods introduced in Humanoid that provide the desired interfaces for the Target class(es). The implementation of these interfaces is often straightforward. We simply call the inherited method defined in Person, which will easily complete the task at hand (but which uses an unacceptable interface to do so). For example, our Humanoid::GetPrimaryName() method simply calls `Person::GetLastName();` to complete the task. However, `GetPrimaryName()` may more so represent appropriate intergalactic lingo than `Person::GetLastName()`. We can see how `Humanoid` is serving as an Adapter for `Person`. We can also see how most of the member functions of the Adapter class, `Humanoid`, use inline functions to simply wrap the `Person` methods with more suitable interfaces while adding no overhead.

Note that it is not necessary to precede the calls to `Person` base class methods within Humanoid methods with `Person::` (except when a `Humanoid` method calls the same named method in `Person`, such as with `GetTitle()`). The scope resolution usage of `Person::` avoids potential recursion in these situations.

We also notice that `Humanoid` introduces an abstract polymorphic method (that is, a pure virtual function) with the specification of `virtual const string &Converse() = 0;`. We have made the design decision that only derived classes of `Humanoid` will be instantiable. Nonetheless, we understand that public descendant classes may still be collected by their base class type of `Humanoid`. Here, `Humanoid` serves primarily as the Adapter class and secondarily as a Target class offering the suite of acceptable interfaces.

Notice that our pure virtual function `virtual const String &Converse() = 0;` includes a default implementation. This is rare but allowed, so long as the implementation is not written inline. Here, we utilize the opportunity to specify a default behavior for `Humanoid::Converse()`, by simply calling `Person::Speak()`.

Deriving concrete classes from the Adapter

Next, let's extend our Adapter (`Humanoid`) and take a look at one of our concrete, derived Target classes, `Orkan`:

```
class Orkan: public Humanoid
{
public:
    Orkan() = default;   // default constructor
    Orkan(const string &n2, const string &n1,
        const string &t): Humanoid(n2, n1, t, "Nanu nanu")
        { }
    // default copy constructor prototype not required
    // Orkan(const Orkan &h) = default;
    // default op= suffices, so commented out below, but
    // let's review how we'd write it if needed
    // note explicit Orkan downcast after calling base
    // class Humanoid::op= to match return type needed here
    // Orkan &operator=(const Orkan &h)
```

```
    //       { return dynamic_cast<Orkan &>
    //                 (Humanoid::operator=(h)); }
    // dtor proto. not required since base dtor is virt.
    // ~Orkan() override = default; // virtual destructor
    const string &Converse() override;
};


// Must override Converse to make Orkan a concrete class
const string &Orkan::Converse()
{
    return Humanoid::Converse(); // use scope resolution to
}                                // avoid recursion
```

In our aforementioned Orkan class, we use public inheritance to derive Orkan from Humanoid. An Orkan *Is-A* Humanoid. As such, all of the public interfaces in Humanoid are available to Orkan instances. Notice that our alternate constructor sets the default greeting message to "Nanu nanu", per the Orkan dialect.

Because we wish O r k a n to be a concrete, instantiable class, we must override Humanoid::Converse() and provide an implementation in the Orkan class. Notice, however, that Orkan::Converse() simply calls Humanoid::Converse();. Perhaps Orkan finds the default implementation in its base class acceptable. Notice that we use the Humanoid:: scope resolution to qualify Converse() within the Orkan::Converse() method to avoid recursion.

Interestingly, had Humanoid not been an abstract class, Orkan would not have had to override Converse() – the default behavior would have automatically been inherited. Yet, with Humanoid defined as abstract, the override of Converse() is necessary within Orkan, otherwise, Orkan will also be also viewed as an abstract class. No worries! We can utilize the benefit of the default behavior of Humanoid::Converse() merely by calling it within Orkan::Converse(). This will satisfy the requirements for making Orkan concrete, while allowing Humanoid to remain abstract, while still providing the rare default behavior for Converse()!

Now, let's take a look at our next concrete Target class, Romulan:

```
class Romulan: public Humanoid
{
public:
    Romulan() = default;    // default constructor
    Romulan(const string &n2, const string &n1,
        const string &t): Humanoid(n2, n1, t, "jolan'tru")
        { }
```

```
    // default copy constructor prototype not required
    // Romulan(const Romulan &h) = default;
    // default op= suffices, so commented out below, but
    // let's review how we'd write it if so needed
    // note explicit Romulan downcast after calling base
    // class Humanoid::op= to match return type needed here
    // Romulan &operator=(const Romulan &h)
    //     { return dynamic_cast<Romulan &>
    //                 (Humanoid::operator=(h)); }
    // dtor proto. not required since base dtor is virt.
    // ~Romulan() override = default;   // virt destructor
    const string &Converse() override;
};


// Must override Converse to make Romulan a concrete class
const string &Romulan::Converse()
{
    return Humanoid::Converse(); // use scope resolution to
}                                // avoid recursion
```

Taking a comparably quick look at the aforementioned Romulan class, we notice that this concrete Target is similar to its sibling class, Orkan. We notice that the default message for the greeting passed up to our base class constructor is "jolan'tru" to reflect Romulan dialect. Though we could have made our implementation of Romulan::Converse() more intricate, we chose not to do so. We can quickly understand the full scope of this class.

Next, let's take a look at our third Target class, Earthling:

```
class Earthling: public Humanoid
{
public:
    Earthling() = default;    // default constructor
    Earthling(const string &n2, const string &n1,
        const string &t): Humanoid(n2, n1, t, "Hello") { }
    // default copy constructor prototype not required
    // Earthling(const Romulan &h) = default;
    // default op= suffices, so commented out below, but
    // let's review how we'd write it if so needed
```

```
    // note explicit Earthling downcast after calling base
    // class Humanoid::op= to match return type needed here
    // Earthling &operator=(const Earthling &h)
    //     { return dynamic_cast<Earthling &>
    //                 (Humanoid::operator=(h)); }
    // dtor proto. not required since base dtor is virt.
    // ~Earthling() override = default; // virt destructor
    const string &Converse() override;
};


// Must override to make Earthling a concrete class
const string &Earthling::Converse()
{
    return Humanoid::Converse(); // use scope resolution to
}                                 // avoid recursion
```

Again, taking another comparably quick look at the aforementioned `Earthling` class, we notice that this concrete Target is similar to its sibling classes, `Orkan` and `Romulan`.

Now that we have defined our Adaptee, Adapter, and multiple Target classes, let's bring the pieces together by examining the Client portion of our program.

Bringing the pattern components together

Finally, let us consider what a sample Client may look like in our overall application. Certainly, it may consist of many files with a variety of classes. In its simplest form, as shown next, our Client will contain a `main()` function to drive the application.

Let's now take a look at our `main()` function to see how our pattern is orchestrated:

```
int main()
{
    list<Humanoid *> allies;
    Orkan *o1 = new Orkan("Mork", "McConnell", "Orkan");
    Romulan *r1 = new Romulan("Donatra", "Jarok",
                                "Romulan");
    Earthling *e1 = new Earthling("Eve", "Xu",
                                "Earthling");
```

```
    // Add each specific type of Humanoid to generic list
    allies.push_back(o1);
    allies.push_back(r1);
    allies.push_back(e1);

    // Process the list of allies (which are Humanoid *'s
    // Actually, each is a specialization of Humanoid!)
    for (auto *entity : allies)
    {
        entity->GetInfo();
        cout << entity->Converse() << endl;
    }
    // Though each type of Humanoid has a default
    // Salutation, each may expand their skills with
    // an alternate language
    e1->SetSalutation("Bonjour");
    e1->GetInfo();
    cout << e1->Converse() << endl; // Show the Earthling's
                                    // revised language capabilities
    delete o1;    // delete the heap instances
    delete r1;
    delete e1;
    return 0;
}
```

Reviewing our aforementioned main() function, we first create an STL list of Humanoid pointers with list<Humanoid *> allies;. We then instantiate an Orkan, Romulan, and an Earthling and add each to the list using allies.push_back(). Again, using the Standard Template Library, we next create a list iterator to walk through the list of pointers to Humanoid instances. As we iterate through our generalized list of allies, we call the approved interfaces of GetInfo() and Converse() on each item in our list (that is, for each specific type of Humanoid).

Next, we specify one specific Humanoid, an Earthling, and change this instance's default greeting by invoking e1->SetSalutation("Bonjour");. By calling Converse() again on this instance (we first did so on this object generically in the aforementioned loop), we can request that the Earthling use "Bonjour" to greet allies instead of "Hello" (the default greeting for Earthling).

Let's take a look at the output for this program:

```
Orkan Mork McConnell
Nanu nanu
Romulan Donatra Jarok
jolan'tru
Earthling Eve Xu
Hello
Earthling Eve Xu
Bonjour
```

In the aforementioned output, notice that the planetary specification for each Humanoid is displayed (Orkan, Romulan, and Earthling), followed by their secondary and primary names. Then, the appropriate greeting is displayed for the particular Humanoid. Notice that Earthling Eve Xu first converses using "Hello" and then later converses using "Bonjour".

An advantage of the preceding implementation (using a private base class to derive Adapter from Adaptee) is that the coding is very straightforward. With this approach, any protected methods in the Adaptee class can easily be carried down to be used within the scope of the Adapter methods. We will soon see that protected members will be an issue should we instead use the association as a means of connecting the Adapter to Adaptee.

A disadvantage of the prior mentioned approach is that it is a C++ specific implementation. Other languages do not support private base classes. Alternatively, using a public base class to define the relationship between Adapter and Adaptee would fail to conceal the unwanted Adaptee interface, and would be a very poor design choice.

Considering an alternate specification of Adaptee and Adapter (association)

Let us now briefly consider a slightly revised version of the aforementioned Adapter pattern implementation. We will instead use an association to model the relationship between the Adaptee and Adapter. The concrete derived classes (Targets) will still be derived from the Adapter as before.

Here is an alternative implementation of our Adapter class, Humanoid, using an association between Adapter and Adaptee. Though we will only review the portion of the code that differs from our initial approach, the full implementation can be found as a complete program in our GitHub repository:

https://github.com/PacktPublishing/Deciphering-Object-Oriented-Programming-with-CPP/blob/main/Chapter18/Chp18-Ex2.cpp

```cpp
// Assume that Person exists mostly as before – however,
// Person::ModifyTitle() must be moved from protected to
// public or be unused if modifying Person is not possible.
```

```cpp
// Let's assume we moved Person::ModifyTitle() to public.

class Humanoid      // Humanoid is abstract
{
private:
    Person *life = nullptr;   // delegate all requests to
                              // the associated object
protected:
    void SetTitle(const string &t)
        { life->ModifyTitle(t); }
public:
    Humanoid() = default;
    Humanoid(const string &, const string &,
            const string &, const string &);
    Humanoid(const Humanoid &h);// we have work for copies!
    Humanoid &operator=(const Humanoid &); // and for op=
    virtual ~Humanoid()  // virtual destructor
        { delete life; life = nullptr; }
    // Added interfaces for the Adapter class
    const string &GetSecondaryName() const
        { return life->GetFirstName(); }
    const string &GetPrimaryName() const
        { return life->GetLastName(); }
    const string &GetTitle() const
        { return life->GetTitle();}
    void SetSalutation(const string &m)
        { life->SetGreeting(m); }
    virtual void GetInfo() const { life->Print(); }
    virtual const string &Converse() = 0; // abstract class
};

Humanoid::Humanoid(const string &n2, const string &n1,
        const string &planetNation, const string &greeting)
{
    life = new Person(n2, n1, ' ', planetNation);
    life->SetGreeting(greeting);
```

```
}

// copy constructor (we need to write it ourselves)
Humanoid::Humanoid(const Humanoid &h)
{   // Remember life data member is of type Person
    delete life;  // delete former associated object
    life = new Person(h.GetSecondaryName(),
                      h.GetPrimaryName(),' ', h.GetTitle());
    life->SetGreeting(h.life->Speak());
}

// overloaded operator= (we need to write it ourselves)
Humanoid &Humanoid::operator=(const Humanoid &h)
{
    if (this != &h)
        life->Person::operator=(dynamic_cast
                                <const Person &>(h));
    return *this;
}

const string &Humanoid::Converse() //default definition for
{                                   // pure virtual fn - unusual
    return life->Speak();
}
```

Notice in the aforementioned implementation of our Adapter class, Humanoid is no longer derived from Person. Instead, Humanoid will add a private data member Person *life;, which will represent an association between the Adapter (Humanoid) and the Adaptee (Person). In our Humanoid constructors, we will need to allocate the underlying implementation of the Adaptee (Person). We will also need to delete the Adaptee (Person) in our destructor.

Similar to our last implementation, Humanoid offers the same member functions within its public interface. However, notice that each Humanoid method delegates, through the associated object, a call to the appropriate Adaptee methods. For example, Humanoid::GetSecondaryName() merely calls life->GetFirstName(); to delegate the request (versus calling the inherited, corresponding Adaptee methods).

As in our initial implementation, our derived classes from Humanoid (Orkan, Romulan, and Earthling) are specified in the same fashion, as is our Client within our main() function.

Choosing the relationship between Adaptee and Adapter

An interesting point to consider when choosing between private inheritance or association as the relationship between Adapter and Adaptee is whether or not the Adaptee contains any protected members. Recall that the original code for `Person` included a protected `ModifyTitle()` method. Should protected members exist in the Adaptee class? The private base class implementation allows those inherited, protected members to continue to be accessed within the scope of the Adapter class (that is, by methods of the Adapter). However, using the association-based implementation, the protected methods in the Adaptee (`Person`) are unusable in the scope of the Adapter. To make this example work, we were required to move `Person::ModifyTitle()` to the public access region. However, modifying the Adaptee class is not always possible, nor is it necessarily recommended. Considering the protected member issue, our initial implementation using a private base class is the stronger implementation, as it does not depend on us modifying the class definition of the Adaptee (`Person`).

Let us now take a brief look at an alternate usage of the Adapter pattern. We will simply be using an Adapter class as a wrapper class. We will add an OO interface to an otherwise loosely arranged set of functions that work well, but lack the desired interface our application (Client) desires.

Using an Adapter as a wrapper

As an alternative usage of the Adapter pattern, we will wrap an OO interface around a grouping of related external functions. That is, we will create a wrapper class to encapsulate these functions.

In our example, the external functions will represent a suite of existing database access functions. We will assume that the core database functionality is well tested for our data type (`Person`) and has been used problem-free. However, these external functions by themselves present an undesirable and unexpected functional interface.

We will instead wrap the external functions by creating an Adapter class to encapsulate their collective functionality. Our Adapter class will be `CitizenDataBase`, representing an encapsulated means for reading and writing `Person` instances from and to a database. Our existing external functions will provide the implementation for our `CitizenDataBase` member functions. Let us assume that the OO interfaces, as defined in our Adapter class, meet the requirements of our OO design.

Let's take a look at the mechanics of our simple wrapper Adapter pattern, beginning by examining external functions providing the database access functionality. This example can be found, as a complete program, in our GitHub repository:

https://github.com/PacktPublishing/Deciphering-Object-Oriented-Programming-with-CPP/blob/main/Chapter18/Chp18-Ex3.cpp

```
// Assume Person class exists with its usual implementation
Person objectRead; // holds the object from current read
```

```
                            // to support a simulation of a DB read
void db_open(const string &dbName)
{    // Assume implementation exists
    cout << "Opening database: " << dbName << endl;
}


void db_close(const string &dbName)
{    // Assume implementation exists
    cout << "Closing database: " << dbName << endl;
}


Person &db_read(const string &dbName, const string &key)
{    // Assume implementation exists
    cout << "Reading from: " << dbName << " using key: ";
    cout << key << endl;
    // In a true implementation, we would read the data
    // using the key and return the object we read in
    return objectRead; // non-stack instance for simulation
}


const string &db_write(const string &dbName, Person &data)
{    // Assume implementation exists
    const string &key = data.GetLastName();
    cout << "Writing: " << key << " to: " <<
            dbName << endl;
    return key;
}
```

In our previously defined external functions, let's assume all functions are well tested and allow Person instances to be read from or written to a database. To support this simulation, we have created an external Person instance with Person objectRead; to provide a brief, non-stack located storage place for a newly read instance (used by db_read()) until the newly read instance is captured as a return value. Keep in mind that the existing external functions do not represent an encapsulated solution.

Now, let's create a simple wrapper class to encapsulate these external functions. The wrapper class, `CitizensDataBase`, will represent our Adapter class:

```
// CitizenDataBase is the Adapter class
class CitizenDataBase   // Adapter wraps undesired interface
{
private:
    string name;
public:
    // No default constructor (unusual)
    CitizenDataBase(const string &);
    CitizenDataBase(const CitizenDataBase &) = delete;
    CitizenDataBase &operator=(const CitizenDataBase &)
                          = delete;   // disallow =
    virtual ~CitizenDataBase();   // virtual destructor
    inline Person &Read(const string &);
    inline const string &Write(Person &);
};

CitizenDataBase::CitizenDataBase(const string &n) : name(n)
{
    db_open(name);   // call existing external function
}

CitizenDataBase::~CitizenDataBase()
{
    db_close(name);   // close database with external
}                     // function

Person &CitizenDataBase::Read(const string &key)
{
    return db_read(name, key);   // call external function
}

const string &CitizenDataBase::Write(Person &data)
{
```

```
    return db_write(name, data);   // call external function
}
```

In our aforementioned class definition for our Adapter class, we simply encapsulate the external database functionality within the CitizenDataBase class. Here, CitizenDataBase is not only our Adapter class but also our Target class, as it contains the interfaces our application at hand (Client) expects. Notice that the CitizenDataBase methods of Read() and Write() have both been inlined in the class definition; their methods merely call the external functions. This is an example of how a wrapper class with inline functions can be a low-cost Adapter class, adding a very minimal amount of overhead (constructors, destructor, and potentially other non-inline methods).

Now, let's take a look at our main() function, which is a streamlined version of a Client:

```
int main()
{
    string key;
    string name("PersonData"); // name of database

    Person p1("Curt", "Jeffreys", 'M', "Mr.");
    Person p2("Frank", "Burns", 'W', "Mr.");
    Person p3;
    CitizenDataBase People(name);    // open Database
    key = People.Write(p1); // write a Person object
    p3 = People.Read(key);   // using a key, retrieve Person
    return 0;
}                              // destruction will close database
```

In the aforementioned main() function, we first instantiate three Person instances. We then instantiate a CitizenDataBase to provide encapsulated access to write or read our Person instances, to or from the database. The methods for our CitizenDataBase constructors call the external function db_open() to open the database. Likewise, the destructor calls db_close(). As expected, our CitizenDataBase methods for Read() and Write() will each, respectively, call the external functions, db_read() or db_write().

Let's take a look at the output for this program:

```
Opening database: PersonData
Writing: Jeffreys to: PersonData
Reading from: PersonData using key: Jeffreys
Closing database: PersonData
```

In the aforementioned output, we can notice the correlation between the various member functions to the wrapped, external functions via construction, a call to write and read, and then the destruction of the database.

Our simple `CitizenDataBase` wrapper is a very straightforward, but reasonable, use of the Adapter pattern. Interestingly, our `CitizenDataBase` also has commonalities with the **Data Access Object pattern**, as this wrapper provides a clean interface to a data storage mechanism, concealing the implementation (access) to the underlying database.

We have now seen three implementations of the Adapter pattern. We have folded the concepts of Adapter, Adaptee, Target, and Client into the framework of classes we are accustomed to seeing, namely `Person`, as well as into descendants of our Adapter (`Orkan`, `Romulan`, and `Earthling`, as in our first two examples). Let us now briefly recap what we have learned relating to patterns before moving forward to our next chapter.

Summary

In this chapter, we have advanced our pursuit to become better C++ programmers through widening our knowledge of design patterns. We have explored the Adapter pattern in both the concept and through multiple implementations. Our first implementation used private inheritance two derive the Adapter from the Adaptee class. We specified our Adapter as an abstract class and then used public inheritance to introduce several Target classes based on the interface provided by our Adapter class. Our second implementation instead modeled the relationship between the Adapter and Adaptee using association. We then looked at an example usage of an Adapter as a wrapper to simply add an OO interface to existing function-based application components.

Utilizing common design patterns, such as the Adapter pattern, will help you more easily reuse existing, well-tested portions of code in a manner understood by other programmers. By utilizing core design patterns, you will be contributing to well-understood and reusable solutions with more sophisticated programming techniques.

We are now ready to continue forward with our next design pattern, in *Chapter 19, Using the Singleton Pattern*. Adding more patterns to our arsenal of programming skills makes us more versatile and valued programmers. Let's continue forward!

Questions

1.   Using the Adapter examples found in this chapter, create a program as follows:

   a.   Implement a `CitizenDataBase` that stores various types of `Humanoid` instances (`Orkan`, `Romulan`, `Earthling`, and perhaps `Martian`). Decide whether you will use the private base class Adapter-Adaptee relationship or the association relationship between the Adapter and Adaptee (hint: the private inheritance version will be easier).

   b.   Noting that the `CitizenDataBase` handles `Person` instances, can this class be used *as-is* to store various types of `Humanoid` instances, or must it be adapted in some way? Recall that `Person` is a base class of `Humanoid` (if you chose this implementation), but also remember that we can never upcast past a non-public inheritance boundary.

2.   What other examples can you imagine that might easily incorporate the Adapter pattern?

Using the Singleton Pattern

This chapter will continue our goal to expand your C++ programming skills beyond core OOP concepts, with the objective of empowering you to solve recurring types of coding conundrums utilizing core design patterns. Utilizing design patterns in coding solutions can not only provide refined solutions but also contribute to easier code maintenance and provide potential opportunities for code reuse.

The next core design pattern that we will learn how to implement effectively in C++ is the **Singleton pattern**.

In this chapter, we will cover the following main topics:

- Understanding the Singleton pattern and how it contributes to OOP

- Implementing the Singleton pattern in C++ (with simple techniques versus a paired-class approach), and using a registry to allow many classes to utilize the Singleton pattern

By the end of this chapter, you will understand the Singleton pattern and how it can be used to ensure that only a single instance of a given type can exist. Adding an additional core design pattern to your knowledge set will further augment your programming skills to help you become a more valuable programmer.

Let's increase our programming skillset by examining another common design pattern, the Singleton pattern.

Technical requirements

Online code for full program examples can be found in the following GitHub URL: `https://github.com/PacktPublishing/Deciphering-Object-Oriented-Programming-with-CPP/tree/main/Chapter19`. Each full program example can be found in the GitHub repository under the appropriate chapter heading (subdirectory) in a file that corresponds to the chapter number, followed by a dash, followed by the example number in the chapter at hand. For example, the first full program in this chapter can be found in the subdirectory `Chapter19` in a file named `Chp19-Ex1.cpp` under the aforementioned GitHub directory.

The CiA video for this chapter can be viewed at: `https://bit.ly/3ThNKe0`.

Understanding the Singleton pattern

The Singleton pattern is a creational design pattern that guarantees only one instance will exist for a class embracing this idiom; two or more instances of the type may simply not exist simultaneously. A class embracing this pattern will be known as a **Singleton**.

A Singleton can be implemented using static data members and static methods. This means that a Singleton will have a global point of access to the instance at hand. This ramification initially seems dangerous; introducing global state information into the code is one criticism that has led the Singleton to sometimes be considered an anti-pattern. However, with the appropriate use of access regions for the static data members defining the Singleton, we can insist that access to the Singleton (other than initialization) only uses the appropriate static methods of the class at hand (and alleviate this potential pattern concern).

Another criticism of the pattern is that it is not thread-safe. There may be race conditions to enter the segment of code where the Singleton instance is created. Without guaranteeing mutual exclusivity to that critical region of code, the Singleton pattern will break, allowing multiple such instances. As such, if multithreaded programming is employed, so must be proper locking mechanisms to protect the critical region of code where the Singleton is instantiated. A Singleton (implemented using static memory) is shared memory between threads in the same process; at times, Singleton can be criticized for monopolizing resources.

The Singleton pattern can utilize several techniques for implementation. Each manner of implementation inevitably will have benefits and drawbacks. We will use a pair of related classes, `Singleton` and `SingletonDestroyer`, to robustly fulfill the pattern. Whereas there are more simple, straightforward implementations (two of which we will briefly review), the simplest techniques leave the possibility that the Singleton will not be adequately destructed. Recall that a destructor may include important and necessary activities.

Singletons tend to be long-lived; it is, therefore, appropriate for a Singleton to be destructed just before the application terminates. Many Clients may have pointers to a Singleton, so no single Client should delete the Singleton. We will see that a `Singleton` will be *self-created*, so it should ideally be *self-destructed* (that is, with the help of its `SingletonDestroyer`). As such, the paired-class approach, though not as simple, will ensure proper `Singleton` destruction. Note that our implementation will also allow the Singleton to be directly deleted; this is rare, but our code will also handle this situation.

The Singleton pattern with the paired-class implementation will include the following:

- A **Singleton** class, which represents the core mechanics needed to implement the concept of a Singleton.

- A **SingletonDestroyer** class, which will serve as a helper class to `Singleton`, ensuring that a given Singleton is properly destructed.

- A class derived from `Singleton` represents a class that we want to ensure can only create a single instance of its type at a given time. This will be our **Target** class.

- Optionally, the Target class may be both derived from `Singleton` and another class, which may represent existing functionality that we would like to specialize in or simply encompass (that is, *mix-in*). In this case, we will multiply inherit from an application-specific class and the Singleton class.

- Optional **Client** classes, which will interact with the Target class(es) to fully define the application at hand.

- Alternatively, the Singleton may also be implemented within a Target class, bundling the class functionalities together in a single class.

- A true Singleton pattern can be expanded to allow for multiple (discrete), but not an undetermined number of instances to be made. This is rare.

We will focus on a traditional Singleton pattern that ensures only a single instance of a class embracing this pattern will exist at a given time.

Let's move forward to first examine two simple implementations, then our preferred paired-class implementation of the Singleton pattern.

Implementing the Singleton pattern

The Singleton pattern will be used to ensure that a given class may only instantiate a single instance of that class. However, a true Singleton pattern will also have expansion capabilities to allow for multiple (but a well-defined number of) instances to be made. This unusual and not well-known caveat of the Singleton pattern is rare.

We will start with two simple Singleton implementations to understand their limitations. We will then progress to the more robust paired-class implementation of the Singleton, with the most common pattern goal of only allowing one Target class instantiation at any given time.

Using a simple implementation

To implement a very simple Singleton, we will use a straightforward single class specification for the Singleton itself. We will define a class, known as `Singleton`, to encapsulate the pattern. We will ensure that our constructor(s) are private so that they cannot be applied more than once. We will also add a static `instance()` method to provide the interface for instantiation of the `Singleton` object. This method will ensure that the private construction occurs exactly once.

Let's take a look at this straightforward implementation, which can be found in our GitHub repository:

https://github.com/PacktPublishing/Deciphering-Object-Orient-
ed-Programming-with-CPP/blob/main/Chapter19/Chp19-Ex1.cpp

```cpp
class Singleton
{
private:
    static Singleton *theInstance;   // initialized below
    Singleton();   // private to prevent multiple
                   // instantiation
public:
    static Singleton *instance(); // interface for creation
    virtual ~Singleton(); // never called, unless you
};                        // delete Singleton explicitly,
                          // which is unlikely and atypical
Singleton *Singleton::theInstance = nullptr; // extern var
                                             // to hold static member
Singleton::Singleton()
{
    cout << "Constructor" << endl;
    // Below line of code is not necessary and therefore
    // commented out - see static member init. above
    // theInstance = nullptr;
}

Singleton::~Singleton()   // the destructor is not called in
{                         // the typical pattern usage
    cout << "Destructor" << endl;
    if (theInstance != nullptr)
    {
        Singleton *temp = theInstance;
        // Remove pointer to Singleton and prevent recursion
        // Remember, theInstance is static, so
        // temp->theInstance = nullptr; would be duplicative
        theInstance = nullptr;
        delete temp;                     // delete the Singleton
```

```
            // Note, delete theInstance; without temp usage
            // above would be recursive
    }
}


Singleton *Singleton::instance()
{
    if (theInstance == nullptr)
        theInstance = new Singleton();// allocate Singleton
    return theInstance;
}


int main()
{
    // create Singleton
    Singleton *s1 = Singleton::instance();
    // returns existing Singleton (not a new one)
    Singleton *s2 = Singleton::instance();
    // note: addresses are the same (same Singleton!)
    cout << s1 << " " << s2 << endl;
    return 0;
}
```

Notice, in the aforementioned class definition, we include data member static Singleton *theInstance; to represent the Singleton instance itself. Our constructor is private so it cannot be used multiple times to create multiple Singleton instances. Instead, we add a static Singleton *instance() method to create the Singleton. Within this method, we check whether data member theInstance is equal to the nullptr and if so, we instantiate the one and only Singleton instance.

Outside of the class definition, we see the external variable (and its initialization) to support the memory requirements of the static data member with the definition of Singleton *Singleton::theInstance = nullptr;. We also see how, in main(), we call the static instance() method to create a Singleton instance using Singleton::instance(). The first call to this method will instantiate a Singleton, whereas subsequent calls to this method will merely return a pointer to the existing Singleton object. We can verify that the instances are the same by printing the address of these objects.

Let's take a look at the output for this simple program:

```
Constructor
0xee1938 0xee1938
```

In the previously mentioned output, we notice something perhaps unexpected – the destructor is not called! What if the destructor had crucial tasks to perform?

Understanding a key deficiency with the simple Singleton implementation

The destructor is not called for our `Singleton` in the simple implementation merely because we have not deleted the dynamically allocated `Singleton` instance through either the `s1` or `s2` identifiers. Why not? There clearly may be multiple pointers (handles) to a `Singleton` object. Deciding which handle should be responsible for removing the `Singleton` is difficult to determine – the handles would minimally need to collaborate or employ reference counting.

Additionally, a `Singleton` tends to exist for the duration of the application. This longevity further suggests that a `Singleton` should be in charge of its own destruction. But how? We soon see an implementation that will allow a `Singleton` to control its own destruction with a helper class. With the simple implementation, however, we might simply throw our hands in the air and suggest that the operating system will reclaim the memory resources when the application terminates – including the heap memory for this small `Singleton`. This is true; however, what if an important task needs to be completed in the destructor? We are running into a limitation within the simple pattern implementation.

If we need the destructor to be called, shall we resort to allowing one of the handles to delete the instance using, for example, `delete s1;`? We have previously reviewed issues regarding whether to allow any one handle to perform the deletion, but let's now additionally examine potential issues within the destructor itself. For example, if our destructor hypothetically only includes `delete theInstance;`, we will have a recursive function call. That is, calling `delete s1;` will invoke the `Singleton` destructor, yet `delete theInstance;` within the destructor body will recognize `theInstance` as a `Singleton` type and again call the `Singleton` destructor – *recursively*.

Not to worry! Our destructor, as shown, instead manages recursion by first checking whether `theInstance` data member is not equal to the `nullptr` and then arranges for `temp` to point to `theInstance` to save a handle to the instance we need to delete. We then make the assignment `temp->theInstance = nullptr;` to prevent recursion when we `delete temp;`. Why? Because `delete temp;` will also call the `Singleton` destructor. Upon this destructor call, `temp` will bind to `this` and will fail the conditional test `if (theInstance != nullptr)` on this first recursive function call, backing us out of continued recursion. Note that our upcoming implementation with a paired-class approach will not have this potential issue.

It is important to note that in an actual application, we would not create a domain-unspecific `Singleton` instance. Rather, we would factor our application into the design to employ the pattern. After all, we want to have a `Singleton` instance of a meaningful class type. To do so using our simple `Singleton` class as a basis, we simply inherit our Target (application-specific) class from `Singleton`. The Target class will also have private constructors – ones that accept the arguments necessary to adequately instantiate the Target class. We will then move the static `instance()` method from `Singleton` to the Target class and ensure that the argument list for `instance()` accepts the necessary arguments to pass to a private Target constructor.

To sum up, our simple implementation has the inherent design flaw that there is no guaranteed proper destruction for the `Singleton` itself. Letting the operating system collect the memory when the application terminates does not call the destructor. Choosing one of many handles for the `Singleton` to delete the memory, though possible, requires coordination and also defeats the usual application of the pattern to allow the `Singleton` to live for the duration of the application.

Let us next consider an alternate simple implementation of using a reference to static local memory, rather than a pointer to heap memory, for our Singleton.

An alternate simple implementation

As an alternative approach for implementing a very straightforward Singleton, we will modify the previous simple class definition. First, we will remove the static pointer data member (which was dynamically allocated within `Singleton::instance()`). Instead of using a static data member within the class, we will use a (non-pointer) static local variable within the `instance()` method to represent the Singleton.

Let's take a look at this alternative implementation, which can be found in our GitHub repository:

https://github.com/PacktPublishing/Deciphering-Object-Orient-ed-Programming-with-CPP/blob/main/Chapter19/Chp19-Ex1b.cpp

```
class Singleton
{
private:
    string data;
    Singleton(string d); // private to prevent multiple
public:                  // instantiation
    static Singleton &instance(string); // return reference
    // destructor is called for the static local variable
    // declared in instance() before the application ends
    virtual ~Singleton();   // destructor is now called
    const string &getData() const { return data; }
};
```

```cpp
Singleton::Singleton(string d): data(d)   // initialize data
{
    cout << "Constructor" << endl;
}

Singleton::~Singleton()
{
    cout << "Destructor" << endl;
}

// Note that instance() takes a parameter to reflect how we
// can provide meaningful data to the Singleton constructor
Singleton &Singleton::instance(string d)
{   // create the Singleton with desired constructor; But,
    // we can never replace the Singleton in this approach!
    // Remember, static local vars are ONLY created and
    // initialized once - guaranteeing one Singleton
    static Singleton theInstance(d);
    return theInstance;
}

int main()
{
    // First call, creates/initializes Singleton
    Singleton &s1 = Singleton::instance("Unique data");
    // Second call returns existing Singleton
    // (the static local declaration is ignored)
    Singleton &s2 = Singleton::instance("More data");
    cout << s1.getData() << " " << s2.getData() << endl;
    return 0;
}
```

Notice, in the aforementioned Singleton class definition, we no longer include a static data member (nor the external static variable declaration to support this data member) to represent the Singleton instance itself. Instead, we have specified the Singleton's implementation using a static local (non-pointer) variable in the static instance() method. Our constructor is private; it can be invoked to initialize this static local variable within a static member function of the class. The local variable, as static (and not a pointer with an allocation), will only be created and initialized once. Its space will be set aside when the application starts and the static variable will be initialized upon the first call to `instance()`. Subsequent calls to `instance()` will not yield a replacement of this `Singleton`; the static local variable declaration will be ignored for anything other than the first call to `instance()`. Notice that the return value of `instance()` is now a reference (&) to this static local `Singleton` instance. Remember, a static local variable will exist for the entire application (it is not stored on the stack with other local variables).

Also, quite importantly, notice that we have passed data to initialize the Singleton to the `instance()` method via the parameter list; this data is passed along to the `Singleton` constructor. The ability to construct the Singleton with appropriate data is quite important. By implementing the Singleton as a static local (non-pointer) variable in the static `instance()` method, we have the opportunity to construct the Singleton within this method. Note that a static pointer data member defined in the class also has this ability, as the allocation (and hence construction, such as in the previous example) is also made within the `instance()` method. However, a non-pointer static data member of the class would not allow the ability to provide meaningful constructor arguments because the instance would be created and initialized at the start of the program before such meaningful initializers would be available (not actually within the `instance()` method). In the latter case, the Singleton would only be returned from `instance()`, not initialized within it.

Now notice, in `main()`, we call the static `instance()` method to create a `Singleton` instance using `Singleton::instance()`. We create an alias, `s1`, using a reference to the Singleton returned from `Singleton::instance()`. The first call to this method will instantiate the `Singleton`, whereas subsequent calls to this method will merely return a reference to the existing `Singleton` object. We can verify that the instance referenced by both aliases (`s1` and `s2`) is the same object by printing the data contained within the Singleton.

Let's take a look at the output for this simple program:

```
Constructor
Unique data
Unique data
Destructor
```

In the previously mentioned output, we notice that the destructor is automatically called to clean up the Singleton before the application ends. We also notice that the attempted creation of the second `Singleton` instance only returns the existing `Singleton`. This is because the static local variable, `theInstance`, is only created and initialized only once per application, no matter how many times `instance()` is invoked (a simple property of static local variables). However, this implementation also has a potential drawback; let's take a look.

Understanding a limitation with the alternate simple Singleton implementation

The implementation using a non-pointer static local variable in `instance()` for the Singleton does not give us the flexibility to change the Singleton. In a function, any static local variable has its memory set aside when the application begins; this memory is only initialized once (on the initial call to `instance()`). The implication is that we always have exactly one `Singleton` in the application. The space for this `Singleton` exists even if we never call `instance()` to initialize it.

Additionally, the `Singleton` in this implementation cannot be exchanged for another `Singleton` object due to the nature of how static local variables are implemented. In some applications, we may want one `Singleton` object at a time, yet also desire the ability to change out one instance of a `Singleton` for another. Imagine, for example, that an organization can have exactly one president; however, it is desirable that the (Singleton) president can be replaced every few years with a different (Singleton) president. The initial simple implementation using a pointer allows for this possibility, yet has the potential deficiency that its destructor is never called. Each of the simple implementations has a potential drawback.

Now, because we understand the limitations of the simple Singleton implementations, we will instead move onward to a preferred paired-class implementation of the Singleton pattern. The paired-class approach will guarantee proper destruction of our `Singleton`, whether the application allows the `Singleton` to be destructed just prior to the application's termination through the deliberate class pairing (the most frequently encountered situation), or in the rare case that a `Singleton` is destroyed prematurely in the application. This approach will also allow us to replace a Singleton with another instance of a Singleton.

Using a more robust paired-class implementation

To implement the Singleton pattern with a paired-class approach in a nicely encapsulated fashion, we will define a Singleton class to purely add the core mechanics of creating a single instance. We will name this class `Singleton`. We will then add a helper class to `Singleton`, known as `SingletonDestroyer`, to ensure that our `Singleton` instance always goes through proper destruction before our application terminates. This pair of classes will be related through aggregation and association. More specifically, the `Singleton` will conceptually contain a `SingletonDestroyer` (aggregate), and the `SingletonDestroyer` will hold an association to the (outer) `Singleton` in which it is conceptually embedded. Because the implementation of the `Singleton` and `SingletonDestroyer` is through static data members, the aggregation is conceptual – static members are stored as external variables.

Once these core classes have been defined, we will consider how we may incorporate the Singleton pattern into a class hierarchy with which we have familiarity. Let's imagine that we would like to implement a class to encapsulate the concept of a *president*. Whether it be a president of a nation or the president of a university, it is important that there be only one president at a given point in time. President will be our Target class; President is a good candidate to utilize our Singleton pattern.

It is interesting to note that, whereas there will only be one president at a given point in time, it is possible to replace a president. For example, the term of a U.S. president is only four years at a time, with possible re-election for one more term. There may be similar conditions for a university president. A president may leave prematurely through resignation, impeachment, or death, or may simply leave upon term expiration. Once a sitting president's existence is removed, it is then acceptable to instantiate a new, Singleton President. Hence, our Singleton pattern allows only one Singleton of the Target class at a given point in time.

Reflecting on how we may best implement a President class, we realize that a President *Is-A* Person and also needs to *mix-in* Singleton capabilities. With this in mind, we now have our design. President will use multiple inheritances to extend the concept of Person and to mix-in the functionality of a Singleton.

Certainly, we could have built a President class from scratch, but why do so when the Person components of the President class are represented in a well-tested and available class? Also, certainly, we could embed the Singleton class information into our President class, rather than inheriting it from a separate Singleton class. Absolutely, this is also an option. However, our application will instead encapsulate each piece of the solution. This will enable easier future reuse. Nonetheless, the design choices are many.

Specifying the Singleton and the SingletonDestroyer

Let's take a look at the mechanics of our Singleton pattern, starting by examining the Singleton and SingletonDestroyer class definitions. These classes work cooperatively to implement the Singleton pattern. This example can be found, as a complete program, on our GitHub:

https://github.com/PacktPublishing/Deciphering-Object-Oriented-Programming-with-CPP/blob/main/Chapter19/Chp19-Ex2.cpp

```
class Singleton;     // Necessary forward class declarations
class SingletonDestroyer;
class Person;
class President;

class SingletonDestroyer
{
private:
```

```
        Singleton *theSingleton = nullptr;
public:
    SingletonDestroyer(Singleton *s = nullptr)
        { theSingleton = s; }
    // disallow copies and assignment
    SingletonDestroyer(const SingletonDestroyer &)
                                        = delete;
    SingletonDestroyer &operator=
        (const SingletonDestroyer &) = delete;
    ~SingletonDestroyer(); // dtor shown further below
    void setSingleton(Singleton *s) { theSingleton = s; }
    Singleton *getSingleton() { return theSingleton; }
};
```

In the aforementioned code segment, we begin with several forward class declarations, such as class Singleton;. These declarations allow references to be made to these data types before their complete class definitions have been seen by the compiler.

Next, let's take a look at our SingletonDestroyer class definition. This simple class contains a private data member, Singleton *theSingleton;, which will be the association to the Singleton that the SingletonDestroyer will one day be responsible for deallocating (we will examine the destructor definition for SingletonDestroyer shortly). Notice that our destructor is not virtual as this class is not meant to be specialized.

Notice that our constructor has a default value of the nullptr specified for the Singleton *, which is an input parameter. SingletonDestroyer also contains two member functions, setSingleton() and getSingleton(), which merely provide the means to *set* and *get* the associated Singleton member.

Also notice that both the use of the copy constructor and the overloaded assignment operator in SingletonDestroyer have been disallowed using =delete in their prototypes.

Before we examine the destructor for this class, let us examine the class definition for Singleton:

```
// Singleton will be mixed-in using inheritance with a
// Target class. If Singleton is used stand-alone, the data
// members would be private. Also be sure to add a
// Static *Singleton instance();
// method to the public access region.
class Singleton
{
```

```
protected:    // protected data members
    static Singleton *theInstance;
    static SingletonDestroyer destroyer;
protected:    // and protected member functions
    Singleton() = default;
    // disallow copies and assignment
    Singleton(const Singleton &) = delete;
    Singleton &operator=(const Singleton &) = delete;
    friend class SingletonDestroyer;
    virtual ~Singleton()
        { cout << "Singleton destructor" << endl; }
};
```

The aforementioned `Singleton` class contains protected data member `static Singleton *theInstance;`, which will represent (when allocated) a pointer to the one and only instance allocated for a class employing the Singleton idiom.

The protected data member `static SingletonDestroyer destroyer;` represents a conceptual aggregate or contained member. The containment is truly only conceptual, as static data members are not stored within the memory layout for any instance; they are instead stored in external memory and *name-mangled* to appear as part of the class. This (conceptual) aggregate subobject, `destroyer`, will be responsible for the proper destruction of the `Singleton`. Recall that the `SingletonDestroyer` has an association with the one and only `Singleton`, representing the outer object in which the `SingletonDestroyer` is conceptually contained. This association is how the `SingletonDestroyer` will access the Singleton.

When the memory for the external variable that implements the static data member `static SingletonDestroyer destroyer;` goes away at the end of the application, the destructor for `SingletonDestroyer` (the static, conceptual, subobject) will be called. This destructor will `delete theSingleton;`, ensuring that the outer `Singleton` object (which was dynamically allocated), will have the appropriate destructor sequence run on it. Because the destructor in `Singleton` is protected, it is necessary that `SingletonDestroyer` is specified as a friend class of `Singleton`.

Notice that both uses of the copy constructor and the overloaded assignment operator in `Singleton` have been disallowed using `=delete` in their prototypes.

In our implementation, we have assumed that `Singleton` will be mixed-in via inheritance to a derived Target class. It will be in the derived class (the one that intends to use the Singleton idiom), that we provide the required static `instance()` method to create the `Singleton` instance. Note that had `Singleton` been used as a standalone class to create Singletons, we would instead add `static Singleton* instance()` to the public access region of `Singleton`.

We would also then move the data members from the protected to the private access region. However, having an application-unspecific Singleton is only of use to demonstrate the concept. Instead, we will apply the Singleton idiom to an actual type requiring the use of this idiom.

With our `Singleton` and `SingletonDestroyer` class definitions in place, let's next examine the remaining implementation necessities for these classes:

```cpp
// External (name mangled) vars to hold static data mbrs
Singleton *Singleton::theInstance = nullptr;
SingletonDestroyer Singleton::destroyer;


// SingletonDestroyer destructor definition must appear
// after class definition for Singleton because it is
// deleting a Singleton (so its destructor can be seen)
// This is not an issue when using header and source files.
SingletonDestroyer::~SingletonDestroyer()
{
    if (theSingleton == nullptr)
        cout << "SingletonDestroyer destructor: Singleton
                has already been destructed" << endl;
    else
    {
        cout << "SingletonDestroyer destructor" << endl;
        delete theSingleton;
    }
}
```

In the aforementioned code fragment, let's first notice the two external variable definitions that provide the memory to support the two static data members within the `Singleton` class – that is, `Singleton *Singleton::theInstance = nullptr;` and `SingletonDestroyer Singleton::destroyer;`. Recall that static data members are not stored within any instance of their designated class. Rather, they are stored in external variables; these two definitions designate the memory. Notice that the data members are both labeled as `protected`. This means that though we may define their outer storage directly in this manner, we may not access these data members other than through static member functions of `Singleton`. This will give us some peace of mind. Though there is a potential global access point to the static data members, their levied protected access region requires appropriate static methods of the `Singleton` class to be used to properly manipulate these important members.

Next, draw your attention to the destructor for `SingletonDestroyer`. This clever destructor first checks whether its association to the `Singleton` for which it is responsible is equal to the `nullptr`. This will be rare and will happen in the very unusual situation when a Client releases the Singleton object directly with an explicit `delete`.

The usual destruction scenario in the `SingletonDestroyer` destructor will be the execution of the `else` clause in which the `SingletonDestructor`, as a static object, will be responsible for the deletion, and hence destruction, of its paired `Singleton`. Remember, there will be a contained `SingletonDestroyer` object within the `Singleton`. The memory for this static (conceptual) subobject will not go away until the application has finished. Recall that static memory is not actually part of any instance. However, the static subobject will be destructed just prior to `main()`'s completion. So, when the `SingletonDestroyer` is destructed, its usual case will be to `delete theSingleton;`, which will release its paired Singleton's memory, allowing the `Singleton` to be properly destructed.

The driving design decision behind the Singleton pattern is that a Singleton is a long-lived object, and its destruction may most often correctly occur near the end of the application. The Singleton is responsible for its own inner Target object creation so that the Singleton should not be deleted (and hence destructed) by a Client. Rather, the preferred mechanism is that the `SingletonDestroyer`, when removed as a static object, deletes its paired `Singleton`.

Nonetheless, occasionally, there are reasonable scenarios for deleting a `Singleton` mid-application. Should a replacement `Singleton` never be created, our `SingletonDestroyer` destructor will still work correctly, identifying that its paired `Singleton` has already been released. However, it is more likely that our `Singleton` will be replaced with another `Singleton` instance somewhere in the application. Recall our application example where a president may be impeached, resign, or die, but will be replaced by another president. In these cases, it is acceptable for a `Singleton` to be deleted directly and a new `Singleton` is then created. In this case, the `SingletonDestroyer` will now reference the replacement `Singleton`.

Deriving a Target class from Singleton

Next, let's take a look at how we can create our Target class, `President`, from `Singleton`:

```
// Assume our Person class is as we are accustomed

// A President Is-A Person and also mixes-in Singleton
class President: public Person, public Singleton
{
private:
    President(const string &, const string &, char,
            const string &);
```

```cpp
public:
    ~President() override;    // virtual destructor
    // disallow copies and assignment
    President(const President &) = delete;
    President &operator=(const President &) = delete;
    static President *instance(const string &,
                      const string &, char, const string &);
};

President::President(const string &fn, const string &ln,
    char mi, const string &t): Person(fn, ln, mi, t),
                                Singleton()
{
}

President::~President()
{
    destroyer.setSingleton(nullptr);
    cout << "President destructor" << endl;
}

President *President::instance(const string &fn,
          const string &ln, char mi, const string &t)
{
    if (theInstance == nullptr)
    {
        theInstance = new President(fn, ln, mi, t);
        destroyer.setSingleton(theInstance);
        cout << "Creating the Singleton" << endl;
    }
    else
        cout << "Singleton previously created.
                Returning existing singleton" << endl;
    // below cast is necessary since theInstance is
    // a Singleton *
    return dynamic_cast<President *>(theInstance);
}
```

In our aforementioned Target class, `President`, we merely inherit `President` from `Person` using public inheritance and then multiply inherit `President` from `Singleton` to *mix-in* the `Singleton` mechanics.

We place our constructor in the private access region. Static method `instance()` will utilize this constructor internally to create the one and only `Singleton` instance permitted, to adhere to the pattern. There is no default constructor (unusual) because we do not wish to allow `President` instances to be created without their relevant details. Recall that C++ will not link in a default constructor if we have provided an alternate constructor interface. As we do not desire copies of a `President` or the assignment of a `President` to another potential `President`, we have disallowed copies and assignments using the `=delete` specification in the prototypes for these methods.

Our destructor for `President` is simple, yet crucial. In the case that our `Singleton` object will be deleted explicitly, we prepare by setting `destroyer.setSingleton(nullptr);`. Recall, `President` inherits the protected `static SingletonDestroyer destroyer;` data member. Here, we are setting the destroyer's associated `Singleton` to the `nullptr`. This line of code in our `President` destructor then enables the destructor in `SingletonDestroyer` to accurately depend on checking for the unusual case of whether its associated `Singleton` has already been deleted before commencing the usual deletion of its `Singleton` counterpart.

Finally, we have defined a static method to provide the creation interface for our `President` as a `Singleton` with `static President *instance(const string &, const string &, char, const string &);`. In the definition of `instance()`, we first check whether the inherited, protected data member `Singleton *theInstance` is equal to the `nullptr`. If we have not yet allocated the `Singleton`, we allocate `President` using the aforementioned private constructor and assign this newly allocated `President` instance to `theInstance`. This is an upcast from a `President *` to a `Singleton *`, which is no problem across a public inheritance boundary. If, however, in the `instance()` method, we find that `theInstance` is not equal to a `nullptr`, we simply return a pointer to the previously allocated `Singleton` object. As users will undoubtedly want to use this object as a `President` to enjoy the inherited `Person` features, we downcast `theInstance` to `President *` for its return value from this method.

Finally, let us consider the logistics of a sample Client in our overall application. In its simplest form, our Client will contain a `main()` function to drive the application and showcase our Singleton pattern.

Bringing the pattern components together within the Client

Let's now take a look at our `main()` function to see how our pattern is orchestrated:

```
int main()
{
    // Create a Singleton President
    President *p1 = President::instance("John", "Adams",
                                    'Q', "President");
```

```
        // This second request will fail, returning
        // the original instance
        President *p2 = President::instance("William",
                          "Harrison", 'H', "President");
        if (p1 == p2)   // Verification there's only one object
            cout << "Same instance (only 1 Singleton)" << endl;
        p1->Print();
        // SingletonDestroyer will release Singleton at end
        return 0;
}
```

Reviewing our main() function in the preceding code, we first allocate a Singleton President using President *p1 = President::instance("John", "Adams", 'Q', "President");. We then try to allocate an additional President on the next line of code using *p2. Because we can only have one Singleton (a President *mixes-in* a Singleton), a pointer is returned to our existing President and stored in p2. We verify that there is only one Singleton by comparing p1 == p2; the pointers indeed point to the same instance.

Next, we take advantage of using our President instance in its intended manner, such as by using some of the inherited member functions from Person. As an example, we invoke p1->Print();. Certainly, our President class could have added specialized functionality that would be appropriate to utilize in our Client as well.

Now, at the end of main(), our static object SingletonDestroyer Singleton::destroyer; will be appropriately destructed before its memory is reclaimed. As we have seen, the destructor for SingletonDestroyer will (most often) issue a delete to its associated Singleton (which is actually a President) using delete theSingleton;. This will trigger our President destructor, Singleton destructor, and Person destructor to each be called and executed (going from most specialized to most general subobjects). As our destructor in Singleton is virtual, we are guaranteed to start at the proper level for destruction and to include all destructors.

Let's take a look at the output for this program:

```
Creating the Singleton
Singleton previously created. Returning existing singleton
Same instance (only 1 Singleton)
President John Q Adams
SingletonDestroyer destructor
President destructor
Singleton destructor
Person destructor
```

In the preceding output, we can visualize the creation of the Singleton `President`, as well as see that the second `instance()` request for a `President` merely returns the existing `President`. We then see the details of the `President` that were printed.

Most interestingly, we can see the destruction sequence for the `Singleton`, which is driven by the static object reclamation of the `SingletonDestroyer`. Through proper deletion of the `Singleton` in the `SingletonDestroyer` destructor, we see that `President`, `Singleton`, and `Person` destructors are each invoked as they contribute to the complete `President` object.

Examining explicit Singleton deletion and its impact on SingletonDestroyer destructor

Let's take a look at an alternate version of the Client with an alternate `main()` function. Here, we force deletion of our `Singleton`; this is rare. In this scenario, our `SingletonDestroyer` will not delete its paired `Singleton`. This example can be found, as a complete program, in our GitHub repository:

https://github.com/PacktPublishing/Deciphering-Object-Oriented-Programming-with-CPP/blob/main/Chapter19/Chp19-Ex3.cpp

```cpp
int main()
{
    President *p1 = President::instance("John", "Adams",
                                        'Q', "President");
    President *p2 = President::instance("William",
                            "Harrison", 'H', "President");
    if (p1 == p2)   // Verification there's only one object
        cout << "Same instance (only 1 Singleton)" << endl;
    p1->Print();
    delete p1;   // Delete the Singleton - unusual.
    return 0;    // Upon checking, the SingletonDestroyer
}   // will no longer need to destroy its paired Singleton
```

In the aforementioned `main()` function, notice that we explicitly deallocate our Singleton `President` using `delete p1;`, versus allowing the instance to be reclaimed via static objection deletion as the program ends. Fortunately, we have included a test in our `SingletonDestroyer` destructor to let us know whether the `SingletonDestroyer` must delete its associated `Singleton` or whether this deletion has already occurred.

Let's take a look at the revised output to notice the differences from our original main():

```
Creating the Singleton
Singleton previously created. Returning existing singleton
Same instance (only 1 Singleton)
President John Q Adams
President destructor
Singleton destructor
Person destructor
SingletonDestroyer destructor: Singleton has already been
destructed
```

In the aforementioned output for our revised Client, we can again visualize the creation of the Singleton President, the *unsuccessful* creation request of a second President, and so on.

Let's notice the destruction sequence and how it differs from our first Client. Here, the Singleton President is explicitly deallocated. We can see the proper deletion of the President through the call and execution of the destructors in President, Singleton, and Person as each is executed. Now, when the application is about to end and the static SingletonDestroyer is about to have its memory reclaimed, we can visualize the destructor called on the SingletonDestroyer. However, this destructor no longer will delete its associated Singleton.

Understanding design advantages and disadvantages

An advantage of the preceding (paired-class) implementation of the Singleton pattern (irrespective of which main() is employed) is that we have guaranteed proper destruction of the Singleton. This happens regardless of whether the Singleton is long-lived and is deleted in its usual fashion by its associated SingletonDestroyer, or whether it is deleted earlier on in the application directly (a rare scenario).

A disadvantage of this implementation is inherited from the concept of the Singleton. That is, there can only be one derived class of Singleton that incorporates the specific mechanics of the Singleton class. Because we have inherited President from Singleton, we are using the Singleton logistics (namely static data members, stored in external variables) for President and President alone. Should another class wish to be derived from Singleton to embrace this idiom, the internal implementation for the Singleton has already been utilized for President. Ouch! That does not seem fair.

Not to worry! Our design can be easily expanded to accommodate multiple classes that wish to use our Singleton base class. We will augment our design to accommodate multiple Singleton objects. We will assume, however, that we still intend to have only one Singleton instance per class type.

Another potential concern is thread safety. For example, if multithreaded programming will be utilized, we need to ensure that our `static President::instance()` method acts as though it is atomic, that is, uninterruptible. We can do this through carefully synchronized access to the static method itself.

Let us now take a brief look at how we may expand the Singleton pattern to solve this issue.

Using a registry to allow many classes to utilize Singleton

Let us more closely examine a shortcoming with our current Singleton pattern implementation. Currently, there can only be one derived class of `Singleton` that can effectively utilize the `Singleton` class. Why is this? `Singleton` is a class that comes with external variable definitions to support the static data members within the class. The static data member representing `theInstance` (implemented using the external variable `Singleton *Singleton::theInstance`) may only be set to one `Singleton` instance. *Not one per class* – there is only one set of external variables creating the memory for the crucial `Singleton` data members of `theInstance` and `destroyer`. Herein lies the problem.

We can, instead, specify a `Registry` class to keep track of the classes applying the Singleton pattern. There are many implementations for a **Registry**, and we will review one such implementation.

In our implementation, the `Registry` will be a class that pairs class names (for classes employing the Singleton pattern) with `Singleton` pointers to the single allowed instance of each registered class. We will still derive each Target class from `Singleton` (and from any other class as deemed appropriate by our design).

Our `instance()` method in each class *derived* from `Singleton` will be revised, as follows:

- Our first check within `instance()` will be to call a `Registry` method (with the derived class' name) asking whether a `Singleton` had previously been created for that class. If the `Registry` method determines a `Singleton` for the requested derived type has previously been instantiated, a pointer to the existing instance will be returned by `instance()`.

- Instead, if the `Registry` provides permission to allocate the `Singleton`, `instance()` will allocate the `Singleton` much as before, setting the inherited protected data member of `theInstance` to the allocated derived `Singleton`. The static `instance()` method will also set the backlink through the inherited protected destroyer data member using `setSingleton()`. We will then pass the newly instantiated derived class instance (which is a `Singleton`) to a `Registry` method to `Store()` the newly allocated `Singleton` within the `Registry`.

We notice that four pointers to the same `Singleton` will exist. One will be the specialized pointer of our derived class type, which is returned from our derived class `instance()` method. This pointer will be handed to our Client for application usage. The second `Singleton` pointer will be the pointer stored in our inherited, protected data member `theInstance`. The third `Singleton` pointer will be the pointer stored in the `SingletonDestroyer`. The fourth pointer to the `Singleton` will be a pointer that is stored in the `Registry`. No problem, we can have multiple pointers to a `Singleton`. This is one reason the `SingletonDestroyer`, used in its traditional destruction capacity, is so important – it will destroy our one and only `Singleton` for each type at the end of the application.

Our `Registry` will maintain a pair for each class employing the `Singleton` pattern, consisting of a class name and the (eventual) pointer to the specific `Singleton` for the corresponding class. The pointer to each specific `Singleton` instance will be a static data member and will additionally require an external variable to garner its underlying memory. The result is one additional external variable per class embracing the Singleton pattern.

The idea of the `Registry` can be expanded further still if we choose to additionally accommodate the rare use of the Singleton pattern to allow multiple (but a finite set of) `Singleton` objects per class type. This rare existence of controlled, multiple singletons is known as the **Multiton pattern**. An example of this extended pattern in action might be that we chose to model a high school that has a single principal, yet multiple vice-principals. `Principal` would be an expected derived class of `Singleton`, yet the multiple vice-principals would represent a fixed number of instances of the `Vice-Principal` class (derived from `Singleton`). Our registry could be expanded to allow up to N registered `Singleton` objects for the `Vice-Principal` type (the multiton).

We have now seen an implementation of the Singleton pattern using a paired-class approach. We have folded the classes and concepts of `Singleton`, `SingetonDestroyer`, Target, and Client into the framework of classes we are accustomed to seeing, namely `Person`, as well as into a descendant class of our `Singleton` and `Person` (`President`). Let's now briefly recap what we have learned relating to patterns before moving forward to our next chapter.

Summary

In this chapter, we have furthered our goal of becoming better C++ programmers by expanding our programming repertoire by embracing another design pattern. We have explored the Singleton pattern by first employing two simple approaches, and then a paired-class implementation using `Singleton` and `SingletonDestroyer`. Our approach uses inheritance to incorporate our Singleton's implementation into our Target class. Optionally, we incorporate a useful, existing base class into our Target class using multiple inheritances.

Making use of core design patterns, such as the Singleton pattern, will help you more easily reuse existing, well-tested portions of code in a manner understood by other programmers. By employing familiar design patterns, you will be contributing to well-understood and reusable solutions with avant-garde programming techniques.

We are now ready to continue onward with our final design pattern in *Chapter 20, Removing Implementation Details Using the pImpl Pattern*. Adding more patterns to our arsenal of programming skills makes us more versatile and valued programmers. Let's continue onward!

Questions

1. Using the Singleton pattern examples found in this chapter, create a program to accomplish the following:

    a. Implement either an interface for a `President` to `Resign()` or implement the interface to `Impeach()` a `President`. Your method should delete the current Singleton `President` (and remove that link from the `SingletonDestroyer`). `SingletonDestroyer` has a `setSingleton()` method that may be useful to aid in removing the backlink.

    b. Noting that the former Singleton `President` has been removed, create a new `President` using `President::instance()`. Verify that the new `President` has been installed.

    c. (Optional) Create a `Registry` to allow `Singleton` to be used effectively in multiple classes (not mutually exclusively, as is the current implementation).

2. Why can you not label the `static instance()` method as virtual in `Singleton` and override it in `President`?

3. What other examples can you imagine that might easily incorporate the Singleton pattern?

20
Removing Implementation Details Using the pImpl Pattern

This chapter will wrap up our quest to expand your C++ programming repertoire beyond core OOP concepts, with the objective of further empowering you to solve recurring types of coding problems, utilizing common design patterns. Incorporating design patterns in your coding can not only provide refined solutions but also contribute to easier code maintenance and provide for potential code reuse.

The next design pattern that we will learn how to implement effectively in C++ is the **pImpl pattern**.

In this chapter, we will cover the following main topics:

- Comprehending the pImpl pattern and how it reduces compile-time dependencies
- Understanding how to implement the pImpl pattern in C++ using association and unique pointers
- Recognizing performance issues relating to pImpl and necessary trade-offs

By the end of this chapter, you will understand the pImpl pattern and how it can be used to separate implementation details from a class interface to reduce compiler dependencies. Adding an additional design pattern to your skillset will help you become a more valuable programmer.

Let's increase our programming skillset by examining another common design pattern, the pImpl pattern.

Technical requirements

Online code for full program examples can be found in the following GitHub URL: `https://github.com/PacktPublishing/Deciphering-Object-Oriented-Programming-with-CPP/tree/main/Chapter20`. Each full program example can be found in the GitHub repository under the appropriate chapter heading (subdirectory) in a file that corresponds to the chapter number, followed by a dash, followed by the example number in the chapter at hand. For example, the first full program in this chapter can be found in the subdirectory `Chapter20` in a file named `Chp20-Ex1.cpp` under the aforementioned GitHub directory. Some programs are in applicable subdirectories as indicated in the examples.

The CiA video for this chapter can be viewed at: `https://bit.ly/3CfQxhR`.

Understanding the pImpl pattern

The **pImpl pattern** (**p**ointer to **Impl**ementation idiom) is a structural design pattern that separates the implementation of a class from its public interface. This pattern was originally known as the **Bridge pattern** by the **Gang of Four** (**GofF**) and is also known as the **Cheshire Cat**, **compiler-firewall idiom**, **d-pointer**, **opaque pointer**, or **Handle pattern**.

The primary purpose of this pattern is to minimize compile-time dependencies. The result of reducing compile-time dependencies is that changes in a class definition (most notably, the private access region) will not send a wave of timely recompilations throughout a developing or deployed application. Instead, the necessary recompiled code can be isolated to the *implementation* of the class itself. The other pieces of the application that depend on the class definition will no longer be affected by recompilation.

Private members within a class definition can affect a class with respect to recompilation. This is because changing the data members can alter the size of an instance of that type. Also, private member functions must be matched by signature to function calls for overloading resolution as well as potential type conversions.

The manner in which traditional header (`.h` or `.hpp`) and source code files (`.cpp`) specify dependencies trigger recompilation. By removing the class inner implementation details from a class header file (and placing these details in a source file), we can remove many dependencies. We can change which header files are included in other header and source code files, streamlining the dependencies and hence the recompilation burden.

The pImpl pattern will compel the following adjustments to a class definition:

- Private (non-virtual) members will instead be replaced by a pointer to a nested class type that includes the former private data members and methods. A forward declaration to the nested class will also be necessary.

- The pointer to the implementation (`pImpl`) will be an association to which method calls of the class implementation will be delegated.

- The revised class definition will exist in a header file for the class embracing this idiom. Any formerly included header files that this header file once depended upon will now be moved to instead be included in the source code file for this class.

- Other classes including the header file of a pImpl class will now not face recompilation should the implementation of the class within its private access region be modified.

- To effectively manage dynamic memory resources of the associated object that represents the implementation, we will use a unique pointer (smart pointer).

The compilation freedom within the revised class definition takes advantage of the fact that pointers only require a forward declaration of the class type of the pointer to compile.

Let's move forward to first examine a basic, and then a refined, implementation of the pImpl pattern.

Implementing the pImpl pattern

In order to implement the pImpl pattern, we will need to revisit the typical header and source file composition. We will then replace the private members in a typical class definition with a pointer to the implementation, taking advantage of an association. The implementation will be encapsulated within a nested class of our target class. Our pImpl pointer will delegate all requests to our associated object that provides the inner class details or implementation.

The inner (nested) class will be referred to as the **implementation class**. The original, now outer, class will be referred to as the **target** or **interface class**.

We will start by reviewing the typical (non-pImpl pattern) file composition containing class definitions and member function definitions.

Organizing file and class contents to apply the pattern basics

Let's first review the organization strategy of the typical C++ class with respect to file placement regarding the class definition and member function definitions. We will next consider the revised organization strategy of a class utilizing the pImpl pattern.

Reviewing typical file and class layout

Let's take a look at a typical class definition and how we previously have organized a class with respect to source and header files, such as in our discussions in *Chapter 5, Exploring Classes in Detail*, and in *Chapter 15, Testing Classes and Components*.

Recall that we organized each class into a header (.h or .hpp) file containing the class definition and inline function definitions, plus a corresponding source code (.cpp) file containing the non-inline member function definitions. Let's review a familiar sample class definition, Person:

```
#ifndef _PERSON_H   // preprocessor directives to avoid
#define _PERSON_H   // multiple inclusion of header
using std::string;

class Person
{
private:
    string firstName, lastName, title;
    char middleInitial = '\0';   // in-class initialization
protected:
    void ModifyTitle(const string &);
public:
    Person() = default;   // default constructor
    Person(const string &, const string &, char,
           const string &);   // alternate constructor
    // prototype not needed for default copy constructor
    // Person(const Person &) = default;   // copy ctor
    virtual ~Person() = default;   // virtual destructor
    const string &GetFirstName() const
        { return firstName; }
    const string &GetLastName() const { return lastName; }
    const string &GetTitle() const { return title; }
    char GetMiddleInitial() const { return middleInitial; }
    virtual void Print() const;
    virtual void IsA() const;
    virtual void Greeting(const string &) const;
    Person &operator=(const Person &);   // overloaded op =
};
#endif
```

In the aforementioned header file (`Person.h`), we have included our class definition for `Person` as well as inline function definitions for the class. Any larger inline function definitions not appearing within the class definition (indicated with the keyword `inline` in the prototype) would also appear in this file, after the class definition itself. Notice the use of preprocessor directives to ensure that a class definition is only included once per compilation unit.

Let's next review the contents of the corresponding source code file, `Person.cpp`:

```
#include <iostream>  // also incl. other relevant libraries
#include "Person.h"  // include the header file
using std::cout;      // preferred to: using namespace std;
using std::endl;
using std::string;


// Include all the non-inline Person member functions
// The alt. constructor is one example of many in the file
Person::Person(const string &fn, const string &ln, char mi,
            const string &t): firstName(fn), lastName(ln),
                              middleInitial(mi), title(t)
{
    // dynamically alloc. memory for any ptr data mbrs here
}
```

In the previously defined source code file, we define all the non-inline member functions for the class, `Person`. Though not all methods are shown, all can be found in our GitHub code. Also, if the class definition contains any static data members, the definition of the external variables designating the memory for these members should be included in the source code file.

Let's now consider how we can remove the implementation details from the `Person` class definition and its corresponding header file, by applying the pImpl pattern.

Applying the pImpl pattern with revised class and file layout

To employ the pImpl pattern, we will reorganize our class definition and its respective implementation. We will add a nested class within our existing class definition to represent the private members of our original class and the core of its implementation. Our outer class will include a pointer of the inner class type, serving as an association to our implementation. Our outer class will delegate all implementation requests to the inner, associated object. We will restructure the placement of classes and source code within the header and source code files.

Let's take a closer look at our revised implementation for our class to understand each new detail required to implement the pImpl pattern. This example, composed of a source file `PersonImpl.cpp` and one header file `Person.h`, can be found in the same directory as a simple driver to test the pattern in our GitHub repository. To make a complete executable, you will need to compile and link together `PersonImp.cpp` and `Chp20-Ex1.cpp` (the driver), found in this same directory. Here is the GitHub repository URL for the driver:

https://github.com/PacktPublishing/Deciphering-Object-Oriented-Programming-with-CPP/blob/main/Chapter20/Chp20-Ex1.cpp

```cpp
#ifndef _PERSON_H        // Person.h header file definition
#define _PERSON_H
class Person
{
private:
    class PersonImpl;   // forward declaration nested class
    PersonImpl *pImpl = nullptr; // ptr to implementation
                                 // of class
protected:
    void ModifyTitle(const string &);
public:
    Person();   // default constructor
    Person(const string &, const string &, char,
           const string &);
    Person(const Person &);  // copy const. will be defined
    virtual ~Person();  // virtual destructor
    const string &GetFirstName() const; // no longer inline
    const string &GetLastName() const;
    const string &GetTitle() const;
    char GetMiddleInitial() const;
    virtual void Print() const;
    virtual void IsA() const;
    virtual void Greeting(const string &) const;
    Person &operator=(const Person &);  // overloaded =
};
#endif
```

In our aforementioned revised class definition for `Person`, notice that we have removed the data members in the private access region. Any non-virtual private methods, had they existed, would have also been removed. Instead, we include a forward declaration to our nested class with `class PersonImpl;`. We also declare a pointer to the implementation using `PersonImpl *pImpl;`, which represents an association to the nested class members encapsulating the implementation. In our initial implementation, we will use a native (raw) C++ pointer to specify the association to the nested class. We will subsequently revise our implementation to utilize a *unique pointer*.

Notice that our public interface for `Person` is much as before. All of our existing public and protected methods exist as expected, interface-wise. We notice, however, that the inline functions (which depend on the implementation of the data members) have been replaced with non-inline member function prototypes.

Let's move forward to see the class definition for our nested class, `PersonImpl`, as well as the placement of the member functions of `PersonImpl` and `Person` in a common source code file, `PersonImpl.cpp`. We will start with the nested `PersonImpl` class definition:

```cpp
// PersonImpl.cpp source code file includes nested class
// Nested class definition supports implementation
class Person::PersonImpl
{
private:
    string firstName, lastName, title;
    char middleInitial = '\0';  // in-class initialization
public:
    PersonImpl() = default;   // default constructor
    PersonImpl(const string &, const string &, char,
               const string &);
    // Default copy ctor does not need to be prototyped
    // PersonImpl(const PersonImpl &) = default;
    virtual ~PersonImpl() = default;  // virtual destructor
    const string &GetFirstName() const
        { return firstName; }
    const string &GetLastName() const { return lastName; }
    const string &GetTitle() const { return title; }
    char GetMiddleInitial() const { return middleInitial; }
    void ModifyTitle(const string &);
    virtual void Print() const;
    virtual void IsA() const { cout << "Person" << endl; }
```

```
        virtual void Greeting(const string &msg) const
            { cout << msg << endl; }
        PersonImpl &operator=(const PersonImpl &);
};
```

In the previously mentioned nested class definition for `PersonImpl`, notice that this class looks surprisingly similar to the original class definition for `Person`. We have private data members and a full host of member function prototypes, even some inline functions written for brevity (which won't actually be inlined because they are virtual). `PersonImpl` represents the implementation for `Person`, so it is crucial that this class can access all data and implement each method fully. Notice that the scope resolution operator (`::`) in the definition of class `Person::PersonImpl` is used to specify that `PersonImpl` is a nested class of `Person`.

Let's continue by taking a look at the member function definitions for `PersonImpl`, which will appear in the same source file `PersonImpl.cpp` as the class definition. Though some methods have been abbreviated, their full online code is available in our GitHub repository:

```
// File: PersonImpl.cpp - See online code for full methods
// Nested class member functions.
// Notice that the class name is Outer::Inner class


// Notice that we are using the system-supplied definitions
// for default constructor, copy constructor and destructor


// alternate constructor
Person::PersonImpl::PersonImpl(const string &fn,
            const string &ln, char mi, const string &t):
            firstName(fn), lastName(ln),
            middleInitial(mi), title(t)
{
}


void Person::PersonImpl::ModifyTitle(const string &newTitle)
{
    title = newTitle;
}


void Person::PersonImpl::Print() const
{   // Print each data member as usual
```

```
}

// Note: same as default op=, but it is good to review what
// is involved in implementing op= for upcoming discussion
Person::PersonImpl &Person::PersonImpl::operator=
                                (const PersonImpl &p)
{
    if (this != &p)   // check for self-assignment
    {
        firstName = p.firstName;
        lastName = p.lastName;
        middleInitial = p.middleInitial;
        title = p.title;
    }
    return *this;   // allow for cascaded assignments
}
```

In the aforementioned code, we see the implementation for the overall Person class using the nested class PersonImpl. We see the member function definitions for PersonImpl and notice that the bodies of these methods are exactly how we previously implemented the methods in our original Person class without the pImpl pattern. Again, we notice the use of the scope resolution operator (::) to specify the class name for each member function definition, such as void Person::PersonImpl::Print() const. Here, Person::PersonImpl indicates the nested class of PersonImpl within the Person class.

Next, let's take a moment to review the member function definitions for Person, our class employing the pImpl pattern. These methods will additionally contribute to the PersonImpl.cpp source code file and can be found in our GitHub repository:

```
// Person member functions - also in PersonImpl.cpp

Person::Person(): pImpl(new PersonImpl())
{ // As shown, this is the complete member fn. definition
}

Person::Person(const string &fn, const string &ln, char mi,
               const string &t):
               pImpl(new PersonImpl(fn, ln, mi, t))
{ // As shown, this is the complete member fn. definition
```

```
}

Person::Person(const Person &p):
                pImpl(new PersonImpl(*(p.pImpl)))
{   // This is the complete copy constructor definition
}   // No Person data members to copy from 'p' except deep
    // copy of *(p.pImpl) to data member pImpl


Person::~Person()
{
    delete pImpl;    // delete associated implementation
}


void Person::ModifyTitle(const string &newTitle)
{   // delegate request to the implementation
    pImpl->ModifyTitle(newTitle);
}


const string &Person::GetFirstName() const
{   // no longer inline in Person;
    // non-inline method further hides implementation
    return pImpl->GetFirstName();
}


// Note: methods GetLastName(), GetTitle(), and
// GetMiddleInitial() are implemented similar to
// GetFirstName(). See online code


void Person::Print() const
{
    pImpl->Print();    // delegate to implementation
}                      // (same named member function)


// Note: methods IsA() and Greeting() are implemented
// similarly to Print() - using delegation. See online code
```

```
Person &Person::operator=(const Person &p)
{   // delegate op= to implementation portion
    pImpl->operator=(*(p.pImpl)); // call op= on impl. piece
    return *this;   // allow for cascaded assignments
}
```

In the aforementioned member function definitions for `Person`, we notice that all methods delegate the required work to the nested class via the associated `pImpl`. In our constructors, we allocate the associated `pImpl` object and initialize it appropriately (using the member initialization list of each constructor). Our destructor is responsible for deleting the associated object using `delete pImpl;`.

Our `Person` copy constructor will set member `pImpl` to the newly allocated memory, while invoking the `PersonImpl` copy constructor for the nested object creation and initialization, passing `*(p.pImpl)` to the nested object's copy constructor. That is, `p.pImpl` is a pointer, so we dereference the pointer using `*` to obtain a referenceable object for the `PersonImpl` copy constructor.

We use a similar strategy in our overloaded assignment operator for `Person`. That is, there are no data members other than `pImpl` to perform a deep assignment, so we merely call the `PersonImpl` assignment operator on associated object `pImpl`, again passing in `*(p.pImpl)` as the right-hand value.

Finally, let us consider a sample driver to demonstrate our pattern in action. Interestingly, our driver will work with either our originally specified non-pattern class (source and header files) or with our revised pImpl pattern-specific source and header files!

Bringing the pattern components together

Let's finally take a look at our `main()` function in our driver source file `Chp20-Ex1.cpp` to see how our pattern is orchestrated:

```
#include <iostream>
#include "Person.h"
using std::cout;   // preferred to: using namespace std;
using std::endl;
constexpr int MAX = 3;

int main()
{
    Person *people[MAX] = { }; // initialized to nullptrs
    people[0] = new Person("Elle", "LeBrun", 'R',"Ms.");
    people[1] = new Person("Zack", "Moon", 'R', "Dr.");
    people[2] = new Person("Gabby", "Doone", 'A', "Dr.");
```

```
    for (auto *individual : people)
       individual->Print();
    for (auto *individual : people)
       delete individual;
    return 0;
}
```

Reviewing our aforementioned `main()` function, we simply dynamically allocate several `Person` instances, call selected `Person` method(s) on the instances (`Print()`), and then delete each instance. We have included the `Person.h` header file, as expected, to be able to utilize this class. From the Client's point of view, everything looks *as usual* and appears pattern unspecific.

Note that we separately compile `PersonImp.cpp` and `Chp20-Ex1.cpp`, linking the object files together into an executable. However, due to the pImpl pattern, if we change the implementation for `Person`, the change will be encapsulated by its implementation in the `PersonImp` nested class. Only `PersonImp.cpp` will require recompilation. The Client will not need to recompile the driver, `Chp20-Ex1.cpp`, because the changes will not have occurred in the `Person.h` header file (which the driver depends on).

Let's take a look at the output for this program:

```
Ms. Elle R. LeBrun
Dr. Zack R. Moon
Dr. Gabby A. Doone
```

In the aforementioned output, we see the expected results of our simple driver.

Let's move forward to consider how we may improve our implementation of the pImpl pattern using a unique pointer.

Improving the pattern with a unique pointer

Our initial implementation using an association with a native C++ pointer relieves many compiler dependencies. This is because the compiler only needs to see a forward class declaration of the pImpl pointer type in order to compile successfully. So far, we have achieved the core goal of using the pImpl pattern – reducing recompilation.

However, there is always criticism of using native or *raw* pointers. We are responsible for managing the memory ourselves, including remembering to delete the allocated nested class type in our outer class destructor. Memory leaks, memory misuse, and memory errors are potential drawbacks for managing memory resources ourselves with raw pointers. For that reason, it is customary to implement the pImpl pattern using **smart pointers**.

We will continue our quest to implement pImpl by examining a key component often used with the pImpl pattern – smart pointers, or more specifically, the `unique_ptr`.

Let's start by understanding smart pointer basics.

Understanding smart pointers

To implement the pImpl pattern customarily, we must first understand smart pointers. A **smart pointer** is a small wrapper class that encapsulates a raw pointer, ensuring that the pointer it contains is automatically deleted when the wrapper object goes out of scope. The class implementing the smart pointer can be implemented using templates to create a smart pointer for any data type.

Here is a very simple example of a smart pointer. This example can be found on our GitHub:

```
https://github.com/PacktPublishing/Deciphering-Object-Orient-
ed-Programming-with-CPP/blob/main/Chapter20/Chp20-Ex2.cpp
```

```cpp
#include <iostream>
#include "Person.h"
using std::cout;   // preferred to: using namespace std;
using std::endl;

template <class Type>
class SmartPointer
{
private:
    Type *pointer = nullptr;   // in-class initialization
public:
    // Below ctor also handles default construction
    SmartPointer(Type *ptr = nullptr): pointer(ptr) { }
    virtual ~SmartPointer();   // allow specialized SmrtPtrs
    Type *operator->() { return pointer; }
    Type &operator*() { return *pointer; }
};

SmartPointer::~SmartPointer()
{
    delete pointer;
    cout << "SmartPtr Destructor" << endl;
}
```

```
int main()
{
    SmartPointer<int> p1(new int());
    SmartPointer<Person> pers1(new Person("Renee",
                            "Alexander", 'K', "Dr."));
    *p1 = 100;
    cout << *p1 << endl;
    (*pers1).Print();    // or use: pers1->Print();
    return 0;
}
```

In the previously defined, straightforward SmartPointer class, we simply encapsulate a raw pointer. The key benefit is that the SmartPointer destructor will ensure that the raw pointer is destructed when the wrapper object is popped off the stack (for local instances) or before the program terminates (for static and extern instances). Certainly, this class is basic, and we must determine the desired behaviors for the copy constructor and the assignment operator. That is, allow shallow copies/assignment, require deep copies/assignment, or disallow all copies/assignment. Nonetheless, we can now visualize the concept of a smart pointer.

Here is the output for our smart pointer example:

```
100
Dr. Renee K. Alexander
SmartPtr Destructor
SmartPtr Destructor
```

The aforementioned output shows that the memory for each object contained within a SmartPointer is managed for us. We can quite easily see with the "SmartPtr Destructor" output strings that the destructor for each object is called on our behalf when the local objects in main() go out of scope and are popped off the stack.

Understanding unique pointers

A **unique pointer**, specified as unique_ptr in the Standard C++ Library, is a type of smart pointer that encapsulates exclusive ownership and access to a given heap memory resource. A unique_ptr cannot be duplicated; the owner of a unique_ptr will have sole use of that pointer. Owners of unique pointers can choose to move these pointers to other resources, but the repercussions are that the original resource will no longer contain the unique_ptr. We must #include <memory> to include the definition for unique_ptr.

> **Additional types of smart pointers**
>
> Other types of smart pointers are available in the Standard C++ Library, in addition to `unique_ptr`, such as `weak_ptr` and `shared_ptr`. These additional types of smart pointers will be explored in *Chapter 21, Making C++ Safer*.

Modifying our smart pointer program to instead utilize `unique_ptr`, we now have the following:

```cpp
#include <iostream>
#include <memory>
#include "Person.h"
using std::cout;   // preferred to: using namespace std;
using std::endl;
using std::unique_ptr;

int main()
{
    unique_ptr<int> p1(new int());
    unique_ptr<Person> pers1(new Person("Renee",
                        "Alexander", 'K', "Dr."));
    *p1 = 100;
    cout << *p1 << endl;
    (*pers1).Print();   // or use: pers1->Print();
    return 0;
}
```

Our output will be similar to the `SmartPointer` example; the difference is that no `"SmartPtr Destructor"` call message will be displayed (as we are using a `unique_ptr` instead). Notice that because we included `using std::unique_ptr;`, we did not need to qualify `unique_ptr` with `std::` in the unique pointer declaration.

With this knowledge, let's add unique pointers to our pImpl pattern.

Adding unique pointers to the pattern

To implement the pImpl pattern using a `unique_ptr`, we will make minimal changes to our previous implementation, starting with our `Person.h` header file. The full program example of our pImpl pattern utilizing a `unique_ptr` can be found in our GitHub repository and will additionally include a revised file for `PersonImpl.cpp`. Here is the URL for the driver, `Chp20-Ex3.cpp`; note the subdirectory, `unique`, in our GitHub repository for this complete example:

https://github.com/PacktPublishing/Deciphering-Object-Orient-
ed-Programming-with-CPP/blob/main/Chapter20/unique/Chp20-Ex3.cpp

```cpp
#ifndef _PERSON_H      // Person.h header file definition
#define _PERSON_H
#include <memory>

class Person
{
private:
    class PersonImpl;  // forward declaration nested class
    std::unique_ptr<PersonImpl> pImpl; //unique ptr to impl
protected:
    void ModifyTitle(const string &);
public:
    Person();   // default constructor
    Person(const string &, const string &, char,
           const string &);
    Person(const Person &);  // copy constructor
    virtual ~Person();  // virtual destructor
    const string &GetFirstName() const; // no longer inline
    const string &GetLastName() const;
    const string &GetTitle() const;
    char GetMiddleInitial() const;
    virtual void Print() const;
    virtual void IsA() const;
    virtual void Greeting(const string &) const;
    Person &operator=(const Person &);  // overloaded =
};
#endif
```

Notice, in the revised aforementioned class definition for `Person`, the unique pointer declaration of `std::unique_ptr<PersonImpl> pImpl;`. Here, we use the `std::` qualifier because the standard namespace has not been explicitly included in our header file. We also `#include <memory>` to gain the definition for `unique_ptr`. The remainder of the class is identical to our initial implementation of pImpl using an association implemented with a raw pointer.

Next, let's understand the extent to which our source code needs to be modified from our initial pImpl implementation. Let's now take a look at the necessary modified member functions in our source file, `PersonImpl.cpp`:

```
// Source file PersonImpl.cpp

// Person destructor no longer needs to delete pImpl member
// and hence can simply be the default destructor!
// Note: prototyped with virtual in header file.
Person::~Person() = default;
// unique_pointer pImpl will delete its own resources
```

Taking a look at the aforementioned member functions requiring modification, we see that it is only the `Person` destructor! Because we are using a unique pointer to implement the association to the nested class implementation, we no longer need to manage the memory for this resource ourselves. That's pretty nice! With these minor changes, our pImpl pattern now features a `unique_ptr` to designate the implementation of the class.

Next, let's examine some of the performance issues relating to using the pImpl pattern.

Understanding pImpl pattern trade-offs

Incorporating the pImpl pattern into production code has both benefits and disadvantages. Let's review each so that we can better understand the circumstances that may warrant deploying this pattern.

The negligible performance issues encompass most of the disadvantages. That is, nearly every request made of the target (interface) class will need to be delegated to its nested implementation class. The only requests that can be handled by the outer class will be those not involving any data members; those circumstances will be extraordinarily rare! Another disadvantage includes slightly higher memory requirements of instances to accommodate the added pointer as part of the pattern implementation. These issues will be paramount in embedded software systems and those requiring peak performance, but relatively minor otherwise.

Maintenance will be a little more difficult for classes employing the pImpl pattern, an unfortunate disadvantage. Each target class is now paired with an extra (implementation) class, including a set of forwarding methods to delegate requests to the implementation.

A few implementation difficulties may also arise. For example, if any of the private members (now in the nested implementation class) need to access any of the protected or public methods of the outer (interface) class, we will need to include a backlink from the nested class to the outer class to access that member. Why? The `this` pointer in the inner class will be of the nested object type. Yet the protected and public methods in the outer object will expect a `this` pointer to the outer object – even if those public methods will then redelegate the request to call a private nested class method for help. This backlink will also be required to call public virtual functions of the interface from the scope of the inner class (implementation). Keep in mind, however, that we impact performance with another added pointer per object and with delegation to call each method in the associated object.

There are several advantages of utilizing the pImpl pattern, offering important considerations. Of most importance, recompile time during the development and maintenance of code decreases significantly. Additionally, the compiled, binary interface of a class becomes independent of the underlying implementation of the class. Changing the implementation of a class only requires the nested implementation class to be recompiled and linked in. Users of the outer class are unaffected. As a bonus, the pImpl pattern provides a way to hide the underlying private details of a class, which may be useful when distributing class libraries or other proprietary code.

An advantage of including a `unique_ptr` in our pImpl implementation is that we have guaranteed proper destruction of the associated implementation class. We also have the potential to save inadvertent programmer-introduced pointer and memory mishaps!

The use of the pImpl pattern is a trade-off. Careful analysis of each class and of the application at hand will help determine whether the pImpl pattern is appropriate for your design.

We have now seen implementations of the pImpl pattern initially using a raw pointer, and then applying a `unique_ptr`. Let us now briefly recap what we have learned relating to patterns before moving to the bonus chapter of our book, *Chapter 21, Making C++ Safer*.

Summary

In this chapter, we have advanced our objective of becoming more indispensable C++ programmers by furthering our programming skills with another core design pattern. We have explored the pImpl pattern with an initial implementation using native C++ pointers and association and then improved our implementation by using a unique pointer. By examining the implementation, we easily understand how the pImpl pattern reduces compile-time dependencies and can make our code more implementation-dependent.

Making use of core design patterns, such as the pImpl pattern, will help you more easily contribute to reusable, maintainable code that is understood by other programmers familiar with common design patterns. Your software solutions will be based on creative and well-tested design solutions.

We have now completed our final design pattern together, wrapping up a long journey of understanding OOP in C++. You now have a multitude of skills, including a deep understanding of OOP, extended language features, and core design patterns, all of which make you a more valuable programmer.

Though C++ is an intricate language with additional features, supplemental techniques, and additional design patterns to discover, you have more than a solid basis and level of expertise to easily navigate and embrace any additional language features, libraries, and patterns you may wish to acquire. You've come a long way; this has been an adventurous journey together! I have enjoyed every minute of our quest and I hope you have as well.

We began by reviewing basic language syntax and understanding the C++ essentials necessary to serve as building blocks for our then-upcoming OOP journey. We then embraced C++ as an OOP language, learning not only essential OO concepts but also how to implement them with either C++ language features, coding techniques, or both. We then extended your skills by adding knowledge of exception handling, friends, operator overloading, templates, STL basics, and testing OO classes and components. We then ventured into sophisticated programming techniques by embracing core design patterns and delving into code by applying each pattern of interest.

Each of these acquired skill segments represents a new tier of C++ knowledge and mastery. Each will help you to create more easily maintainable and robust code. Your future as a well-versed, skilled OO programmer in C++ awaits. Now, let's move on to our bonus chapter, and then, let's get programming!

Questions

1. Modify the pImpl pattern example in this chapter, which uses a unique pointer to additionally introduce unique pointers within the implementation of the nested class.

2. Revise your Student class from a previous chapter solution to simply inherit from the Person class in this chapter that embraces the pImpl pattern. What difficulties, if any, do you have? Now, modify your Student class to additionally utilize the pImpl pattern with a unique pointer. A suggested Student class is one that includes an association with a Course. Now, what difficulties, if any, do you have?

3. What other examples can you imagine that might reasonably incorporate the pImpl pattern for relative implementation independence?

Part 5: Considerations for Safer Programming in C++

The goal of this part is to understand what can be done as a programmer to make C++ a safer language, which in turn will help make our programs more robust. At this point, we will have learned a lot about C++, from language essentials to implementing OO designs in C++. We will have added additional skills to our repertoire, such as using friends and operator overloading, exception handling, templates, and the STL. We will have even looked in depth at a handful of popular design patterns. We will know that we can do nearly anything in C++, but we will have also seen that having so much power can leave room for cavalier programming and grave errors, which can lead to unwieldy code that is difficult to maintain.

In this section, we will review what we have learned throughout the book with a keen eye toward understanding how we can work to make our code bulletproof. We will work toward a set of core programming guidelines to follow with one goal in mind: to make our programs safe!

We will revisit and expand upon our knowledge of smart pointers (unique, shared, and weak) as well as introduce a complimentary idiom, RAII. We will review what we have seen along the way relating to safety issues with native C++ pointers and sum up our safety concerns with a programming guideline: always prefer smart pointers in newly created C++ code.

We will review modern programming features, such as range-based `for` loops and for-each style loops to understand how these simple constructs can help us avoid common errors. We will revisit `auto` instead of explicit typing to add safety to our code. We will revisit using well-tested STL types to ensure our code is not error-prone with ad hoc containers. We will revisit how the `const` qualifier can add safety to our code in a variety of ways. By reviewing specific language features used throughout the book, we will revisit how each of these features can add safety to our code. We will also consider thread safety and how various topics we have seen throughout the book relate to thread safety.

Finally, we will discuss core programming guidelines, such as preferring initialization over assignment, or using one of `virtual`, `override`, or `final` to specify polymorphic operations and their methods. We will understand the importance of adopting a programming guideline and see the resources available to support programming safely in C++.

This part comprises the following chapter:

- *Chapter 21, Making C++ Safer*

21
Making C++ Safer

This bonus chapter will add insight into what we can do as C++ programmers to make the language as safe as possible in our everyday usage. We have progressed from basic language features to our core interest of OO programming with C++, to additional useful language features and libraries (exceptions, operator overloading, templates, and STL), to design patterns to give us a knowledge base to solve recurring types of OO programming problems. At every point along the way, we've seen that C++ requires extra care on our part to avoid tricky and potentially problematic programming situations. C++ is a language that will allow us to do anything, but with this power comes the need for guidelines to ensure our programming follows safe practices. After all, our goal is to create programs that will run successfully without errors and, additionally, be easy to maintain. The ability of C++ to do anything needs to be paired with sound practices to simply make C++ safer.

The goal of this chapter is to revisit topics that we have covered in previous chapters, reviewing them with an eye toward safety. We will also incorporate topics strongly related to ones we have seen previously. This chapter is not meant to cover wholly new topics or previous topics in great depth, but to provide a grouping of safer programming practices and the encouragement to seek further information on each topic as needed. Some of these topics can encompass entire chapters (or books) themselves!

In this bonus chapter, we will cover selected popular programming conventions to meet our safety challenge:

- Revisiting smart pointers (unique, shared, and weak), as well as a complementary idiom (RAII)

- Using modern `for` loops (range-based, for-each) to avoid common errors

- Adding type safety: usage of `auto` instead of explicit typing

- Preferring usage of STL types for simple containers (`std::vector`, and so on)

- Utilizing `const` appropriately to ensure non-modification of select items

- Understanding thread safety issues

- Considering core programming guideline essentials, such as preferring initialization to assignment, or choosing only one of `virtual`, `override`, or `final`

- Adopting C++ core programming guidelines for safety (build and assemble one, if necessary)
- Understanding resources for programming safety in C++

By the end of this chapter, you will understand some of the current industry standards and concerns for programming safely in C++. This chapter is not meant to be a comprehensive list of all safety concerns and practices in C++, but to showcase the types of issues you will need to become mindful of as a successful C++ programmer. In some cases, you may desire to investigate a topic more deeply to gain a more thorough level of competence and proficiency. Adding safety to your C++ programming will make you a more valuable programmer, as your code will be more reliable and have more longevity and success.

Let's round out our programming skillset by considering how we can make C++ safer.

Technical requirements

Online code for full program examples can be found in the following GitHub URL: `https://github.com/PacktPublishing/Deciphering-Object-Oriented-Programming-with-CPP/tree/main/Chapter21`. Each full program example can be found in the GitHub repository under the appropriate chapter heading (subdirectory) in a file that corresponds to the chapter number, followed by a dash, followed by the example number in the chapter at hand. For example, the first full program in this chapter can be found in the subdirectory `Chapter21` in a file named `Chp21-Ex1.cpp` under the aforementioned GitHub directory. Some programs are in applicable subdirectories as indicated in the examples.

The CiA video for this chapter can be viewed at: `https://bit.ly/3wpOG6b`.

Revisiting smart pointers

Throughout the book, we have developed a reasonable understanding of how to use raw or native C++ pointers, including the associated memory allocation and deallocation for heap instances. We have persevered through native C++ pointers because they are pervasive in existing C++ code. Having knowledge of how to properly utilize native pointers is essential in working with the volume of existing C++ code currently in use. But, for newly created code, there is a safer way to manipulate heap memory.

We have seen that dynamic memory management with native pointers is a lot of work! Especially when there may be multiple pointers to the same chunk of memory. We've talked about reference counting to shared resources (such as heap memory) and mechanisms for deleting memory when all instances are done with the shared memory. We also know that memory deallocation can easily be overlooked, leading to memory leakage.

We have also seen, firsthand, that errors with native pointers can be costly. Our programs can end abruptly when we dereference memory we don't intend to access, or when we dereference uninitialized native pointers (interpreting the memory to contain a valid address and meaningful data at that address—neither of which are actually valid). Pointer arithmetic to walk through memory can be laden with errors by an otherwise adept programmer. When a memory error is made, pointer or heap memory misuse are often the culprits.

Certainly, using references can ease the burden of many errors with native pointers. But references can still point to dereferenced heap memory that someone forgets to deallocate. For these and many other reasons, smart pointers have become popular in C++ with the primary purpose of making C++ safer.

We've talked about smart pointers in previous chapters and have seen them in action with our pImpl pattern (using `unique_ptr`). But there are more types of smart pointers for us to review in addition to unique: shared and weak. Let's also make a programming premise (a future style guide addition) to prefer smart pointers in our newly created code to native pointers for the purpose and value of pointer safety.

Recall that a **smart pointer** is a small wrapper class that encapsulates a raw or native pointer, ensuring that the pointer it contains is automatically deleted when the wrapper object goes out of scope. The Standard C++ Library implementations of *unique*, *shared*, and *weak* smart pointers use templates to create a specific category of smart pointer for any data type.

Though we could devote an entire chapter to each type of smart pointer in depth, we will review each type briefly as a starting point to encourage their usage in newly created code to support our goal of making C++ safer.

Now, let's revisit each type of smart pointer, one by one.

Using smart pointers – unique

Recall that a **unique pointer**, specified as `unique_ptr` in the Standard C++ Library, is a type of smart pointer that encapsulates exclusive ownership and access to a given heap memory resource. A `unique_ptr` cannot be duplicated; the owner of a `unique_ptr` will have sole use of that pointer. Owners of unique pointers can choose to move these pointers to other resources, but the repercussions are that the original resource will no longer contain `unique_ptr`. Recall that we must use `#include <memory>` to include the definition for `unique_ptr`.

Here is a very simple example illustrating how to create unique pointers. This example can be found in our GitHub repository:

https://github.com/PacktPublishing/Deciphering-Object-Orient-ed-Programming-with-CPP/blob/main/Chapter21/Chp21-Ex1.cpp

```cpp
#include <iostream>
#include <memory>
#include "Person.h"
using std::cout;    // preferred to: using namespace std;
using std::endl;
using std::unique_ptr;

// We will create unique pointers, with and without using
// the make_unique (safe wrapper) interface

int main()
{
    unique_ptr<int> p1(new int(100));
    cout << *p1 << endl;

    unique_ptr<Person> pers1(new Person("Renee",
                            "Alexander",'K', "Dr."));
    (*pers1).Print();       // or use: pers1->Print();

    unique_ptr<Person> pers2; // currently uninitialized
    pers2 = move(pers1);    // take over another unique
                            // pointer's resource
    pers2->Print();         // or use: (*pers2).Print();

    // make_unique provides a safe wrapper, eliminating
    // obvious use of heap allocation with new()
    auto pers3 = make_unique<Person>("Giselle", "LeBrun",
                                    'R', "Ms.");
    pers3->Print();
    return 0;
}
```

First, notice that because we included using `std::unique_ptr;`, we did not need to qualify `unique_ptr` or `make_unique` with `std::` in the unique pointer declarations. In this small program, we create several unique pointers, starting with one to point to an integer, `p1`, and one to point to an instance of a `Person`, `pers1`. Each of these variables has exclusive use of the heap memory each points to because we are using unique pointers.

Next, we introduce a unique pointer, `pers2`, that takes over the memory originally allocated and linked to `pers1` using `pers2 = move(pers1);`. The original variable no longer has access to this memory. Note that though we could have allocated `pers2` to have its own, unique heap memory, we instead chose to demonstrate how to allow one unique pointer to relinquish its memory to another unique pointer using `move()`. Changing the ownership of unique pointers with `move()` is typical, as unique pointers cannot be copied (because that would allow two or more pointers to share the same memory and, therefore, not be unique!)

Finally, we create another unique pointer, `pers3`, that utilizes `make_unique` as a wrapper to allocate the heap memory for the unique pointer that `pers3` will represent. The preference for using `make_unique` is that the call to `new()` will be made internally, on our behalf. Additionally, any exceptions thrown during the construction of the object will be handled for us, as will any call to `delete()`, should the underlying `new()` not complete successfully and a call to `delete()` is then warranted.

The heap memory will be managed for us automatically; this is one of the benefits of using a smart pointer.

Here is the output for our `unique_ptr` example:

```
100
Dr. Renee K. Alexander
Dr. Renee K. Alexander
Ms. Giselle LeBrun
Person destructor
Person destructor
```

Under the hood, the destructor will automatically be called for each object pointed to by a smart pointer, when the memory is no longer utilized. In the case of this example, the destructor for each `Person` object is called on our behalf when the local objects in `main()` go out of scope and are popped off the stack. Note that our `Person` destructor contains a `cout` statement so that we can visualize that there are only two `Person` objects destructed. Here, the destructed `Person` objects represent the instance taken over by `pers2` (from `pers1`) via the `move()` statement, and the `pers3` object that was created using the `make_unique` wrapper.

Next, let's add examples using shared and weak smart pointers.

Using smart pointers – shared

A **shared pointer**, specified as `shared_ptr` in the Standard C++ Library, is a type of smart pointer that permits shared ownership of and access to a given resource. The last shared pointer to the resource in question will trigger the destruction and memory deallocation of the resource. Shared pointers can be used in multithreaded applications; however, race conditions may occur if non-constant member functions are used to modify the shared resource. Since shared pointers only provide reference counting, we will need to enlist additional library methods to solve these issues (alleviating race conditions, synchronizing access to critical regions of code, and so on). The Standard C++ Library, for example, provides overloaded atomic methods to lock, store, and compare the underlying data pointed to by a shared pointer.

We have seen many example programs that could take advantage of shared pointers. For example, we utilized associations between the `Course` and `Student` classes – a given student is associated with many courses and a given course is associated with many students. Clearly, multiple `Student` instances may point to the same `Course` instance, and vice versa.

Previously, with raw pointers, it was the programmer's responsibility to employ reference counting. In contrast, using shared pointers, the internal reference counter is atomically incremented and decremented in support of both pointer and thread safety.

Dereferencing a shared pointer is nearly as fast as dereferencing a raw pointer; however, because a shared pointer represents a wrapped pointer in a class, constructing and copying a shared pointer is more expensive. However, we are interested in making C++ safer, so we will simply note this very minor performance expense and move forward.

Let's take a look at a very simple example using `shared_ptr`. This example can be found in our GitHub repository:

https://github.com/PacktPublishing/Deciphering-Object-Oriented-Programming-with-CPP/blob/main/Chapter21/Chp21-Ex2.cpp

```
#include <iostream>
#include <memory>
#include "Person.h"
using std::cout;    // preferred to: using namespace std;
using std::endl;
using std::shared_ptr;

int main()
{
    shared_ptr<int> p1 = std::make_shared<int>(100);
```

```
// alternative to preferred, previous line of code:
// shared_ptr<int> p1(new int(100));

shared_ptr<int> p2;// currently uninitialized (caution)
p2 = p1; // p2 now shares the same memory as p1
cout << *p1 << " " << *p2 << endl;

shared_ptr<Person> pers1 = std::make_shared<Person>
                    ("Gabby", "Doone", 'A', "Miss");
// alternative to preferred, previous lines of code:
// shared_ptr<Person> pers1(new Person("Gabby",
//                          "Doone",'A', "Miss"));

shared_ptr<Person> pers2 = pers1;  // initialized
pers1->Print();   // or use: (*pers1).Print();
pers2->Print();
pers1->ModifyTitle("Dr."); // changes shared instance
pers2->Print();

cout << "Number of references: " << pers1.use_count();
return 0;
}
```

In the aforementioned program, we create four shared pointers – two to point to the same integer (p1 and p2) and two to point to the same instance of Person (pers1 and pers2). Each of these variables may change the specific shared memory they point to because we are using shared pointers (which allow such a reassignment). A change to the shared memory through pers1, for example, will be reflected should we then review the (shared) memory through pointer pers2; both variables point to the same memory location.

The heap memory will again be managed for us automatically as a benefit of using smart pointers. In this example, the memory will be destructed and deleted when the last reference to the memory is removed. Notice that reference counting is done on our behalf and that we can access this information using use_count().

Let us notice something interesting about the previous example. Notice the mixed use of -> and . notation with shared pointer variables pers1 and pers2. For example, we utilize pers1->Print(); and yet we also utilize pers1.use_count(). This is no mistake and reveals the wrapper implementation of the smart pointer. With that in mind, we know that use_count() is a method of shared_ptr. Our shared pointers pers1 and pers2 are each declared as instances of shared_ptr (definitely not using raw C++ pointers with the symbol *). Hence, dot notation is appropriate to access method `use_count()`. Yet, we are using `->` notation to access `pers1->Print();`. Here, recall that this notation is equivalent to `(*pers1).Print();`. Both `operator*` and `operator->` in the `shared_ptr` class are overloaded to delegate to the wrapped, raw pointer contained within the smart pointer. Hence, we may utilize standard pointer notation to access `Person` methods (through the safely wrapped raw pointer).

Here is the output for our `shared_ptr` pointer example:

```
100 100
Miss Gabby Doone
Miss Gabby Doone
Dr. Gabby Doone
Number of references: 2
Person destructor
```

Shared pointers seem like a wonderful way to ensure that memory resources pointed to by multiple pointers are properly managed. Overall, this is true. However, there are situations with circular dependencies such that shared pointers simply cannot release their memory – another pointer is always pointing to the memory in question. This happens when a cycle of memory is orphaned; that is, when no outside shared pointers point *into* the circular connection. In such unique cases, we might actually (and counterintuitively) mismanage memory with shared pointers. In these situations, we can elicit help from a weak pointer to help us break the cycle.

With that in mind, let's next take a look at weak smart pointers.

Using smart pointers – weak

A **weak pointer**, specified by `weak_ptr` in the Standard C++ Library, is a type of smart pointer that does not take ownership of a given resource; instead, the weak pointer acts as an observer. Weak pointers can be used to help break a circular connection that may exist between shared pointers; that is, situations where the destruction of a shared resource would otherwise never occur. Here, a weak pointer is inserted into the chain to break the circular dependency that shared pointers alone might otherwise create.

As an example, imagine our `Student` and `Course` dependencies from our initial programming examples utilizing association, or from our more complex program illustrating the Observer pattern. Each contains pointer data members of the associated object types, effectively creating a potential circular dependency. Now, should an outside (from the circle) shared pointer exist, such as an external list of courses or an external list of students, the exclusive circular dependency scenario may not arise. In this case, for example, the master list of courses (the external pointer, separate from any circular dependency existing between the associated objects) will provide the means to cancel a course, leading to its eventual destruction.

Likewise in our example, an external set of students comprising the university's student body can provide an external pointer to the potentially circular shared pointer scenario resulting from the association between Student and Course. Yet in both of these cases, work will need to be done to remove a canceled course from a student's course list (or remove a dropped student from a course's student list). The removal of the associations in this scenario reflects accurately managing a student's schedule or a course's attendance list. Nonetheless, we can imagine scenarios where a circular connection may exist without an outside handle to the links (unlike the aforementioned scenario, which has outside links into the circle).

In the case where a circular dependency exists (with no outside influences), we will need to downgrade one of the shared pointers to a weak pointer. A weak pointer will not control the lifetime of the resource that it points to.

A weak pointer to a resource cannot access the resource directly. This is because operators * and -> are not overloaded in the weak_ptr class. You will need to convert the weak pointer to a shared pointer in order to access methods of the (wrapped) pointer type. One way to do this is to apply the lock() method to a weak pointer, as the return value is a shared pointer whose contents are locked with a semaphore to ensure mutual exclusivity to the shared resource.

Let's take a look at a very simple example using weak_ptr. This example can be found on our GitHub:

https://github.com/PacktPublishing/Deciphering-Object-Orient-ed-Programming-with-CPP/blob/main/Chapter21/Chp21-Ex3.cpp

```
#include <iostream>
#include <memory>
#include "Person.h"
using std::cout;   // preferred to: using namespace std;
using std::endl;
using std::weak_ptr;
using std::shared_ptr;

int main()
{
    // construct the resource using a shared pointer
    shared_ptr<Person> pers1 = std::make_shared<Person>
                            ("Gabby", "Doone", 'A', "Miss");
    pers1->Print(); // or alternatively: (*pers1).Print();

    // Downgrade resource to a weak pointer
    weak_ptr<Person> wpers1(pers1);
```

```
        // weak pointer cannot access the resource;
        // must convert to a shared pointer to do so
        // wpers1->Print();    // not allowed! operator-> is not
                               // overloaded in weak_ptr class


        cout << "# references: " << pers1.use_count() << endl;
        cout << "# references: " << wpers1.use_count() << endl;


        // establish a new shared pointer to the resource
        shared_ptr<Person> pers2 = wpers1.lock();
        pers2->Print();


        pers2->ModifyTitle("Dr.");    // modify the resource
        pers2->Print();


        cout << "# references: " << pers1.use_count() << endl;
        cout << "# references: " << wpers1.use_count() << endl;
        cout << "# references: " << pers2.use_count() << endl;
        return 0;
}
```

In the aforementioned program, we allocate our resource using a shared pointer in `pers1`. Now, let us imagine we had a reason in our program to downgrade our resource to a weak pointer – perhaps we would like to insert a weak pointer to break an otherwise cycle of shared pointers. Using `weak_ptr<Person> wpers1(pers1);`, we establish a weak pointer to this resource. Notice that we cannot use `wpers1` to call `Print();`. This is because `operator->` and `operator*` have not been overloaded in the `weak_ptr` class.

We print out `use_count()` for each of `pers1` and `wpers1` to notice that each shows a value of 1. That is, there is only one non-weak pointer controlling the resource in question (the weak pointer may temporarily hold the resource, but cannot modify it).

Now, imagine that we would like to convert the resource pointed to by `wpers1` on-demand to another shared pointer, so that we may access the resource. We can do so by first gaining a lock on the weak pointer; `lock()` will return a shared pointer whose contents are protected by a semaphore. We assign this value to `pers2`. We then call `pers2->ModifyTitle("Dr.");` on the resource using the shared pointer.

Finally, we print out `use_count ()` from the perspective of each of `pers1`, `wpers1`, and `pers2`. In each case, the reference count will be 2, as there are two non-weak pointers referencing the shared resource. The weak pointer does not contribute to the reference count of that resource, which is exactly how weak pointers can help break a chain of circular dependencies. By inserting a weak pointer into the dependency loop, the reference count to the shared resource will not be affected by the weak pointer's presence. This strategy allows the resource to be deleted when only the weak pointer to the resource remains (and the reference count is 0).

The heap memory will again be managed for us automatically as a benefit of using smart pointers. In this example, the memory will be destructed and deleted when the last reference to the memory is removed. Again, note that the weak pointer did not contribute a reference to this count. We can see from the `cout` statement in the `Person` destructor that only one instance was destructed.

Here is the output for our `weak_ptr` pointer example:

```
Miss Gabby Doone
# references: 1
# references: 1
Miss Gabby Doone
Dr. Gabby Doone
# references: 2
# references: 2
# references: 2
Person destructor
```

In this section, we've reviewed and added to the basics regarding smart pointers. However, there could be a chapter easily spent on each type of smart pointer. Nonetheless, hopefully, you have enough comfort with the essentials to begin to include a variety of smart pointers in your code and investigate each type further as your need arises.

Exploring a complementary idea – RAII

A programming idiom that complements smart pointers (as well as other concepts) is **RAII (Resource Acquisition Is Initialization)**. RAII binds the life cycle of a (potentially shared) resource to the lifetime of an object by requiring the resource to be acquired before use. This concept can help control the life cycle of a shared resource. RAII can be applied to concepts we have previously seen, such as allocated heap memory (through the usage of smart pointers), or to the concept of reference counting that we covered in *Chapter 10, Implementing Association, Aggregation, and Composition*. This programming technique is also applicable to multithreaded programming (accessing shared resources with a mutex lock), as well as to coordinate access to other shared resources such as sockets, files, or databases. Ownership of a resource can be transferred from one object to another safely using a `move ()` operation.

Many C++ class libraries follow RAII for resource management, such as `std::string` and `std::vector`. These classes follow the idiom in that their constructors acquire the necessary resources (heap memory), and release the resources automatically in their destructors. The user of these classes is not required to explicitly release any memory for the container itself. In these class libraries, RAII as a technique is used to manage these resources, even though the heap memory is not managed using smart pointers. Instead, the concepts of RAII are encapsulated and hidden within the class implementations themselves.

When we implemented our own smart pointers in *Chapter 20, Removing Implementation Details Using the pImpl Pattern*, we used RAII, without knowing it, to ensure the allocation of the heap resource within our constructor and the release of the resource in our destructor. The smart pointers implemented in the Standard C++ Library (`std::unique_ptr`, `std::shared_ptr`, and `std::weak_ptr`) embrace this idiom as well. Embracing RAII by using classes employing this idiom (or by adding it yourself to classes when this is not possible), can help ensure code is safer and easier to maintain. Because of the safety and robustness that this idiom adds to code, savvy developers urge us to embrace RAII as one of the most important practices and features available in C++.

Next in our effort to make C++ safer, let's consider several easy C++ features we can easily embrace to ensure our coding is more robust.

Embracing additional C++ features promoting safety

As we have seen through 20 previous chapters of programming, C++ is an extensive language. We know that C++ has great power and that we can do nearly anything in C++. As object-oriented C++ programmers, we have seen how to adopt OO designs, with the goal of making our code more easily maintainable.

We have also gained a lot of experience utilizing raw (native) pointers in C++, primarily because raw pointers are very pervasive in existing code. You truly need experience and facility in using native pointers for when the need arises. In gaining this experience, we have seen firsthand the pitfalls we may encounter with mismanagement of heap memory – our programs may have crashed, we may have leaked memory, overwritten memory accidentally, left dangling pointers, and so on. Our first order of business in this chapter was to prefer using smart pointers in newly created code – to promote safety in C++.

Now, we will explore other areas of C++ that we can similarly employ safer features. We have seen these various features throughout the book; it is important to establish a guideline that select language features promote safety in C++. Just because we can do anything in C++ doesn't mean that we should routinely include features in our repertoire that have a high level of misuse associated with them. Applications that continually crash (or crash even once) are unacceptable. Certainly, we have noted no-nos throughout the book. Here, let's point out language features that are worth embracing to further our goal of making C++ safer, leaving our applications more robust and more easily maintainable.

Let's start by reviewing simple items we can incorporate into our everyday code.

Revisiting range for loops

C++ has a variety of looping constructs that we have seen throughout the book. One common error that occurs when processing a complete set of items is correctly keeping track of how many items are in the set, especially when this counter is used as a basis to loop through all items in the set. Processing too many elements when our set is stored as an array, for example, could lead our code to raise an exception unnecessarily (or worse, could lead our program to crash).

Rather than relying on a MAX value to conduct our looping for all elements in a set, it is more desirable to loop through every item in the set in a way that doesn't count on the programmer correctly remembering this upper loop value. Instead, for each item in the set, let's do some sort of processing. A for-each loop answers this need quite nicely.

One common error that occurs when processing a non-complete set of items is correctly keeping track of how many items are currently in the set. For example, a Course may have a maximum number of students permitted. Yet, as of today, only half of the potential Student slots are filled. When we peruse the list of students enrolled in the course, we need to ensure we are processing only the filled student spots (that is, the current number of students). Processing all maximum student spots would clearly be an error and could lead our program to crash. In this scenario, we must use care to iterate only over the currently utilized Student spots in the Course, either through using logic to exit a loop when appropriate or by selecting a container type whose current size represents the complete size of the set to be iterated upon (with no empty *to be filled* spots); the latter scenario making a for-each loop an ideal choice.

Also, what if we rely on looping based upon a currentNumStudents counter? This may be better than a MAX value in cases as previously illustrated, but what if we've not kept that counter correctly updated? We're subject to an error on this as well. Again, combining a container class where the number of entries represents the current number of entries with a for-each type of loop can ensure that we process the complete, current grouping in a less error-prone manner.

Now that we have revisited modern and more safe looping styles, let's embrace auto to ensure type safety. We will then see an example incorporating these collective features.

Using auto for type safety

Many situations arise in which using auto makes coding easier with respect to variable declarations, including loop iterators. Additionally, using auto instead of explicit typing can ensure type safety.

Choosing to use auto is a simple way to declare a variable that has a complicated type. Using auto can also ensure that the best type is chosen for a given variable and that implicit conversion will not occur.

We can use auto as a placeholder for types in a variety of situations, allowing the compiler to deduce what is needed in a particular situation. We can even use auto as a return type for a function in many cases. Using auto allows our code to appear more generic and can complement templates as an alternative to genericizing a type. We can pair auto with const, and also pair these qualifiers with references; note that these qualifiers *combined* cannot be extrapolated with `auto` and must be specified individually by the programmer. Additionally, `auto` cannot also be used with qualifiers augmenting a type, such as `long` or `short`, nor can it be used with `volatile`. Though outside the scope of our book, `auto` can be utilized with lambda expressions.

Of course, using `auto` has a few drawbacks. For example, if the programmer doesn't understand the type of object being created, the programmer may anticipate the compiler to select a certain type, and yet another (unexpected) type is deduced. This may create subtle errors in your code. For example, if you have overloaded functions for both the type you think `auto` will select and for the type the compiler actually deduces the `auto` declaration to be, you may call a different function than anticipated! Certainly, this may mostly be due to the programmer not fully understanding the context of usage at hand when inserting the `auto` keyword. Another drawback is when the programmer uses `auto` just to force the code to compile, without truly working through the syntax at hand and thinking about how the code should be written.

Now that we have revisited adding `auto` to our code, let's revisit embracing STL in our everyday code. We will then see an example incorporating these collective features.

Preferring STL for simple containers

The Standard Template Library, as we've seen in *Chapter 14, Understanding STL Basics*, includes a very complete and robust set of container classes that are widely utilized in C++ code. Using these well-tested components instead of native C++ mechanisms (such as an array of pointers) to collect like items can add robustness and reliability to our code. The memory management is eased on our behalf (eliminating many potential errors).

The STL, by using templates to implement its large variety of container classes, allows its containers to be used generically for any data type our programs may encounter. By comparison, had we utilized native C++ mechanisms, it is likely that we may have tied our implementation to a specific class type, such as an array of pointers to `Student`. Certainly, we could have implemented an array of pointers to a templatized type, but why do so when such a nice variety of well-tested containers are readily available for our use?

STL containers also avoid using `new()` and `delete()` for memory management, choosing to use allocators to improve efficiency for STL's underlying memory management. For example, a vector, stack, or queue may grow and shrink in size. Rather than allocating the maximum number of elements you may anticipate (which may be both difficult to estimate or inefficient to over-allocate for typical usage that does not reach the maximum), a certain buffer size or a number of elements may be allocated under the hood up front. This initial allocation allows multiple additions to the container without a resize necessary for each new addition to the set (as might otherwise be done to avoid over-allocation). Only when the underlying container's internal allocation (or buffer) size exceeds the pre-allocated amount will an internal reallocation be necessary (unknown to the user of the container). The expense of an internal reallocation, or a *move*, is the allocation of a larger piece of memory, copying from the original memory to the larger piece, and then the release of the original memory. The STL works to fine-tune, under the hood, the internal allocations to balance typical usage needs versus costly reallocation that might otherwise be performed.

Now that we have revisited preferring STL in our code, let's revisit applying const when necessary to ensure code isn't modified unless we so intend it to be. We will wrap up this section with an example illustrating all of the key safety points featured in this section.

Applying const as needed

Applying the const qualifier to objects is an easy way to indicate that instances that should not be modified are not, in fact, modified. We may recall that const instances may only call const member functions. And that const member functions may not modify any part of the object calling the method (this). Remembering to utilize this simple qualifier can ensure that this chain of checkpoints occurs for objects that we truly do not intend to modify.

With that in mind, remember that const can be utilized in parameter lists, to qualify objects and methods. Using const adds readability to the objects and methods it qualifies as well as adding the valuable enforcement of read-only objects and methods. Let's remember to use const as needed!

Now, let's take a look at how we can use each of these easily added C++ features that contribute to safer programming. This example revisits preferred looping styles, using auto for type safety, using the STL for simple containers, and applying const as appropriate. This example can be found in our GitHub repository:

https://github.com/PacktPublishing/Deciphering-Object-Orient-ed-Programming-with-CPP/blob/main/Chapter21/Chp21-Ex4.cpp

```cpp
#include <vector>
using std::vector;
// Assume additional #include/using as typically included
// Assume classes Person, Student are as typically defined

// In this const member function, no part of 'this' will
// be modified. Student::Print() can be called by const
// instances of Student, including const iterators
void Student::Print() const
{   // need to use access functions as these data members
    // are defined in Person as private
    cout << GetTitle() << " " << GetFirstName() << " ";
    cout << GetMiddleInitial() << ". " << GetLastName();
    cout << " with id: " << studentId << " GPA: ";
```

```
        cout << setprecision(3) <<  " " << gpa;
        cout << " Course: " << currentCourse << endl;
}


int main()
{   // Utilize STL::vector instead of more native C++ data
    // structures (such as an array of pointers to Student)
    // There's less chance for us to make an error with
    // memory allocation, deallocation, deep copies, etc.
    vector<Student> studentBody;
    studentBody.push_back(Student("Hana", "Sato", 'U',
                         "Miss", 3.8, "C++", "178PSU"));
    studentBody.push_back(Student("Sam", "Kato", 'B',
                         "Mr.", 3.5, "C++", "272PSU"));
    studentBody.push_back(Student("Giselle", "LeBrun", 'R',
                         "Ms.", 3.4, "C++", "299TU"));


    // Notice that our first loop uses traditional notation
    // to loop through each element of the vector.
    // Compare this loop to next loop using an iterator and
    // also to the preferred range-for loop further beyond
    // Note: had we used MAX instead of studentBody.size(),
    // we'd have a potential error - what if MAX isn't the
    // same as studentBody.size()?
    for (int i = 0; i < studentBody.size(); i++)
        studentBody1[i].Print();


    // Notice auto keyword simplifies iterator declaration
    // However, an iterator is still not the most
    // preferred looping mechanism.
    // Note, iterator type is: vector<Student>::iterator
    // the use of auto replaces this type, simplifying as:
    for (auto iter = studentBody.begin();
             iter != studentBody.end(); iter++)
        (*iter).EarnPhD();
```

```cpp
    // Preferred range-for loop
    // Uses auto to simplify type and const to ensure no
    // modification. As a const iterator, student may only
    // call const member fns on the set it iterates thru
    for (const auto &student : studentBody)
        student.Print();
    return 0;
}
```

In the aforementioned program, we initially notice that we have included the use of `std::vector` from C++'s STL. Further in `main()`, we notice the instantiation of a vector using `vector<Student> studentBody;`. Utilizing this well-tested container class certainly adds robustness to our code versus managing a dynamically sized array ourselves.

Next, notice the specification of a constant member function `void Student::Print() const;`. Here, the `const` qualification ensures that no part of the object invoking this method (`this`) will be able to be modified. Furthermore, should any `const` instances exist, they will be able to invoke `Student::Print()` as the `const` qualification guarantees this method to be safe (that is, read-only) for `const` instances to utilize.

Next, we notice three looping styles and mechanisms, progressing from least to most safe in style. The first loop cycles through each element in the loop with a traditional style `for` loop. What if we had used `MAX` for the looping condition instead of `studentBody.size()`? We might have tried to process more elements than are currently in the container; this type of oversight can be error-prone.

The second loop utilizes an iterator and the `auto` keyword to make the type specification easier (and hence safer) for the iterator itself. Iterators, though well defined, are still not the preferred looping mechanism. A subtlety from the increment in the second statement in the `for` statement can also lead to inefficiency. Consider, for example, the pre versus post increment in the statement that is executed before the loop condition is retested (that is, `++iter`). Had this been `iter++`, the code would be less efficient. This is because `iter` is an object and the pre-increment returns a reference to the object, whereas the post-increment returns a temporary object (what is created and destroyed with each loop iteration). The post-increment also utilizes an overloaded function, so the compiler cannot optimize its usage.

Finally, we see the preferred and safest looping mechanism, featuring a range-for loop combined with `auto` for the iterator specification (to simplify the type declaration). The use of `auto` replaces `vector<Student>::iterator` as the type for `iter`. Any time there is an ease in notation, there is also less room for error. Also, notice the use of `const` added to the iterator declaration to ensure that the loop will only call non-modifiable methods on each instance iterated upon; this is an example of an additional, appropriate safety feature we can employ in our code.

Here is the output for our aforementioned program:

```
Miss Hana U. Sato with id: 178PSU GPA:  3.8 Course: C++
Mr. Sam B. Kato with id: 272PSU GPA:  3.5 Course: C++
Ms. Giselle R. LeBrun with id: 299TU GPA:  3.4 Course: C++
Everyone to earn a PhD
Dr. Hana U. Sato with id: 178PSU GPA:  3.8 Course: C++
Dr. Sam B. Kato with id: 272PSU GPA:  3.5 Course: C++
Dr. Giselle R. LeBrun with id: 299TU GPA:  3.4 Course: C++
```

We have now revisited several straightforward C++ language features that can easily be embraced to promote safety in our everyday coding practices. Using range-for loops provides code simplification and removes dependencies from often incorrect upper limits of loop iteration. Embracing `auto` simplifies variable declarations, including within loop iterators, and can help ensure type safety versus explicit typing. Using well-tested STL components can add robustness, reliability, and familiarity to our code. Finally, applying `const` to data and methods is an easy way to ensure data is not modified unintentionally. Each of these principles is easy to employ and adds value to our code by adding to its overall safety.

Next, let's consider how understanding thread safety can contribute to making C++ safer.

Considering thread safety

Multithreaded programming in C++ is an entire book unto itself. Nonetheless, we have mentioned several situations throughout the book that potentially require the consideration of thread safety. It is worth re-iterating these topics to provide an overview of the issues you may encounter in various niches of C++ programming.

A program may be comprised of multiple threads, each of which may potentially compete against one another to access a shared resource. For example, a shared resource could be a file, socket, region of shared memory, or output buffer. Each thread accessing the shared resource needs to have carefully coordinated (known as mutually exclusive) access to the resource.

Imagine, for example, if two threads wanted to write output to your screen. If each thread could access the output buffer associated with `cout` without waiting for the other to complete a cohesive statement, the output would be a garbled mess of random letters and symbols. Clearly, synchronized access to a shared resource is important!

Thread safety involves understanding atomic actions, mutual exclusion, locks, synchronization, and so on—all of which are aspects of multithreaded programming.

Let's begin with an overview of threads and multithreaded programming.

Multithreaded programming overview

A **thread** is a separate flow of control within a process, conceptually working like a subprocess (or further subdivision of a process) within a given process. Threads are sometimes referred to as *threads of control*. Applications that have many threads of control are known as **multithreaded applications**.

In uniprocessor environments, threads give the appearance that multiple tasks are running concurrently. Just as with processes, threads are swapped in and out of the CPU quickly to appear to the user that they are being processed simultaneously (though they aren't). In a shared, multiprocessor environment, the use of threads within an application can significantly speed up processing and allow parallel computing to be realized. Even in a uniprocessor system, threads can actually (and perhaps counterintuitively) speed up a process, in that one thread may run while waiting for the I/O of another thread to complete.

Threads related by the tasks they are performing may find themselves in similar methods of a class simultaneously. If each thread is working on a distinct dataset (such as a distinct this pointer, even if working within the same method), there is generally no need to synchronize access to those methods. For example, imagine s1.EarnPhd(); and s2.EarnPhD();. Here, two separate instances are in the same method (possibly concurrently). However, the datasets worked upon in each method differ – in the first scenario, s1 will bind to this; in the second scenario, s2 will bind to this. There is most likely no overlap in shared data between the two instances. However, if these methods are accessing static data (that is shared by all instances of a given class, such as a numStudents data member), synchronization to the critical pieces of code accessing the shared memory regions will be required. Traditionally, system-dependent locks or semaphores are added around data or functions that require mutual exclusivity to critical regions of code.

Multithreaded programming in C++ is available through a variety of commercial or public domain multithreading libraries. Additionally, the Standard C++ Library features thread support in a variety of capacities including using std::condition_variable for thread synchronization, std::mutex to ensure mutual exclusivity of critical resources (by avoiding race conditions), and std::semaphore to model resource counting. By instantiating a std::thread object and becoming proficient with the aforementioned features, we can add multithreaded programming using an established C++ library. Additionally, the std::atomic template can be added to a type to establish it as an atomic type and ensure type-safe synchronization. The std::exception_ptr type allows the transport of exceptions between coordinating threads. Overall, there are many thread library features to consider; this is a vast topic.

The details for multithreaded programming are beyond the scope of this book; however, we can discuss scenarios within this book that may be augmented to require the knowledge of using threads. Let's revisit some of those situations.

Multithreaded programming scenarios

There are many programming scenarios that can benefit from the use of multithreaded programming. We will just mention a few that extend the ideas we have covered in this book.

The Observer pattern may certainly be employed in multithreaded programming scenarios! In these instances, care must be used in the `Update()` and `Notify()` methods of `Observer` and `Subject`, respectively, to add synchronization and locking mechanisms.

Smart pointers, such as `shared_ptr` and `weak_ptr`, can be used in multithreaded applications and already include the means to lock and synchronize access to shared resources via reference counting (and with the use of atomic library methods).

Objects related through association may arise with multithreaded programming or through shared memory regions. Any time access is conducted through a shared resource using multithreaded programming, mutexes (locks) should be employed to ensure mutual exclusivity to those shared resources.

Objects throwing exceptions that need to communicate with one another will need to include synchronization within catcher blocks or delegate exceptions to the `main()` program thread. Employing worker threads to communicate with the `main()` program thread is a typical design model. Utilizing shared memory is a means to store the data that will need to be shared between threads coordinating with a `throw` and `catch` of the exception itself. An instance of `std::exception_ptr` can be utilized with `std::current_exception()` to store an instance needing to be shared. This shared instance (between threads) can be rethrown to a participating thread using `std::rethrow_exception()`.

Multithreaded programming is a fascinating topic unto itself and requires in-depth understanding to utilize safely in C++. We've revisited a few areas in which thread safety considerations may complement areas we have covered in this book. It is highly recommended to delve deeply into thread safety in C++ before embarking on adding multithreaded programming to your code.

Next, let's move forward to investigate how programming guidelines can add a necessary level of safety to C++ programming.

Utilizing core programming guidelines

Programming guidelines are much more than a set of conventions to indicate how many spaces to indent or naming conventions for variables, functions, classes, data members, and member functions. A modern programming guideline is a covenant between programmers within an organization to create code adhering to specific standards, with the largest goal to provide robust and easily extensible code by following these common standards. The bottom line is that most of the conventions contained within a programming guideline are simply to make programming in C++ safer.

The consensus of what comprises a C++ programming guideline may vary from organization to organization, but there are many resources available (including from standards committees) to provide examples and direction.

Let's move forward to examine a sampling of programming guide essentials and then discuss adopting a core set of guidelines, as well as understanding resources widely available for programming safely in C++.

Examining guideline essentials

Let's start by examining a sampling of meaningful conventions to follow from a typical C++ programming guideline. We have examined many of these programming issues throughout the book, yet it is useful to review a few items to provide a starting point for choosing conventions to promote C++ safety.

Preferring initialization over assignment

Always choose initialization, whenever possible, over assignment. It's simply more efficient and safer! Use in-class initialization or the member initialization list. When assignment is used after initialization, it can be less efficient. Imagine, for example, a member object that is default constructed, only to quickly overwrite its values with more suitable values via assignment in the body of the constructor. It would have been more efficient to utilize the member initialization list to initialize this member object via an alternate constructor.

Also, neglecting to give each piece of memory an initial value can cost us dearly in terms of safety – memory in C++ is not clean, so it is truly inappropriate to interpret whatever is in an uninitialized variable (or data member) as valid. Accessing an uninitialized value is an undefined behavior. We truly never know what is lurking in uninitialized memory, but we know it is never the correct value to be used as an initializer!

Let's review preferred initialization with a small program. This example can be found in our GitHub:

https://github.com/PacktPublishing/Deciphering-Object-Orient-ed-Programming-with-CPP/blob/main/Chapter21/Chp21-Ex5.cpp

```
class Person
{
private:
    string firstName; // str mbrs are default constructed so
    string lastName;  // we don't need in-class initializers
    char middleInitial = '\0';  // in-class initialization
    string title;
protected:
    void ModifyTitle(const string &);
```

```
public:
    Person() = default;    // default constructor
    Person(const string &, const string &, char,
           const string &);
    // use default copy constructor and default destructor
    // inline function definitions
    const string &GetFirstName() const { return firstName; }
    const string &GetLastName() const { return lastName; }
    const string &GetTitle() const { return title; }
    char GetMiddleInitial() const { return middleInitial; }
};


// With in-class initialization, it often not necessary to
// write the default constructor yourself - there's often
// nothing remaining to initialize!

// alternate constructor
// Note use of member init list to initialize data members
Person::Person(const string &fn, const string &ln, char mi,
               const string &t): firstName(fn),
               lastName(ln), middleInitial(mi), title(t)
{
    // no need to assign values in body of method -
    // initialization has handled everything!
}
```

Examining the preceding code, we notice that the `Person` class uses in-class initialization to set the `middleInitial` data member to the null character (`'\0'`). For each instance of `Person`, `middleInitial` will be set to the null character prior to any constructor call that further initializes the instance in question. Notice that the other data members in the class are all of type `string`. Because `string` is a class itself, these data members are actually member objects of type `string` and will be default constructed, appropriately initializing each of these string members.

Next, notice that we opted not to provide a default (no argument) constructor, allowing the system-supplied default constructor to be linked in for us. In-class initialization, coupled with the appropriate member object initialization of the `string` members, left no additional initialization necessary for new `Person` instances, and hence no need for a programmer-specified default constructor.

Finally, notice our use of the member initialization list in the alternate constructor for `Person`. Here, each data member is set with an appropriate value from the parameter list of this method. Notice that every data member is set via initialization, leaving no assignments necessary in the body of the alternate constructor.

Our preceding code follows the popular code guideline: whenever possible, always opt to set values via initialization versus assignment. Knowing that each data member has an appropriate value during construction leads us to provide a safer code. Initialization is also more efficient than assignment.

Now, let's consider another core C++ guideline relating to virtual functions.

Choosing one of virtual, override, or final

Polymorphism is a wonderful concept that C++ easily supports with the use of virtual functions. We learned in *Chapter 7, Utilizing Dynamic Binding through Polymorphism,* that the keyword `virtual` is used to indicate a polymorphic operation – an operation that may be overridden by derived classes with a preferred method. Derived classes are not obligated to override a polymorphic operation (virtual function) by providing a new method, but may find it meaningful to do so.

When a derived class chooses to override a virtual function introduced by a base class with a new method, the overridden method may use both the keywords `virtual` and `override` in the signature of the method. However, it is a convention to use only `override` at this overridden (derived class) level.

When a virtual function is introduced in the hierarchy, it may be desirable at some point to indicate that a certain method is the *final* implementation of this operation. That is, the operation in question may no longer be overridden. We know that it is appropriate to apply the `final` specifier to the virtual function at this level of the hierarchy to indicate that a given method may no longer be overridden. Though we may also include the keyword `virtual` at this level as well, it is recommended to only utilize `final`.

To sum up, when specifying a virtual function, only choose one label at each level: `virtual`, `override`, or `final` – even though the keyword `virtual` can be added to complement `override` and `final`. By doing so, it will be much clearer if the virtual function at hand is newly introduced (`virtual`), an overridden method of a virtual function (`override`), or the final method of a virtual function (`final`). Clarity causes fewer errors to occur and that helps make C++ safer.

Let's review the preferred keyword usage with virtual functions with a program segment. The complete example can be found in our GitHub repository:

```
https://github.com/PacktPublishing/Deciphering-Object-Orient-
ed-Programming-with-CPP/blob/main/Chapter21/Chp21-Ex6.cpp
```

```
class Person
{
private:
```

```
    string firstName;
    string lastName;
    char middleInitial = '\0';  // in-class initialization
    string title;  // Mr., Ms., Mrs., Miss, Dr., etc.
protected:
    void ModifyTitle(const string &);
public:
    Person() = default;   // default constructor
    Person(const string &, const string &, char,
          const string &);
    virtual ~Person();  // virtual destructor
    const string &GetFirstName() const
        { return firstName; }
    const string &GetLastName() const { return lastName; }
    const string &GetTitle() const { return title; }
    char GetMiddleInitial() const { return middleInitial; }
    virtual void Print() const; // polymorphic operations
    virtual void IsA() const;   // introduced at this level
    virtual void Greeting(const string &) const;
};
// Assume the non-inline member functions for Person
// follow and are as we are accustomed to seeing

class Student: public Person
{
private:
    float gpa = 0.0;   // in-class initialization
    string currentCourse;
    const string studentId;
    static int numStudents;  // static data member
public:
    Student();  // default constructor
    Student(const string &, const string &, char,
            const string &, float, const string &,
            const string &);
    Student(const Student &);  // copy constructor
```

```
    ~Student() override;   // virtual destructor
    void EarnPhD();
    // inline function definitions
    float GetGpa() const { return gpa; }
    const string &GetCurrentCourse() const
        { return currentCourse; }
    const string &GetStudentId() const
        { return studentId; }
    void SetCurrentCourse(const string &); // proto. only

    // In the derived class, keyword virtual is optional,
    // and not currently recommended. Use override instead.
    void Print() const final; // override is optional here
    void IsA() const override;
    // note, we choose not to redefine (override):
    // Person::Greeting(const string &) const
    static int GetNumberStudents(); // static mbr. function
};
// definition for static data member
int Student::numStudents = 0;   // notice initial value of 0

// Assume the non-inline, non-static member functions for
// Students follow and are as we are accustomed to seeing
```

In the preceding example, we see our Person class that we have carried forward throughout the book. As a base class, notice that Person specifies polymorphic operations of Print(), IsA(), and Greeting(), as well as the destructor using the virtual keyword. These operations are intended to be overridden by a derived class with more suitable methods (not including the destructor), but are not required to be overridden should the derived class find the base class implementation suitable.

In the derived class, Student, we override IsA() with a more suitable method. Notice that we use override in the signature of this function, though we could have also included virtual. Next, notice that we have chosen not to override Greeting() at the Student level; we can assume that Student finds the implementation in Person acceptable. Also notice that the destructor is overridden to provide the entry point to the destruction chain. Recall with a destructor that not only is the derived class destructor called, but the base class destructor will also be called (implicitly as the last line of code in the derived class destructor), allowing the object's full destruction sequence to properly commence.

Finally, notice that `Print ()` has been overridden as `final` in `Student`. Though we could have added `override` to the signature of this function as well, we choose to only utilize `final` per the recommended coding convention.

Now, let's look at another typical element in a typical C++ programming set of guidelines, relating to smart pointers.

Preferring smart pointers in new code

We have utilized many native (raw) C++ pointers in this book, as you will undoubtedly be asked to immerse yourself in existing code in which they are plentiful. Having native pointer experience and facility will make you a safer programmer when asked to step into situations that use native pointers.

However, for safety's sake, most programming guides will recommend using smart pointers exclusively in newly created code. After all, their use adds little overhead and can help eliminate many of the potential pitfalls of managing heap memory by the programmer. Smart pointers also aid in exception safety. For example, exception handling implies that the expected flow of code may be interrupted at nearly any time, leading to potential memory leaks with traditional pointer usage. Smart pointers can alleviate some of this burden and provide for exception safety.

Using smart pointers is so important in original code that this point is worth repeating: choosing smart pointers over native pointers leads to safer and far easier to maintain code in C++. The code will also be easier to write, eliminating the need for many destructors, automatically blocking undesired copies and assignment (`unique_ptr`), and so on. With that in mind, whenever possible, choose smart pointers in newly created code.

We've seen smart pointers in this book as well as native pointers. Now, you can choose to use smart pointers in the new code that you create – this is highly recommended. Certainly, there may be some scenarios when this is not possible; perhaps you are creating new code that interfaces heavily with existing native pointer code and need to utilize the same data structures. Nonetheless, you can strive to use smart pointers, when possible, yet you have the flexibility and experience to understand the vast amounts of existing code, libraries, and online examples that exist utilizing native pointers.

What could be better for safety than to have the facility of smart pointers for your original code, paired with the knowledge of native pointers to use only when necessary?

There are many examples of programming guidelines that can be easily followed to make your code safer. The aforementioned examples are just a few of many to illustrate the types of practices you will expect to see in a set of essential C++ programming guidelines.

Now, let's consider how we can assemble or adopt core programming guidelines to help make our code safer.

Adopting a programming guideline

Whether you build or assemble a set of programming guidelines yourself or adhere to a set governed by an organization you are a member of, adopting a core set of C++ programming guidelines is crucial to ensure your code is as safe and robust as possible, translating to more easily maintainable code.

Guidelines should always remain fluid as the language evolves. Let's next consider resources for finding core C++ programming guidelines to either follow directly or to revisit incrementally to improve the accepted guidelines within your organization.

Understanding resources for programming safely in C++

There are many online resources for programming guidelines in C++. The essential resource, however, is the *ISO C++ Core Guidelines*, assembled primarily by Bjarne Stroustrup and Herb Sutter, which can be found at the following GitHub URL: `https://github.com/isocpp/CppCoreGuidelines/blob/master/CppCoreGuidelines.md`. Their collective goal is to help programmers use modern C++ safely and more effectively.

Selected market sectors may have guidelines imposed upon them to obtain or ensure certification within an industry. For example, **MISRA** is a set of C++ coding standards for the **Motor Industry Software Reliability Association**; MISRA has also been adopted as a standard across other industries, such as for medical systems. Another coding standard, developed for embedded systems, is **CERT**, developed at **Carnegie Mellon University (CMU)**. Once an acronym for **Computer Emergency Response Team**, CERT is now a registered trademark of CMU. CERT has been adopted in many financial sectors as well. **JSF AV C++ (Joint Strike Fighter Air Vehicle C++)** is a C++ coding standard used in the aerospace engineering domain, developed by Lockheed Martin, to ensure error-free code for safety-critical systems.

Undoubtedly, each organization you join as a contributor will have a base set of programming guidelines for all programmers in the group to follow. If not, a wise move will be to suggest employing a core set of C++ programming guidelines. After all, you will need to help maintain your own code as well as the code of your colleagues; a uniform and expected set of standards will make this endeavor manageable for everyone involved.

Summary

In this bonus chapter, we have added to our objective of becoming indispensable C++ programmers by understanding the importance of programming safely in C++. After all, our primary goal is to create robust and easily maintainable code. Incorporating safe programming practices will help us achieve this goal.

We have reviewed concepts seen throughout the book, as well as related ideas that culminate in adopting a set of core programming guidelines to ensure safer coding practices.

First, we reviewed smart pointers, examining three types from the Standard C++ Library, namely `unique_ptr`, `shared_ptr`, and `weak_ptr`. We understand that these classes employ the RAII idiom by providing wrappers to allocate and deallocate heap memory safely on our behavior in well-tested standard library classes. We put forth a guideline: always prefer smart pointers in newly created code.

Next, we reiterated a variety of programming practices that we have seen throughout the book that we can employ to make our coding safer overall. For example, preferring for-each style loops and using `auto` for type safety. Also, using STL containers versus less robust native mechanisms, and also adding the `const` qualifier for data and methods to ensure read-only access when so needed. These practices are examples (among many) that can help ensure our code is as safe as possible.

Next, we introduced multithreaded programming in C++ and reviewed programming scenarios we have seen previously that may benefit from the use of threads. We also took a look ahead at the classes available in the Standard C++ Library in support of multithreaded programming, including those that provide synchronization, mutex locks, semaphores, and creating atomic types.

Finally, we examined programming guideline essentials to better understand rules that may be beneficial in a C++ core programming guide. For example, we reviewed preferring initialization over the assignment, virtual function usage with regard to the keywords `virtual`, `override`, and `final`, as well as previously examined topics from this chapter. We talked about the importance of adopting a comprehensive set of core programming guidelines for C++ as well as resources to find sample guidelines used as industry standards.

Understanding how to make C++ safer as you apply the many features covered in the book will undoubtedly make you a more valuable programmer. You now have core language skills plus a very solid understanding of OOP in C++ (essential concepts and how to implement them in C++ with either direct language support or using programming techniques). We have augmented your skills with knowledge of exception handling, friends, operator overloading, templates, STL basics, and testing OO classes and components. We have also embraced core design patterns, delving into each pattern with comprehensive programming examples. Finally in this chapter, we have reviewed how to safely put together the knowledge you have learned by choosing to employ safer programming practices at each available opportunity.

As we wrap up our bonus chapter together, you are now ready to journey further on your own, applying C++ to many new and existing applications. You are ready to create safe, robust, and easy to maintain code. I am sincerely hopeful that you are as intrigued by C++ as I am. Once again, let's get programming!

Assessments

The programming solution for each chapter's questions can be found in our GitHub repository at the following URL: `https://github.com/PacktPublishing/Deciphering-Object-Oriented-Programming-with-CPP/tree/main`. Each full program solution can be found in our GitHub repository in the subdirectory Assessments , and then under the appropriate chapter heading (subdirectory, such as `Chapter01`), in a file that corresponds to the chapter number, followed by a dash, followed by the solution number in the chapter at hand. For example, the solution for *Question 3* in *Chapter 1, Understanding Basic C++ Assumptions*, can be found in the subdirectory `Assessments/Chapter01` in a file named `Chp1-Q3.cpp` under the aforementioned GitHub directory.

The written responses for non-programming questions can be found in the following sections, organized by chapter, as well as in the aforementioned GitHub in the appropriate Assessments subdirectory for a given chapter. For example, `Assessments/Chapter01/Chp1-WrittenQs.pdf` will contain the answers to the non-programming solutions for *Chapter 1, Understanding Basic C++ Assumptions*. Should an exercise have a programming portion and a follow-up question to the program, the answer to the follow-up question may be found both in the next sections (as well as in the aforementioned .pdf file) and in a comment at the top of the programming solution in GitHub (as it may be appropriate to review the solution in order to fully understand the answer to the follow-up question).

Chapter 1, Understanding Basic C++ Assumptions

1. A `flush` may be useful, rather than and `endl`, for clearing the contents of a buffer associated with `cout` for the situations where you do not wish the cursor to be advanced to the next line for output. Recall, an `endl` manipulator is merely a newline character plus a buffer flush.

2. Choosing a pre versus a post increment for a variable, such as `++i` (versus `i++`) will have an impact on the code when used in conjunction with a compound expression. A typical example would be `result = array[i++];` versus `result = array[++i];`. With the post-increment (`i++`) the contents of `array[i]` will be assigned to `result` and then `i` is incremented. With the pre-increment, `i` is first incremented and then `result` will have the value of `array[i]` (that is, using the new value of `i` as an index).

3. Please see `Assessments/Chapter01/Chp1-Q3.cpp` in the GitHub repository.

Chapter 2, Adding Language Necessities

1. The signature of a function is the function's name plus its type and number of arguments (no return type). This relates to name mangling as the signature helps the compiler provide a unique, internal name for each function. For example, `void Print(int, float);` may have mangled name of `Print_int_float();`. This facilitates overloaded functions by giving each function a unique name so that when a call is made, it is evident by the internal function name as to which function is being invoked.

2. a – d. Please see `Assessments/Chapter02/Chp2-Q2.cpp` in the GitHub repository.

Chapter 3, Indirect Addressing: Pointers

1. a – f. Please see `Assessments/Chapter03/Chp3-Q1.cpp` in the GitHub repository.

   d. (follow-up question) `Print(Student)` is less efficient than `Print(const Student *)` as the initial version of this function passes an entire object on the stack, whereas the overloaded version passes only a pointer on the stack.

2. Assuming we have an existing pointer to an object of type `Student`, such as: `Student *s0 = new Student;` (this `Student` is not yet initialized with data)

   a. `const Student *s1;` (does not require initialization)

   b. `Student *const s2 = s0;` (requires initialization)

   c. `const Student *const s3 = s0;` (also requires initialization)

3. Passing an argument of type `const Student *` to `Print()` would allow a pointer to a `Student` to be passed into `Print()` for speed, yet the object pointed to could not be dereferenced and modified. Yet passing a `Student * const` as a parameter to `Print()` would not make sense because a copy of the pointer would be passed to `Print()`. Marking that copy additionally as `const` (meaning not allowing changing where the pointer points) would then be meaningless, as disallowing a *copy* of a pointer to be changed has no effect on the original pointer itself. The original pointer was never in jeopardy of its address being changed within the function.

4. There are many programming situations that might use a dynamically allocated 3-D array. For example, if an image is stored in a 2-D array, a collection of images might be stored in a 3-D array. Having a dynamically allocated 3-D array allows for any number of images to be read in from a filesystem and stored internally. Of course, you'd need to know how many images you'll be reading in before making the 3-D array allocation. For example, a 3-D array might hold 30 images, where 30 is the third dimension to collect the images in a set. To conceptualize a 4-D array, perhaps you would like to organize sets of the aforementioned 3-D arrays.

For example, perhaps you have a set of 31 images for the month of January. That set of January images is a 3-D array (2-D for the image and the third dimension for the set of 31 images comprising January). You may wish to do the same for every month. Rather than having separate 3-D array variables for each month's image set, we can create a fourth dimension to collect the years' worth of data into one set. The fourth dimension would have an element for each of the 12 months of the year. How about a 5-D array? You can extend this image idea by making the fifth dimension a way to collect various years of data, such as collecting images for a century (fifth dimension). Now we have images organized by century, then organized by year, then month, then by image (the image requiring the first two dimensions).

Chapter 4, Indirect Addressing: References

1.   a – c. Please see `Assessments/Chapter04/Chp4-Q1.cpp` in the GitHub repository.

   c. (follow-up question) Pointer variables need not only call the version of `ReadData(Student *)` that accepts a pointer to a `Student` and reference variables need not only call the version of `ReadData(Student &)` that accepts a reference to a `Student`. For example, a pointer variable may be dereferenced with `*` and then call the version that accepts a reference. Likewise, a reference variable may have its address taken using `&` and then call the version that accepts a pointer (though this is less commonly done). You simply need to make the data types match with respect to what you are passing and what the function expects.

Chapter 5, Exploring Classes in Detail

1.   a – e. Please see `Assessments/Chapter05/Chp5-Q1.cpp` in the GitHub repository.

Chapter 6, Implementing Hierarchies with Single Inheritance

1.   a – d. Please see `Assessments/Chapter06/Chp6-Q1.cpp` in the GitHub repository.

2.   a – c. (Optional) Please see `Chapter06/Assessments/Chp6-Q2.cpp` in the GitHub repository.

Chapter 7, Utilizing Dynamic Binding through Polymorphism

1.   a – e. Please see `Assessments/Chapter07/Chp7-Q1.cpp` in the GitHub repository.

Chapter 8, Mastering Abstract Classes

1. a – d. Please see `Assessments/Chapter08/Chp8-Q1.cpp` in the GitHub repository.

   e. Depending on your implementation, your `Shape` class may or may not be considered an interface class. If your implementation is an abstract class that contains no data members and only abstract methods (pure virtual functions), your `Shape` implementation is considered an interface class. If your `Shape` class, however, stores `area` as a data member once it has been calculated by the overridden `Area()` method in the derived classes, it is then just an abstract base class.

Chapter 9, Exploring Multiple Inheritance

1. Please see `Assessments/Chapter09/Chp9-Q1.cpp` in the GitHub repository.

   a. There is one `LifeForm` subobject.

   b. The `LifeForm` constructor and destructor are each invoked once.

   c. The default constructor for `LifeForm` would be invoked if the specification of an alternate constructor of `LifeForm(1000)` was removed from the member initialization list of the `Centaur` constructor.

2. Please see the `Assessments/Chapter09/Chp9-Q2.cpp` in the GitHub repository.

   a. There are two `LifeForm` sub-objects.

   b. The `LifeForm` constructor and destructor are each invoked twice.

Chapter 10, Implementing Association, Aggregation, and Composition

1. Please see `Assessments/Chapter10/Chp10-Q1.cpp` in the GitHub repository.

   (follow-up question) Once you have overloaded a constructor that accepts a `University` `&` as a parameter, this version may be invoked using a `University` `*` by first dereferencing the pointer to the `University` within the constructor call (to make a referenceable object).

2. a – f. Please see `Assessments/Chapter10/Chp10-Q2.cpp` in the GitHub repository.

3. a – b. (optional) Please see `Assessments/Chapter10/Chp10-Q3.cpp` in the GitHub repository.

Chapter 11, Handling Exceptions

1. a – c. Please see `Assessments/Chapter11/Chp11-Q1.cpp` in the GitHub repository.

Chapter 12, Friends and Operator Overloading

1. Please see `Assessments/Chapter12/Chp12-Q1.cpp` in the GitHub repository.
2. Please see `Assessments/Chapter12/Chp12-Q2.cpp` in the GitHub repository.
3. Please see `Assessments/Chapter12/Chp12-Q3.cpp` in the GitHub repository.

Chapter 13, Working with Templates

1. a – b. Please see `Assessments/Chapter13/Chp13-Q1.cpp` in the GitHub repository.
2. Please see `Assessments/Chapter13/Chp13-Q2.cpp` in the GitHub repository.

Chapter 14, Understanding STL Basics

1. a – b. Please see `Assessments/Chapter14/Chp14-Q1.cpp` in the GitHub repository.
2. Please see `Assessments/Chapter14/Chp14-Q2.cpp` in the GitHub repository.
3. Please see `Assessments/Chapter14/Chp14-Q3.cpp` in the GitHub repository.
4. Please see `Assessments/Chapter14/Chp14-Q4.cpp` in the GitHub repository.

Chapter 15, Testing Classes and Components

1. a. Your classes follow orthodox canonical class form if they each include a (user-specified) default constructor, copy constructor, overloaded assignment operator, and a virtual destructor. Your classes additionally follow extended canonical class form if they also include a move copy constructor and an overloaded move assignment operator.

   b. Your class will be considered robust if it follows canonical class form and ensures that all instances of a class have the means to be fully constructed. Testing a class can ensure robustness.

2. a – c. Please see `Assessments/Chapter15/Chp15-Q2.cpp` in the GitHub repository.
3. Please see `Assessments/Chapter15/Chp15-Q3.cpp` in the GitHub repository.

Chapter 16, Using the Observer Pattern

1. a – b. Please see `Assessments/Chapter16/Chp16-Q1.cpp` in the GitHub repository.
2. Other examples which may easily incorporate the Observer pattern include any application requiring customers to receive notification of backordered products that they desire. For example, many people may wish to receive the Covid-19 vaccine and wish to be on a waiting list at a vaccine distribution site. Here, a `VaccineDistributionSite` (the subject of interest) can be inherited from `Subject` and contain a list of `Person` objects, where `Person` inherits from

Observer. The Person objects will contain a pointer to the VaccineDistributionSite. Once enough supply for the vaccine exists at a given VaccineDistributionSite (that is, a distribution event has occurred), Notify() can be called to update the Observer instances (people on the waitlist). Each Observer will be sent an Update(), which will be the means to allow that person to schedule an appointment. If the Update() returns success and the Person has been scheduled for an appointment, the Observer can release itself from the waiting list with the Subject.

Chapter 17, Applying the Factory Pattern

1. a – b. Please see Assessments/Chapter17/Chp17-Q1.cpp in the GitHub repository.

2. Other examples which may easily incorporate the Factory Method pattern include many types of applications in which various derived classes may need to be instantiated based upon the specific values provided at construction. For example, a payroll application may require various types of Employee instances, such as Manager, Engineer, Vice-President, and so on. A factory method can provide a way to instantiate the various types of Employee based on the information provided when the Employee is hired. The Factory Method pattern is a pattern that can be applied to many types of applications.

Chapter 18, Applying the Adapter Pattern

1. a – b. Please see Assessments/Chapter18/Chp18-Q1.cpp in the GitHub repository.

2. Other examples which may easily incorporate the Adapter pattern include many examples of repurposing existing, well tested non-OO code to provide an OO interface (that is, a wrapper type of Adapter). Other examples include creating an Adapter to convert a formerly used class into a currently needed class (again with the idea of reusing previously created and well-tested components). An example is to adapt a Car class that has been previously used to represent gasoline engine cars into a class that models an ElectricCar.

Chapter 19, Using the Singleton Pattern

1. a – c. Please see Assessments/Chapter19/Chp19-Q1.cpp in the GitHub repository.

2. We cannot label the static instance() method as virtual in Singleton and override it in President, simply because static methods can never be virtual. They are statically bound, and never receive a this pointer. Also, the signature may need to be different (and no one likes an un-intentional function hiding situation).

3. Other examples which may easily incorporate the Singleton pattern include creating a Singleton CEO of a company, a Singleton TreasuryDepartment for a country, or a Singleton Queen of a nation. Each of these Singleton instances offers the opportunity to establish a registry to keep track of multiple Singleton objects. That is, many countries may have a single Queen.

In this case, the registry would allow not just one Singleton per object type, but one Singleton per other qualifiers, such as *nation*. This is an example of the rare case in which more than one Singleton object of a given type can occur (but always a controlled number of such objects).

Chapter 20, Removing Implementation Details Using the pImpl Pattern

1. Please see `Assessments/Chapter20/Chp20-Q1.cpp` in the GitHub repository.

2. Please see `Assessments/Chapter20/Chp20-Q2.cpp` in the GitHub repository.

    (follow-up question) In this chapter, simply inheriting `Student` from the `Person` class that embraces the pImpl pattern presents no logistical difficulties. Additionally, modifying the `Student` class to also employ the pImpl pattern and utilize a unique pointer is more challenging. Various approaches may run across various difficulties, including dealing with inline functions, down-casting, avoiding explicit calls to the underlying implementation, or requiring back pointers to help invoke virtual functions. Please see the online solution for details.

3. Other examples which may easily incorporate the pImpl pattern for relative implementation independence include creating generic GUI components, such as for `Window`, `Scrollbar`, `Textbox`, and so on, for various platforms (derived classes). The implementation details can easily be hidden. Another example could be proprietary commercial classes in which the developer wishes to hide the implementation details that might otherwise be seen in a header file.

Index

S

Packt.com

Subscribe to our online digital library for full access to over 7,000 books and videos, as well as industry leading tools to help you plan your personal development and advance your career. For more information, please visit our website.

Why subscribe?

- Spend less time learning and more time coding with practical eBooks and Videos from over 4,000 industry professionals

- Improve your learning with Skill Plans built especially for you

- Get a free eBook or video every month

- Fully searchable for easy access to vital information

- Copy and paste, print, and bookmark content

Did you know that Packt offers eBook versions of every book published, with PDF and ePub files available? You can upgrade to the eBook version at packt.com and as a print book customer, you are entitled to a discount on the eBook copy. Get in touch with us at customercare@packtpub.com for more details.

At www.packt.com, you can also read a collection of free technical articles, sign up for a range of free newsletters, and receive exclusive discounts and offers on Packt books and eBooks.

Other Books You May Enjoy

If you enjoyed this book, you may be interested in these other books by Packt:

C++20 STL Cookbook

Bill Weinman

ISBN: 978-1-80324-871-4

- Understand the new language features and the problems they can solve
- Implement generic features of the STL with practical examples
- Understand standard support classes for concurrency and synchronization
- Perform efficient memory management using the STL
- Implement seamless formatting using std::format
- Work with strings the STL way instead of handcrafting C-style code

CMake Best Practices

Rafał Świdziński

ISBN: 978-1-80107-005-8

- Understand best practices for building C++ code
- Gain practical knowledge of the CMake language by focusing on the most useful aspects
- Use cutting-edge tooling to guarantee code quality with the help of tests and static and dynamic analysis
- Discover how to manage, discover, download, and link dependencies with CMake
- Build solutions that can be reused and maintained in the long term
- Understand how to optimize build artifacts and the build process itself

C++ High Performance

Björn Andrist, Viktor Sehr

ISBN: 978-1-83921-654-1

- Write specialized data structures for performance-critical code
- Use modern metaprogramming techniques to reduce runtime calculations
- Achieve efficient memory management using custom memory allocators
- Reduce boilerplate code using reflection techniques
- Reap the benefits of lock-free concurrent programming
- Gain insights into subtle optimizations used by standard library algorithms
- Compose algorithms using ranges library
- Develop the ability to apply metaprogramming aspects such as constexpr, constraints, and concepts
- Implement lazy generators and asynchronous tasks using C++20 coroutines

CMake Best Practices

Fedor G. Pikus

ISBN: 978-1-80020-811-7

- Discover how to use the hardware computing resources in your programs effectively
- Understand the relationship between memory order and memory barriers
- Familiarize yourself with the performance implications of different data structures and organizations
- Assess the performance impact of concurrent memory accessed and how to minimize it
- Discover when to use and when not to use lock-free programming techniques
- Explore different ways to improve the effectiveness of compiler optimizations
- Design APIs for concurrent data structures and high-performance data structures to avoid inefficiencies

Packt is searching for authors like you

If you're interested in becoming an author for Packt, please visit `authors.packtpub.com` and apply today. We have worked with thousands of developers and tech professionals, just like you, to help them share their insight with the global tech community. You can make a general application, apply for a specific hot topic that we are recruiting an author for, or submit your own idea.

Share Your Thoughts

Now you've finished *Deciphering Object-Oriented Programming with C++*, we'd love to hear your thoughts! Scan the QR code below to go straight to the Amazon review page for this book and share your feedback or leave a review on the site that you purchased it from.

`https://packt.link/r/1-804-61390-8`

Your review is important to us and the tech community and will help us make sure we're delivering excellent quality content.

www.ingramcontent.com/pod-product-compliance
Lightning Source LLC
Chambersburg PA
CBHW081449050326
40690CB00015B/2737